Springer Series in Computational Physics

Editors
H. Cabannes   M. Holt
H. B. Keller   J. Killeen   S. A. Orszag

Springer Series in Computational Physics

Editors: H. Cabannes, M. Holt, H. B. Keller, J. Killeen, S. A. Orszag

M. Kubíček
M. Marek

Computational Methods in Bifurcation Theory and Dissipative Structures

With 91 Illustrations

Springer-Verlag
New York Berlin Heidelberg Tokyo

M. Kubíček
M. Marek
Prague Institute of Chemical Technology
Department of Chemical Engineering
166 28 Praha 6
Suchbátarova 1903
Czechoslovakia

Editors

H. Cabannes
Mécanique Théoretique
Université Pierre et Marie Curie
F–75005 Paris
France

M. Holt
College of Mechanical Engineering
University of California
Berkeley, CA 94720
U.S.A.

H. B. Keller
California Institute of Technology
Pasadena, CA 91125
U.S.A.

J. Killeen
Lawrence Livermore Laboratory
Livermore, CA 94551
U.S.A.

S. A. Orszag
Massachusetts Institute of Technology
Cambridge, MA 02139
U.S.A.

Library of Congress Cataloging in Publication Data
Kubíček, Milan.
  Computational methods in bifurcation theory and dissipative structures.
  (Springer series in computational physics) Bibliography: p.
  1. Bifurcation theory. 2. Stability. 3. Mathematical physics. I. Marek, M.
II. Title. III. Title: Dissipative structures. IV. Series.
QC20.7.B54K8 1983   530.1'55353   82-19630

© 1983 by Springer-Verlag New York Inc.
Softcover reprint of the hardcover 1st edition 1983

Typeset by Composition House Ltd., Salisbury, England.

9 8 7 6 5 4 3 2 1

ISBN-13: 978-3-642-85959-5   e-ISBN-13: 978-3-642-85957-1
DOI: 10.1007/978-3-642-85957-1

Preface

"Dissipative structures" is a concept which has recently been used in physics to discuss the formation of structures organized in space and/or time at the expense of the energy flowing into the system from the outside. The space-time structural organization of biological systems starting from the subcellular level up to the level of ecological systems, coherent structures in laser and plasma physics, problems of elastic stability in mechanics, instability in hydrodynamics leading to the development of turbulence, behavior of electrical networks and chemical reactors form just a short list of problems treated in this framework.

Mathematical models constructed to describe these systems are usually nonlinear, often formed by complicated systems of algebraic, ordinary differential, or partial differential equations and include a number of characteristic parameters. In problems of theoretical interest as well as engineering practice, we are concerned with the dependence of solutions on parameters and particularly with the values of parameters where qualitatively new types of solutions, e.g., oscillatory solutions, new stationary states, and chaotic attractors, appear (bifurcate).

Numerical techniques to determine both bifurcation points and the dependence of steady-state and oscillatory solutions on parameters are developed and discussed in detail in this text. The text is intended to serve as a working manual not only for students and research workers who are interested in dissipative structures, but also for practicing engineers who deal with the problems of constructing models and solving complicated nonlinear systems. Development of computer capacity and speed make possible (and will continue to facilitate) the use of the more complicated computational algorithms described herein.

The text is organized into five chapters and three appendices. In Chapter 1, Section 1.2, examples of models of nonlinear systems used herein for illustration of numerical techniques are discussed. These are drawn from the areas of chemical kinetics, fluid mechanics, morphogenesis, reaction-diffusion systems, and chemical and biological reactors.

Section 1.3 defines the basic concepts in the text. The terms multiplicity of solutions, stability of solutions, limit cycle, strange (chaotic) attractor, solution diagram, bifurcation diagram, and evolution diagram are explained and discussed.

In Chapter 2 we discuss procedures for investigating lumped parameter systems (usually described by sets of ordinary differential equations). We describe in detail techniques for determining branch points (both limit and bifurcation) and for continuation along arcs of solutions. We also present numerical methods for solving systems of nonlinear equations and systems of ordinary differential equations. We provide detailed examples illustrating these techniques which the reader can use to formulate algorithms for his own problems. In the final portion of Chapter 2, we discuss chaotic attractors and numerical methods used in their characterization.

In Chapter 3 we consider distributed parameter systems described mostly by systems of nonlinear partial differential equations. Procedures for computing branch points and the dependence of the solution on parameters are examined. The methods are still under development, particularly for cases where oscillatory solutions appear. One of the detailed examples involves the construction of a bifurcation diagram for the reaction-diffusion system "Brussellator", originally used by Prigogine's group to describe space-time organization of concentrations in reacting media far from equilibrium.

In Chapter 4 we use the results of the previous chapters to discuss the solutions of evolution equations. A sequence of structures which appear when one of the parameters slowly varies in time is interpreted on the basis of the bifurcation diagram. The procedure for distributed systems in Gierer–Meinhardt's example of a reaction-diffusion model of morphogenesis is illustrated; here increasingly more complicated spatial morphogen concentration profiles appear as a result of changing characteristic dimensions of the systems.

In Chapter 5 we summarize numerical techniques recently proposed by others to solve the problems treated here. Finally, because this text appears to be the first of its kind, and the subject is still in an early stage of development, possible future developments are explored. Appendices A and B containing algorithms in the form of FORTRAN routines should be useful to the reader.

Recently, a number of new results in the qualitative theory of differential equations, dynamic systems, and bifurcation theory have appeared in the literature. Some of them are not very familiar to students, research workers, and engineers. Hence we have included discussions of such concepts as invariant manifold, normal form, codimension of bifurcation, and other terms in Appendix C. Characteristic types of bifurcations are illustrated by simple examples. We hope that the appendices will serve to familiarize the reader with the current literature on the theory of bifurcations needed to interpret results in complex systems. However, the text is not dependent on the appendices, which may be omitted in the first reading.

The text is based on results obtained in our research group by the combined efforts of applied mathematicians and engineers over the last 10 years. We would like to express our sincere thanks to our colleagues, Alois Klíč, Martin

Holodniok, Vladimír Hlaváček, Igor Schreiber, and to the many students who participated in developing the subject.

We are thankful to Professor Beiglböck and Professor Keller for their interest, meaningful suggestions and criticism, and to the staff of Springer-Verlag, New York for their invaluable help with the transformation of the manuscript into the final form of this book.

We are indebted to Marie Růžková, Dana Suchanová, and Zdenka Svobodová for their assistance in producing the manuscript.

Table of Contents

CHAPTER 1

Introduction

1.1 General Introduction

Evolutionary processes leading to diversification and increasing complexity are common in physical, chemical, and biological systems. One of the main aspects of the study of such phenomena is the problem of describing self-organization, i.e., detailed study of stationary and/or time-dependent states evolving with changes of characteristic parameters. We can recognize two main approaches in the description of such systems—deterministic and stochastic models. The deterministic models can also give solutions with stochastic character (chaotic solutions). This work is related mainly to the study of deterministic models, but the described numerical procedures can also be used in the development of basic numerical procedures applicable to stochastic systems.

General features of the studied dissipative systems are:

1. The description consists of a set of nonlinear equations, which can be in the form of algebraic, ordinary differential, partial differential, difference, or integral equations.
2. Through the variation of a parameter (one or more), new steady-state, time-periodic, or chaotic solutions—dissipative structures—appear (bifurcate) at certain critical (bifurcation) values of the parameters. These solutions appear, become more complex, and disappear with the change of parameters.

In principle, the term dissipative structures means that the structures are maintained at the expense of energy flowing into the system from the outside. Hence, we shall deal with systems which generally are far from equilibrium. The new structurally stable structures which evolve can possess either coherent or chaotic behavior.

The formation of dissipative structures has long been known to occur in hydrodynamics. We can mention, for example, the problems of Bénard instability—the appearance of regular cells in a layer of liquid heated from below. The bifurcations in hydrodynamics are usually called transitions. The occurrence of multiple steady-state solutions, oscillations, concentration and temperature waves in chemically reacting systems, including combustion, also belong to the same category. Space-time structural organization of biological systems starting from the subcellular level up to the level of ecological systems, the problems of structural stability in mechanics, coherent structures in laser and plasma physics, earth boundary-layer stability, cycling behavior of economic systems are all problems which can be treated in the framework of dissipative structures.

In the study of dissipative structures, we can recognize three different types of approaches. In the first one, the most general and qualitative, topological techniques are used to classify topologically distinct types of behavior possible in the given model. Catastrophe theory and singularity theory [1.1, 1.2] belong to this category. Mathematical foundations of this approach have been built up with increasing intensity during the last 15 years (cf., [1.3, 1.4]).

In the second approach, approximate analytical techniques using linearization in the neighborhood of singular points, methods of small parameters, asymptotic expansions, and averaging are used to construct local descriptions of dissipative structures. However, in most situations, we can thus obtain only special solutions (e.g., solutions close to bifurcation points, solutions for extreme values of parameters). An advantage of this approach is that the description of the behavior of solutions is explicit (even though approximate).

Where we would like to compare the predictions of the model with experimental data, we are generally interested in the full solution diagram,[1] i.e., the dependence of the solution on parameters over the entire region of their physical significance. Here the third type of approach—direct numerical solution of the model equations—usually has to be used.

For example, we may wish to solve a problem for some given values of the parameters, or we may have to generate the solution as a chosen parameter varies over the region of interest.

This work is devoted to the description of efficient numerical techniques for examining the dependence of solutions on parameters. Full solution diagrams can be constructed in this way.

Since this text is intended as a working manual of numerical techniques in the study of dissipative structures, discussion of the relevant theory is kept to a minimum. However, in the bibliography, the reader is offered a variety of references, including both theory and applications.

---

[1] By a "solution diagram" we mean a diagram of the dependence on a parameter of a given characteristic of the solution (e.g., norm). The term "*bifurcation diagram*" is used in the above sense by some authors but we shall define bifurcation diagram as the graph showing the dependence of branch (bifurcation) points on a parameter, cf., Section 2.4.

1.2 Dissipative Structures in Physical, Chemical, and Biological Systems

We shall illustrate the wide variety of dissipative systems with examples from various branches of science and engineering, and discuss some problems encountered in an analysis of the dependence on parameters.

1.2.1 The problems of elastic stability

There is extensive literature on elastic stability problems [1.5, 1.6]. We shall briefly discuss some examples.

One of the best known is the buckling of a circular plate. The edge of a circular elastic plate is clamped and compressed by a uniform, radial edge thrust. For sufficiently small thrusts, the plate deforms by uniformly contracting in its plane to a circle of smaller radius. This equilibrium state is called the unbuckled state. When the thrust exceeds a critical value, called the buckling load, the unbuckled state becomes unstable. Then the plate deforms by deflecting out of its plane.

Let us assume that the nonlinear-plate theory of von Kármán [1.7] describes the nonlinear deformations of the plate. If we denote the dimensionless radial variable as x and the polar angle as Θ, then the dimensionless lateral deflection of the plate $w(x, \Theta)$ and the dimensionless excess stress function $f(x, \Theta)$ are governed by the following boundary-value problem:

$$\Delta^2 f = -\tfrac{1}{2}[w, w], \tag{1.1}$$

$$\Delta^2 w = [f, w] - \lambda w_{xx}, \tag{1.2}$$

for $0 \le x \le 1$ and $0 \le \Theta \le 2\pi$.

Corresponding boundary conditions are in the form

$$w(1, \Theta) = w_x(1, \Theta) = f(1, \Theta) = f_x(1, \Theta) = 0, \tag{1.3}$$

where $x = 1$ is the edge of the plate.

In Eqs. (1.1) and (1.2), Δ is the Laplacian in polar coordinates and the nonlinear operator $[f, w]$ is defined by

$$[f, w] = \frac{1}{x} \left\{ f_{xx}\left(w_x + \frac{1}{x} w_{\Theta\Theta}\right) + w_{xx}\left(f_x + \frac{1}{x} f_{\Theta\Theta}\right) - 2x\left(\frac{w_\Theta}{x}\right)_x \left(\frac{f_\Theta}{x}\right)_x \right\}. \tag{1.4}$$

(a subscript denotes differentiation with respect to that variable). The parameter λ is proportional to the edge thrust ($\lambda > 0$ corresponds to compressive thrust).

The unbuckled state is given by $w = f = 0$ and is a solution of Eqs. (1.1) and (1.2) for all values of λ. For increasing values of λ, beginning at $\lambda = \lambda_c$, different buckled states appear. Hence analyzing the solutions' dependence on the value of λ is the main task in the theory of nonlinear buckling. Both axially symmetric and asymmetric solutions appear for increasing values of λ; the first ones arise through primary bifurcations and the second ones through secondary bifurcations. The unbuckled state is stable for $\lambda < \lambda_c$ and unstable for $\lambda \geq \lambda_c$. Both approximate analytical techniques and numerical procedures have been used to construct bifurcated buckled states [1.8–1.13]. No complete continuous dependence of the solutions on the parameter λ has appeared in the literature so far.

We describe the buckling of a rectangular simply supported plate that is deformed by uniform compressive thrusts applied to the two edges normal to the x-axis (the other two edges are free to expand in the y direction) by the set of differential equations

$$\Delta^2 f = -\tfrac{1}{2}[w, w], \tag{1.5}$$

$$\Delta^2 w = [f, w] - \lambda w_{xx}. \tag{1.6}$$

Equations (1.5) and (1.6) are defined on the rectangle $0 \leq x \leq l, 0 \leq y \leq 1$, subject to the boundary conditions

$$w = \Delta w = f = \Delta f = 0. \tag{1.7}$$

Here Δ is the two-dimensional Cartesian Laplacian, $l > 0$ is the aspect ratio of the plate, λ is a parameter proportional to the thrust, and the nonlinear operator $[f, w]$ is defined as

$$[f, w] = f_{yy} w_{xx} + f_{xx} w_{yy} - 2 f_{xy} w_{xy}. \tag{1.8}$$

The unbuckled state $w = f = 0$ is a solution of Eqs. (1.6)–(1.8) for all values of λ and l. For $\lambda < \lambda_c(l)$, the unbuckled state is the unique solution of Eq. (4.1). For $\lambda \geq \lambda_c(l)$, primary buckled states branch off from the unbuckled state. They branch off in pairs, but if the eigenvalues of the corresponding linearized problems are multiple, then more pairs can branch off at the critical eigenvalue (e.g., four pairs branch off at the double eigenvalue [1.15, 1.16]). It has been observed experimentally [1.17] that, as λ increases, the primary bifurcated state (buckled state) may become unstable and the plate jumps to another equilibrium state; this may occur several times as λ increases. This phenomenon is called mode jumping. Again, both analytical and numerical techniques have been used to find primary and secondary bifurcated solutions [1.7, 1.14, 1.15, 1.18, 1.19, 1.20]. Symmetric and asymmetric buckled states have been found, but, as yet, no complete solution diagram containing a full picture of all solutions and the connections between them has been published. Neither has the simulation of mode jumping, i.e., of the evolution of the sequence of buckled states which appear when the compressive thrust (proportional to λ) is changed slowly (quasi-stationary situation) been discussed up until now.

Both solution-diagram and evolution problems can be solved by techniques described in Chapters 3 and 4.

An application of topological techniques to nonlinear buckling problems has been discussed, e.g., by Chillingworth [1.21, 1.22].

1.2.2 Bifurcations to divergence and flutter in flow-induced oscillations

The problem of nonlinear oscillations of a fluttering plate (flow-induced oscillations) is a well-known problem in aeroelasticity [1.23]. The equation of motion of a thin panel, fixed at both ends and undergoing "cylindrical" bending between $z = 0$ and $z = 1$, can be written in terms of the lateral deflection $v = v(z, t)$ as

$$\alpha \dot{v}'''' + v''' - \left\{ \Gamma + \varkappa \int_0^1 (v'(\xi))^2 \, d\xi + \sigma \int_0^1 (v'(\xi)\dot{v}'(\xi)) d\xi \right\} v''$$

$$+ \rho v' + \sqrt{\rho} \, \delta \dot{v} + \ddot{v} = 0. \tag{1.9}$$

Here $\cdot = \partial/\partial t$, $' = \partial/\partial z$, and α, σ describe viscoelastic structural damping terms, $\sqrt{\rho} \delta$ represents aerodynamic damping, $\varkappa$ represents nonlinear (membrane) stiffness, ρ is dynamical pressure and Γ is in-plane tensile load. Either supported,

$$v = (\dot{v} + \alpha v)'' = 0, \tag{1.10}$$

or clamped,

$$v = v' = 0, \tag{1.11}$$

boundary conditions are usually considered. If α, σ, δ, $\varkappa$ are fixed and positive, and solutions of Eq. (1.9) are followed as functions of $\mu = \{(\rho, \Gamma)\}$, $(\rho \geq 0)$, Hopf bifurcation can occur and oscillatory (periodic) solutions appear. Finite-dimensional approximations of Eq. (1.9) have been studied numerically, cf., [1.23] and, recently, center manifold theory has been applied to the study of the problem by Holmes and Marsden [1.24]. However, comprehensive global results have yet to be obtained.

1.2.3 Wheelset nonlinear hunting problem

The analysis of nonlinear oscillations of a railroad vehicle wheelset can be based on the following set of equations:

$$M\ddot{y} + 2f_L\left(\frac{\dot{y}}{v} - \psi\right) + K_y y + F_g(y - \delta) = 0, \tag{1.12}$$

$$I\ddot{\psi} + 2lf_T\left(\frac{l\dot{\psi}}{v} + \frac{r_1 - r_2}{2r_0}\right) + K_\psi \psi = 0. \tag{1.13}$$

The relative position of the wheelset y with respect to the rails determines the lateral force F_g due to wheel-rail contact; the terms involving f_L and f_T are

forces and moments due to creepage (slip) between wheels and rails. The term
$(r_1 - r_2)/2r_0$ describes conicity. Both periodic solutions arising through Hopf
bifurcations and large excitations following small random input are observed
in the system [1.25, 1.26, 1.27].

1.2.4 Buckling of a shallow elastic arch

Consider a shallow elastic arch subjected to a single concentrated load. The
unloaded configuration $y_0(x)$ is assumed to be circular with height c. Under
load p at location x_p, the equilibrium shape $y(x)$ is described by the equation

$$y'''' - y_0'''' + \eta(y)y'' = -p\,\delta(x - x_p), \tag{1.14}$$

where δ is the Dirac function and the axial thrust $\eta(y)$ is given by

$$\eta(y) = 2\int_0^1 [(y_0')^2 - (y')^2]dx. \tag{1.15}$$

The boundary conditions are

$$y(0) = 0, \quad y'(0) = y_0'(0), \quad y(1) = 0, \quad y''(1) = y_0''(1). \tag{1.16}$$

The load-deflection equilibrium paths are monotonically increasing for
sufficiently small c. For higher c, snap-through instability occurs as the load
is increased, either at a limit (maximum) point or at a bifurcation point. The
snap-through corresponds to sudden buckling into an inverted (or partially
inverted) configuration [1.28]. Similar problems have been analyzed by
Zeeman [1.29] in terms of catastrophe theory and by Golubitsky and Schaeffer
[1.30] in terms of singularity theory. Other more complicated problems have
been reviewed by Plant [1.28].

1.2.5 Dissipative structures in fluid mechanics

Observations of dissipative structures and bifurcations in fluid dynamics are
probably the most common among the various examples of dissipative
structures in physical systems. We shall present formulations of several of the
most commonly studied problems here: the problem of equilibrium shapes of
rotating fluid masses held together by gravitation, and the Taylor and Bénard
problems. There exists extensive literature on the subject, cf., e.g.,
[1.31, 1.32, 1.33]. Recently, a rejuvenation of inviscid vortex theory, kinematics
of vortex interactions, and exploration of the concepts of coherent structures
generally have become a central subject of interest in the study of turbulence
[1.34].

The stability of rotating liquid masses
Consider a homogeneous incompressible fluid of given mass and volume,
rotating rigidly round a fixed axis with an angular momentum J. Let the only

acting forces be gravitational and centrifugal forces and let the fluid occupy a connected volume bounded by the surface

$$S(\mathbf{x}) = 0.$$

($S(\mathbf{x}) > 0$ inside the fluid).

In the co-rotating system, the equations of hydrostatic equilibrium hold [1.35].

$$\nabla(p - \phi) = 0. \tag{1.17}$$

Here $p(\mathbf{x})$ is the pressure and

$$\phi(\mathbf{x}) = \phi_g(\mathbf{x}) + \phi_c(\mathbf{x}) \tag{1.18}$$

is the sum of the gravitational and centrifugal potentials. The gravitational potential satisfying Poisson's equation

$$\Delta\phi_g = -\rho, \quad \rho(\mathbf{x}) = \begin{cases} 1, & S(x) \geq 0, \\ 0, & S(x) < 0, \end{cases} \tag{1.19}$$

and the boundary condition $\phi_g(x) \to 0, |x| \to \infty$, is given by

$$\phi_g(S(\mathbf{x})) = \frac{1}{4\pi} \int_{S(y) \geq 0} \frac{dy}{|x - y|}. \tag{1.20}$$

The centrifugal potential can be defined as

$$\phi_c(S(\mathbf{x}, J^2)) = \frac{J^2}{2I^2} (x_1^2 + x_2^2), \tag{1.21}$$

where

$$I(S) = \frac{15}{8\pi} \int_{S(\mathbf{x}) \geq 0} (x_1^2 + x_2^2) dx \tag{1.22}$$

is the corresponding moment of inertia.

Equation (1.17) must be integrated subject to the boundary condition

$$p(\mathbf{x})\bigg|_{S(\mathbf{x}) = 0} = 0. \tag{1.23}$$

Hence, it follows that the fluid boundary is an equipotential:

$$\phi(S(\mathbf{x}, J^2))\bigg|_{S(\mathbf{x}) = 0} = \text{const.} \tag{1.24}$$

This is an equation for the function S which determines the equilibrium boundary that depends on the parameter J^2. For $0 < J^2 < 0.384436$, the solutions of Eq. (1.24)—equilibrium solutions—are in the form of ellipsoids (Maclaurin ellipsoids). As J^2 increases beyond the critical value $J^2 = 0.384436$, new solutions appear. The first one is a set of Jacobi ellipsoids. Further on, other solutions bifurcating both from the Maclaurin and Jacobi sequences arise. The relations between individual solutions are still areas of active research, although more than 100 years have passed since the problem's formulation. The problem was studied, for example, by Jacobi, Poincaré, and Lyapunov, and it is one of the basic problems in astrophysics.

The Taylor problem

Taylor [1.36] discovered that the viscous flow between two infinitely long coaxial cylinders rotating in the same direction may have different stationary states. Let us consider [1.33, 1.37] two coaxial circular cylinders of infinite length with radii r_1' and $r_2'(r_1' < r_2')$, rotating with constant angular velocities ω_1 and ω_2. An incompressible fluid rotates in the gap between the cylinders. Let $\lambda = r_1'\omega_1(r_2' - r_1')/v$ be the Reynolds number and $r_i = r_i'/(r_2' - r_1')$, $i = 1, 2$ denote a dimensionless radius.

The Couette flow is a solution of the Navier–Stokes equations independent of λ, given by

$$\mathbf{v}^0 = (0, v_\varphi^0, 0), \quad p^0(r) = \int_{r_1}^{r} \frac{v_\varphi^0(\rho)^2}{\rho}\, d\rho + \text{const},$$

$$v_\varphi^0 = ar + \frac{b}{r}, \qquad a = \frac{1}{r_1\omega_1} \frac{\omega_2 r_2^2 - \omega_1 r_1^2}{r_2^2 - r_1^2},$$

$$b = \frac{1}{r_1\omega_1} \frac{(\omega_1 - \omega_2)r_1^2 r_2^2}{r_2^2 - r_1^2}. \tag{1.25}$$

Here, in the coordinates r, φ, and z, $\mathbf{v}$ denotes velocity $\mathbf{v} = (v_r, v_\varphi, v_z)$ and p the pressure.

For small λ, the solution (1.25) is unique and asymptotically stable. As λ increases, various types of fluid motions are observed; the simplest solutions are periodic in z and independent of φ (Taylor vortex flow).

Let us consider φ independent solutions and let

$$\mathbf{v} = \mathbf{V} + \mathbf{u}, \quad p = P + q,$$

$$D_t = \frac{\partial}{\partial t}, \quad \nabla = \left(\frac{\partial}{\partial r}, 0, \frac{\partial}{\partial z}\right), \quad \Delta = \frac{\partial^2}{\partial r^2} + \frac{1}{r}\frac{\partial}{\partial r} + \frac{\partial^2}{\partial z^2},$$

$$\tilde{\Delta}_{ik} = \left(\Delta - \frac{1 - \delta_{i3}}{r^2}\right)\delta_{ik},$$

$$L_{ik}^0(\mathbf{V}) = \frac{-2V_\varphi \delta_{i1}\delta_{2k}}{r} + \frac{(V_r\delta_{2k} + V_\varphi\delta_{1k})\delta_{i2}}{r},$$

$$L(\mathbf{V})\mathbf{u} = L^0(\mathbf{V})\mathbf{u} + (\mathbf{V}\cdot\nabla)\mathbf{u} + (\mathbf{u}\cdot\nabla)\mathbf{V},$$

$$Q(u)_i = -\frac{u_\varphi^2\delta_{i1}}{r} + \frac{u_\varphi u_r\delta_{i2}}{r},$$

$$N(\mathbf{u}) = (\mathbf{u}\cdot\nabla)\mathbf{u} + Q(\mathbf{u}), \qquad i, k = 1, 2, 3, \tag{1.26}$$

where $i = 1, 2, 3$ corresponds to r, φ, z. Then the corresponding Navier–Stokes equations can be written in the form

$$D_t\mathbf{u} - \tilde{\Delta}\mathbf{u} + \lambda L(\mathbf{V})\mathbf{u} + \lambda\nabla q = -\lambda N(\mathbf{u}),$$

$$\nabla\cdot\mathbf{u} = 0, \quad \mathbf{u}|_{r=r_1, r_2} = \mathbf{0}, \quad \mathbf{u}|_{t=0} = \mathbf{u}^0. \tag{1.27}$$

Furthermore, we consider the solution $\mathbf{v}$ that is invariant under translations $z \to z + 2\pi/\sigma$, $\varphi \to \varphi + 2\pi$, $\sigma > 0$. For higher values of the Reynolds number λ, the Couette flow becomes unstable, and a new rotationally symmetric stationary flow appears. The flow is periodic in the axial direction, and the trajectories are found on tori forming a system of closed vortices (Taylor vortices). These vortices become unstable again for even higher λ, and a "wavy" vortex pattern appears with its wave number depending on λ. A further increase in λ generates more complex flow patterns.

The Bénard problem
A viscous fluid in the layer between two horizontal planes moves under the influence of viscosity and buoyancy forces, where the latter is caused by heating the lower plane. The constant temperature of the upper plane is T_1 and that of the lower plane T_0 ($T_0 > T_1$). When $T_0 - T_1$ exceeds a certain critical value, convective motion is observed. The convection takes place in a regular pattern of closed cells having the form of rolls or hexagons. Let us use the Boussinesq approximation, i.e., neglect the variations of density except for the buoyant-force term [1.33]. Let α, h, g, v, ρ, and $\varkappa$ denote the coefficient of volume expansion, the thickness of the layer, gravity, the kinematic viscosity, the density, and the coefficient of thermal conductivity, respectively. Let the space vector $\mathbf{x} = (x_1, x_2, x_3)$ and the velocity vector $\mathbf{v} = (v_1, v_2, v_3)$ be given in Cartesian coordinates with the x_3-axis pointing in the direction opposite to that of the effect of the gravitational force, where θ denotes temperature and p the pressure. The basic stationary solution $\mathbf{v}_0$, θ_0, p_0 has the following form:

$$\mathbf{v}_0 = 0, \quad p_0(x_3) = -\frac{gh^3}{v^2}(x_3 + \alpha(T_0 - T_1)), \quad \theta_0(x_3) = -x_3. \qquad (1.28)$$

Let us define

$$\mathbf{v} = \mathbf{V} + \mathbf{u}, \quad P = P + q, \quad \tilde{\theta} = T + \theta,$$

$$\mathbf{w} = (\mathbf{u}, \theta),$$

$$\lambda = \frac{\alpha g(T_0 - T_1)h^3}{v^2} \quad \text{(Grashof number)},$$

$$\text{Pr} = \frac{\varkappa}{v} \quad \text{(Prandtl number)},$$

$$\mathbf{V} = \left(\frac{\partial}{\partial x_1}, \frac{\partial}{\partial x_2}, \frac{\partial}{\partial x_3}, 0\right),$$

$$\tilde{\Delta}_{ik} = \left(\frac{\partial^2}{\partial x_1^2} + \frac{\partial^2}{\partial x_2^2} + \frac{\partial^2}{\partial x_3^2}\right)\left(\delta_{ik} + \frac{1}{\text{Pr}}\delta_{i4}\right),$$

$$L_{ik}^0 = -\delta_{i3}\delta_{k4} - \frac{1}{\text{Pr}}\delta_{i4}\delta_{k3}, \quad i, k = 1, \dots, 4,$$

$$L(\mathbf{V})\mathbf{w} = L^0\mathbf{w} + (\mathbf{V} \cdot \mathbf{V})\mathbf{w} + (\mathbf{u} \cdot \mathbf{V})\mathbf{V},$$

$$N(\mathbf{w}) = (\mathbf{u} \cdot \mathbf{V})\mathbf{w}. \qquad (1.29)$$

Then the velocity and temperature distribution can be described by the following initial-value problem

$$D_t \mathbf{w} - \tilde{\Delta}\mathbf{w} + \lambda L(V)\mathbf{w} + \nabla q = -N(\mathbf{w}),$$

$$\nabla \mathbf{w} = 0, \quad \mathbf{w}|_{x_3 = 0, 1} = 0,$$

$$\mathbf{w}|_{t=0} = \mathbf{w}^0. \tag{1.30}$$

Both Taylor and Bénard problems can be formulated as one general problem of evolution in a suitable Hilbert space. Later on, we shall formulate the problem of the description of flow between two rotating discs and discuss some computational results obtained by the techniques described in Chapters 2 and 3.

1.2.6 Biological systems

Most of the processes occurring in living systems can be described in terms of dissipative structures; like problems which arise, for example, in the study of the formation and functioning of subcellular structures, membranes, biochemical processes in cells, tissues, organs, behavior of populations of microorganisms, and, finally, ecological problems including dynamics of interacting populations. The description of transports (e.g., diffusion, convection, ionic migration) and kinetics (enzymatic reactions) is the basis of the description of the formation and development of dissipative structures on the molecular level. Hence, here we are generally dealing with reaction-diffusion problems in heterogeneous systems. The individual systems are organized in a hierarchy, and we usually assume that the dissipative structure on a lower level of the hierarchy is in a quasi-stationary state when higher levels are studied. Thus, although some of the parameters vary slowly and we observe different stationary states which subsequently appear, the development of biological systems over time is one of the basic problems in the description of the functioning of living systems. This is why the techniques of dissipative structures described in this text can be particularly useful in the future. Vast literature on the subject exists. Here we shall discuss several examples where mathematical techniques have already been applied to the study of dissipative structures.

The electrophysics of a nerve fiber
In 1952, Hodgkin and Huxley [1.38] introduced a phenomenological expression for the ion current density, J_i, flowing through a squid axon membrane in the following form:

$$J_i = -\left(C\frac{\partial V}{\partial t} + D\frac{\partial^2 V}{\partial x^2}\right) = G_K n^4(v_{12} - V_K) + G_{Na} m^3 h(v_{12} - V_{Na})$$

$$+ G_L(v_{12} - V_L),$$

$$\frac{dn}{dt} = -\frac{n - n_0}{\tau_n}, \quad \frac{dm}{dt} = -\frac{m - m_0}{\tau_m}, \quad \frac{dh}{dt} = -\frac{h - h_0}{\tau_h}. \tag{1.31}$$

Here G_K and G_{Na} are the maximum potassium and sodium conductances per unit area, respectively, and G_L is the constant leakage conductance. The phenomenological variables n, m, and h lie between zero and unity, n_0, m_0, and h_0 are their equilibrium values and τ_n, τ_m, and τ_h are their characteristic times. The spatial and time changes of the potential V (C and D are constants) can have traveling-wave characteristics, i.e., periodic solutions exist. An extensive review of different approaches to the derivation of the nonlinear diffusion equation which governs propagation of the action potential has been given by Scott [1.39]. The empirical Hodgkin–Huxley theory for the squid axon has accounted for a large number of experimental facts in a quantitative way. Since the equations are complicated and nonlinear, mostly numerical results have been obtained. However, no full solution diagram (continuous dependence of solutions on parameters) has been given so far.

Fitzhugh [1.40] has constructed a modified relaxation oscillator as a conceptual model for nerve-membrane excitability:

$$\frac{\partial v}{\partial t} = \frac{\partial^2 v}{\partial x^2} - f(v) - w,$$

$$\frac{\partial w}{\partial t} = b(v - dw). \tag{1.32}$$

Here t is dimensionless time, x is dimensionless distance along the "nerve", v is analogous to membrane potential, w is the recovery variable, and b and d are positive parameters. Usually, the cubic function

$$f(v) = v(1 - v)(a - v), \qquad 0 < a < \tfrac{1}{2}, \tag{1.33}$$

is chosen for nonlinearity. Nagumo et al. [1.41] has also studied the cubic case with $a = 0$. The Fitzhugh–Nagumo nerve model equations (1.32) and (1.33), together with the original Hodgkin–Huxley equations (1.31), have been the subject of much interest in the literature on theoretical biology and applied mathematics, cf., [1.42–1.47]. Both numerical techniques and approximate analytical methods have been used to determine the regions of existence (in parameter space) and the stability of periodic (traveling-wave) solutions.

Far more complicated dynamics can be observed in the studies of nonlinear integro-differential equations describing the interactions of neurons, e.g., in the cerebral cortex [1.48].

A model of the immune system defense against cancer

Lefever and Garay [1.49] have constructed a model of the cancer-onset mechanism based on the following assumptions:

1. Some normal cells are continuously transformed into cancerous ones (concentration X) at a very small rate (A).
2. The transformed cells have a higher proliferation rate and the exponential growth rate of their population is characterized by the constant λ.

3. The cytotoxic cells (not bound, M_0; bound, M_1) limit the size of the population of transformed cells by recognizing and destroying them.
   A simplified kinetic scheme can be written in the form

$$\text{normal cells} \xrightarrow{A} X$$

$$X \xrightarrow{\lambda} 2X$$

$$X + M_0 \xrightarrow{k_1} M_1 \xrightarrow{k_2} M_0 + P. \tag{1.34}$$

Here P stands for the degradation products of X, and kinetic constants, k_1 and k_2, characterize the binding and destruction of the malignant cells by the cytotoxic cells.

4. In the absence of transformed cells, it is assumed that the local fluxes of cytotoxic cells are Fickian with a diffusion coefficient μ. In the presence of a sufficient number of transformed cells, lymphokines, which slow down the motion of M_0 and M_1, are released. The time and space evolutions of the cellular populations (X, M_0, M_1) are described by the following set of differential equations:

$$\frac{\partial X}{\partial t} = (N - X)(A + \lambda X) - k_1 M_0 X,$$

$$\frac{\partial M_0}{\partial t} = -k_1 M_0 X + k_2 M_1 + \bar{\mu} \frac{\partial^2}{\partial r^2} \left(\frac{M_0}{1 + KX} \right),$$

$$\frac{\partial M_1}{\partial t} = k_1 M_0 X - k_2 M_1 + \bar{\mu} \frac{\partial^2}{\partial r^2} \left(\frac{M_1}{1 + KX} \right). \tag{1.35}$$

Here N is the total number of normal and malignant cells, and K describes the magnitude of the effect of lymphokines on the diffusion slowdown. Multiple stationary solutions arise due to the diffusion effect in Eq. (1.35).

Bifurcation behavior of periodic solutions of a set of ordinary differential equations describing immunity of an organism from infection has also been studied by Pimbley [1.50].

Population biology
Most of the problems in dissipative structures have also been studied by using mathematical models of interacting populations, cf., [1.51–1.54].
  A classical problem in mathematical ecology is the problem of two interacting populations. In the prey–predator system, a general spatially homogeneous model is described by [1.55]

$$\frac{dN}{dt} = h(N, P)N,$$

$$\frac{dP}{dt} = k(N, P)P, \tag{1.36}$$

where N and P represent the population densities of a prey species and of its predator, respectively. Different forms of the functions h, k have been studied [1.51, 1.52]. For the well-known Lotka–Volterra system,

$$h(N, P) = a - \alpha P,$$

$$k(N, P) = -b + \beta N, \tag{1.37}$$

Another form of functions h, k is

$$h(N, P) = r\left(1 - \frac{N}{K}\right) - \frac{RP}{N + D}, \tag{1.38}$$

$$k(N, P) = -s\left(1 - \frac{\gamma P}{N}\right).$$

The parameters a, b, α, β, r, s, K, and D in relations (1.37) and (1.38) determine the time course of population densities. Oscillations (i.e., dissipative structures over time) can be observed for certain combinations of parameters.

We can also study the effects of migration on populations. Usually, migration is described in terms of Fickian diffusion, and we then obtain the following balance equations:

$$\frac{\partial N}{\partial t} = D_N \frac{\partial^2 N}{\partial x^2} + h(N, P)N, \tag{1.39}$$

$$\frac{\partial P}{\partial t} = D_P \frac{\partial^2 P}{\partial x^2} + k(N, P)P. \tag{1.40}$$

Both uniform spatial patterns and nonhomogeneous stationary patterns (called "patchiness" in ecology) can be observed for properly chosen functions h and k. Periodic solutions (traveling waves of population densities) can also be observed.

Generally, both N and P can be vectors and the description of a certain tropical level can be part of a both vertically and horizontally organized food web. Thus, we are again dealing with flows in organized hierarchical structures.

Glycolytic oscillations in a nonhomogeneous system
There is extensive literature on the subject of biochemical oscillations. It includes oscillations in solution, on carriers (membranes, immobilized enzymes), in cell extracts, cell populations, tissues, etc. [1.56–1.58]. Let us discuss one specific example: glycolytic oscillations under nonhomogeneous conditions.

A simplified model system of equations was used by Hess [1.59]. Let us consider a two-dimensional medium where concentrations of substrate (α) and product (γ) vary due to reaction and transport (described by a Fickian diffusion relation with diffusion coefficients D_α and D_γ). The time development of spatial concentrations in two dimensions (spatial coordinates x and y) can be described by the following two coupled partial differential equations:

$$\frac{\partial \alpha}{\partial t} = D_\alpha \left(\frac{\partial^2 \alpha}{\partial x^2} + \frac{\partial^2 \alpha}{\partial y^2} \right) - v + A\sigma_1, \tag{1.41}$$

$$\frac{\partial \gamma}{\partial t} = D_\gamma \left(\frac{\partial^2 \gamma}{\partial x^2} + \frac{\partial^2 \gamma}{\partial y^2} \right) + v - A\sigma_2 \gamma, \tag{1.42}$$

$$v = v_m \frac{(1 + \alpha/E)(1 + \gamma)^2}{L(1 + c\alpha)^2 + (1 + \gamma)^2 + (1 + \alpha/E)^2}. \tag{1.43}$$

Here σ_1 and σ_2 are rates of source and sink, respectively, and A, v_m, E, c, and L are positive constants.

For different boundary conditions (both defined flux and zero flux at the boundaries), numerical computations have yielded traveling waves, standing waves, and temporal oscillations of concentrations of α and γ over the entire area. The results are in good agreement with experimental observations in yeast extracts as reported by Boiteaux and Hess [1.60].

Instabilities in physiological control systems
A large number of human diseases are characterized by changes in the qualitative dynamics of physiological control systems. Systems can oscillate, stop oscillating, or change patterns of oscillation. Glass and Mackey [1.61] discuss several models of a so-called dynamic disease. They start from the differential equation

$$\frac{dx}{dt} = \lambda - \gamma x, \tag{1.44}$$

where x is the variable of interest (e.g., cell-type concentration). λ is the production rate for x, and γ is the destruction rate of x. For example, mathematical models of respiratory disorders are used in the following forms:

$$\frac{dx}{dt} = - \frac{\alpha V_{max} x_\tau^n x}{\Theta^n + x_\tau^n}. \tag{1.45}$$

Here the ventilation V at time t depends on the carbon dioxide concentration $x_\tau = x(t - \tau)$ at time $(t - \tau)$; V_{max} is the maximum ventilation and n, Θ, and α are positive experimental parameters. The time-delay differential equation (1.45) can describe various patterns of oscillations.

Two models of haematopoiesis are given in the form

$$\frac{dx}{dt} = \frac{\lambda_0 \Theta^n}{\Theta^n + x_\tau^n} - \gamma x, \tag{1.46}$$

$$\frac{dx}{dt} = \frac{\lambda_1 \Theta^n x_\tau}{\Theta^n + x_\tau^n} - \gamma x. \tag{1.47}$$

Here $x(t)$ denotes the concentration of circulation cells, $x_\tau = x(t - \tau)$, and λ_0, λ_1, Θ, and n are experimental parameters. Both periodic and chaotic oscillations are observed for certain values of parameters.

1.2.7 Reaction-diffusion problems

The existence of dissipative structures (multiple steady states, oscillations) in reaction-diffusion systems (including heat transport) and in common types of chemical reactors has been known for more than a hundred years. Both experimental and mathematical results are still being reported at a high rate, and a good historical survey of these developments can be found in recent review articles [1.62–1.67].

Throughout this text we shall use a number of examples to illustrate different computational procedures. Here we shall indicate the best-known problems and the most important results. Some of the problems will be reformulated as examples which will serve to illustrate computational techniques discussed in the text.

The simplest type of chemical reactor which exhibits both multiple steady states and oscillations is the continuous stirred tank reactor. For a single first-order irreversible reaction, the mass and enthalpy balances can be written in the form

$$\frac{dx_1}{dt} = -x_1 + \mathrm{Da}(1 - x_1) \exp\left(\frac{x_2}{1 + x_2/\gamma}\right) = f_1(x_1, x_2), \tag{1.48}$$

$$\mathrm{Lw}\frac{dx_2}{dt} = -x_2 + B\,\mathrm{Da}(1 - x_1) \exp\left(\frac{x_2}{1 + x_2/\gamma}\right) - \beta(x_2 - x_{2c})$$

$$= f_2(x_1, x_2), \tag{1.49}$$

where x_1 denotes the conversion of reactant and x_2 is the dimensionless temperature [1.68]. The system parameters are as follows: B—dimensionless heat of reaction; γ—dimensionless activation energy; β—dimensionless heat-transfer coefficient; x_{2c}—dimensionless coolant temperature; Lw—Lewis number; and Da—dimensionless ratio of reactor residence time to reaction time. This system of equations has been studied for more than 20 years. Let us define $\bar{x}_1, \bar{x}_2$ as the values of the variables in steady state ($f_1(\bar{x}_1, \bar{x}_2) = f_2(\bar{x}_1, \bar{x}_2) = 0$) and $J = \{\partial f_i/\partial x_j\}_{\bar{x}_1, \bar{x}_2}$. The branch points $\bar{x}_1 = m_1(\mathrm{Da}_1)$, $\bar{x}_1 = m_2(\mathrm{Da}_2)$ correspond to points where the determinant of the Jacobian matrix J vanishes. It is possible to show that $\det J > 0$ (and, hence, only unique solutions exist) unless the condition

$$B > \frac{4[1 + \beta + (\beta x_{2c}/\gamma)]^2}{1 + \beta - (4/\gamma)[1 + \beta + \beta x_{2c}/\gamma]} \tag{1.50}$$

is satisfied. The local stability, as well as the existence of bifurcating periodic solutions, is determined by the roots of the characteristic equation of J:

$$\lambda^2 - (\text{tr } J)\lambda + \det J = 0. \tag{1.51}$$

Stability requires that $\det J > 0$, $\text{tr } J < 0$, while the bifurcation of periodic orbits occurs when $\text{tr } J = 0$, $\det J > 0$. The lines $\text{tr } J = 0$, condition (1.50), $\det J = 0$, and $\text{tr } J = \det J = 0$ determine such regions in parameter space which have qualitatively different dynamic behavior due to the placement of bifurcation points. Perturbation analysis at these bifurcation points can be used to determine the directions of bifurcation of the limit cycles (and, hence, their stability). Numerical computations are then used to trace the periodic orbits. Isolated branches of solutions (isolas) are of particular interest cf., [1.69–1.71].

Singularity theory has recently been applied to the problem of determination of possible sequences of static bifurcation problems [1.72].

Theoretical and experimental studies of systems involving more than one phase, more than one reaction, reactors in series, i.e., the cases where the number of dependent variables increases, have revealed more complex behavior, including a large number of steady states. For example, as many as 25 different steady states were encountered in a system of equations describing multiphase polymerization [1.73]. A review of theoretical and experimental work in this field has been given by Schmitz [1.74]. Multiple steady states can also be observed if DNA polymerization is followed in a continuous stirred tank reactor [1.75] experimentally. Hence, multiple steady states can be significant in the formation of the genetic code.

Heat and mass transport in a catalyst particle
Mass- and heat-transport effects in porous catalyst particles where a first-order irreversible reaction occurs can be described by two coupled parabolic partial differential equations:

$$\frac{\partial C_s}{\partial t} = \frac{\partial^2 C_s}{\partial x^2} + \frac{a}{x}\frac{\partial C_s}{\partial x} - \phi^2 C_s \exp\left[\gamma\left(1 - \frac{1}{T_s}\right)\right], \tag{1.52}$$

$$\text{Lw}\frac{\partial T_s}{\partial t} = \frac{\partial^2 T_s}{\partial x^2} + \frac{a}{x}\frac{\partial T_s}{\partial x} + \beta'\phi^2 C_s \exp\left[\gamma\left(1 - \frac{1}{T_s}\right)\right], \tag{1.53}$$

where $a = 0, 1, 2$ for slab, cylindrical, and spherical geometry of the particle, respectively.

Boundary conditions can be written in the form

$$x = 0: \frac{\partial C_s}{\partial x} = \frac{\partial T_s}{\partial x} = 0 \quad \text{(for symmetrical profiles only)},$$

$$x = 1: C_s = C_f - \frac{1}{\text{Sh}}\frac{\partial C_s}{\partial x}, \tag{1.54}$$

$$T_s = T_f - \frac{1}{\text{Nu}}\frac{\partial T_s}{\partial x}. \tag{1.55}$$

In the above balance equations, dependent variables C_s and T_s denote concentration in the pores of the particle and temperature of the solid phase, respectively, C_f and T_f are concentration and temperature in the bulk of the flowing phase, respectively, ϕ is the Thiele modulus, γ is dimensionless activation energy; Lw is the Lewis number, β' is the dimensionless heat-evolution parameter, Sh is the Sherwood number, and Nu is the Nusselt number.

Both multiple steady states and oscillations were predicted by numerical and approximate analytic solutions of the above equations and were also observed experimentally, cf., the comprehensive two-volume work by Aris [1.76]. We shall discuss some of the problems later.

Tubular and fixed-bed reactors
Most mathematical models used in the studies of multiple steady states and stability of tubular or fixed-bed reactors use balance equations of the fluid phase in the form:

$$\frac{\partial C_f}{\partial t} = \frac{1}{\mathrm{Pe}_C} \frac{\partial^2 C_f}{\partial z^2} - \frac{\partial C_f}{\partial z} - R_1, \tag{1.56}$$

$$\mathrm{Lw} \frac{\partial T_f}{\partial t} = \frac{1}{\mathrm{Pe}_T} \frac{\partial^2 T_f}{\partial z^2} - \frac{\partial T_f}{\partial z} + R_2 - \alpha(T_f - T_c). \tag{1.57}$$

The forms of the rate functions R_1 and R_2 depend on whether the model is "homogeneous" or "heterogeneous" (multiple phase). For the irreversible first-order kinetics with Arrhenius-type dependence of the reaction rate on temperature, we have, for homogeneous models,

$$R_1 = \mathrm{Da} \, C_f \exp\left[\gamma\left(1 - \frac{1}{T_f}\right)\right], \tag{1.58}$$

$$R_2 = BR_1. \tag{1.59}$$

For the same kinetics, in heterogeneous models [1.74],

$$R_1 = k_1 \phi^2 \eta C_f \exp\left[\gamma\left(1 - \frac{1}{T_f}\right)\right], \tag{1.60}$$

$$R_2 = k_2 \frac{\mathrm{Nu}}{\mathrm{Lw}} (L - 1)(T_s - T_f). \tag{1.61}$$

The Danckwerts boundary conditions are usually used in the form

$$z = 0 : C_f = 1 + \frac{1}{\mathrm{Pe}_C} \frac{\partial C_f}{\partial z}, \tag{1.62}$$

$$T_f = 1 + \frac{1}{\mathrm{Pe}_T} \frac{\partial T_f}{\partial z}, \tag{1.63}$$

$$z = 1 : \frac{\partial C_f}{\partial z} = \frac{\partial T_f}{\partial z} = 0. \tag{1.64}$$

In the above equations, the dependent variables C_f and T_f denote concentration and temperature of the fluid phase, respectively, z is the axial variable, Pe_C and Pe_T are Péclet numbers for mass and heat transport, respectively, in the axial direction, L is the dimensionless heat-capacity factor, α is the heat-transport coefficient, T_c is the temperature of the coolant, Da is the Damköhler number, γ is the dimensionless activation energy, and B is the dimensionless heat-evolution parameter. The constants k_1 and k_2 describe packed-bed reactor arrangements, ϕ is the Thiele modulus, η is the effectiveness factor of a catalyst particle, T_s is the temperature of the solid catalyst, and Nu and Lw are dimensionless Nusselt and Lewis numbers, respectively. As many as five steady states together with oscillations can be predicted by the model under common operating conditions. Several detailed reviews of the problem, including discussions of more complicated models, may be found, for example, in [1.77–1.81].

Some of the above briefly discussed examples of nonlinear problems from different branches of science and engineering possess certain common features: They are described by a set of coupled nonlinear (partial or ordinary) differential equations. The equations usually contain several physical parameters which may vary continuously in the positive region (the regions of parameter variations very often extend over several orders of magnitude).

With varying parameters, multiple steady-state, oscillatory (periodic), or chaotic (aperiodic) solutions appear. Usually, solutions for isolated values of parameters have been obtained. Numerical methods developed for fast computers, together with advances in the theory of nonlinear dynamic systems (bifurcation theory), now provide us with efficient procedures for computing continuous dependence of solutions on parameters, and with special methods for locating bifurcation points on computed branches of solutions. These methods will be discussed in Chapters 2–4. First, we shall discuss some basic concepts used in the analysis of nonlinear systems. Later, we shall introduce physical examples which will illustrate these techniques in more detail.

1.3 Basic Concepts and Properties of Nonlinear Systems

Behavior of dynamical (evolutionary) systems can usually be described by the differential equation

$$\frac{d\mathbf{X}}{dt} = \mathbf{F}(\mathbf{X}, \mathbf{A}), \tag{1.65}$$

where $\mathbf{X} \in B$, B is a suitably defined Banach space, and $\mathbf{A}$ is a parameter which can also belong to a suitable (e.g., Banach) space. We shall deal mainly with autonomous systems, i.e., systems where the time variable t does not appear in the right-hand side (RHS) of the differential equation (1.65) explicitly.

In applied science it is customary to divide dynamical systems of the type (1.65) into two groups; first, where the space B is of finite dimension, and second, where B is of infinite dimension. When $B = R^n$ (n-dimensional Euclidean space), we speak about lumped-parameter systems (LPS, systems with concentrated parameters); then A also belongs to a finite-dimensional space (even though an example can be constructed where A belongs to an infinite-dimensional space).

When B is an infinite-dimensional space, e.g., a space of functions $C^2_{[a, b]}$, we speak about the dynamical system with distributed parameters (distributed-parameter systems, DPS).

1.3.1 Steady-state solutions

The steady-state solution X of the dynamical system (1.65) is defined by the equation

$$F(X, A) = 0, \tag{1.66}$$

where the value of parameter A is known. Types of dynamical and stationary systems are listed in Table 1.1. The LPS are usually described by systems of ordinary differential equations, while the DPS are described by partial differential equations of the parabolic or hyperbolic type. Examples of various dynamical systems will be given in Section 1.4, where Examples 1–5 correspond to LPS and Examples 6–11 to DPS.

Now we shall briefly discuss some of the concepts used in this text. The terms will be defined heuristically. More precise definitions are contained either in Appendix C or in other parts of the text.

It is often necessary to study the dependence of the steady-state solutions of system (1.65) on the values of the parameter A. Let us assume now that $A \in R^1$ and set

$$A = \alpha \in R^1. \tag{1.67}$$

Table 1.1 Typical examples of dynamical (1.65) and stationary (1.66) systems.

	Dynamical System (1.65)	Stationary State (1.66)
LPS	System of first-order ordinary differential equations	System of nonlinear algebraic equations
DPS	System of parabolic partial differential equations	Nonlinear boundary value problem for ordinary differential equations or partial differential equations of an elliptic type

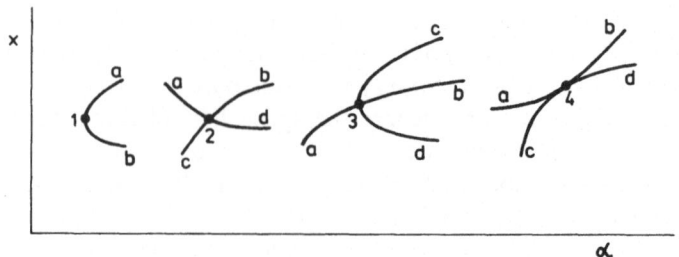

Figure 1.1 Types of branching point. 1—limit point; 2—bifurcation point; 3—bifurcation-limit point; 4—cusp point; a, b, c, d—branches of solutions.

We will define a branch of solutions as a continuous and uniquely dependent $X(\alpha)$. By uniqueness we mean that for every fixed $\alpha \in (\alpha_0, \alpha_1)$ we can determine $\varepsilon > 0$ such that no other solution $X_1(\alpha)$ of Eq. (1.66) satisfying

$$\|X_1(\alpha) - X(\alpha)\| < \varepsilon$$

exists. The branch of solutions can be continued in both directions until certain limit values of the parameter α, say (α_0, α_1) are reached. The uniqueness assumption no longer holds for these values. Such critical points will be called branch points. In principle, we may encounter two types of branch points, limit points (which are sometimes called turning points) and bifurcation points. Different types of branch points are shown schematically in Fig. 1.1.

To illustrate the above definitions, we shall discuss the one-dimensional case in greater detail, i.e., where $X = x \in R^1$. See Iooss and Joseph [1.82]. In the one-dimensional case, Eq. (1.66) becomes

$$f(x, \alpha) = 0. \tag{1.68}$$

We assume that f has continuous first and second derivatives. Points (x, α) satisfying (1.68) can be classified as follows [1.82]:

1. A regular point (x, α) is one at which the implicit function theorem applies,

$$\frac{\partial f}{\partial x} \neq 0 \quad \text{or} \quad \frac{\partial f}{\partial \alpha} \neq 0. \tag{1.69}$$

If (1.69) holds, we can find a unique curve $x = x(\alpha)$—the branch of solutions— or $\alpha = \alpha(x)$ passing through the point.

2. A regular turning point (limit point) is a point where $\partial \alpha/\partial x$ changes sign (and $\partial f/\partial \alpha \neq 0$). In other words, two branches which are joined at this point have a limiting "tangent" $d\alpha/dx = 0$ while $d\alpha/dx$ has opposite signs on either side of this point, cf., point 1 in Fig. 1.1. Evidently, $\partial f/\partial x = 0$ at this point.

3. A singular point (x, α) is a point where

$$\frac{\partial f}{\partial x} = 0, \quad \frac{\partial f}{\partial \alpha} = 0. \tag{1.70}$$

4. A double point (bifurcation point) of the curve satisfying Eq. (1.68) is a singular point at which two (and only two) curves (1.68) possessing distinct (finite) tangents $dx/d\alpha$ cross, cf., point 2 in Fig. 1.1. Four branches of solutions emanate from this point forming the two pairs, each of which has a common tangent $dx/d\alpha$ (and thus form a smooth curve passing through the double point). Two such branches which have a common tangent at the double point are sometimes considered as one branch (computation of the tangents will be discussed in Section 2.4).

5. A singular turning (double) point (bifurcation-limit point) of the curve (1.68) is a double point at which two of the four existing branches have a limiting "tangent" $d\alpha/dx = 0$ and different signs of $d\alpha/dx$ in the neighborhood of the point, cf., point 3 in Fig. 1.1.

6. A cusp point of the curve (1.68) is a point of second-order contact between two curves (1.68). All four branches have the same limiting tangent at the cusp point, cf., point 4 in Fig. 1.1.

7. A higher-order singular point of the curve (1.68) is a singular point at which all three second derivatives of $f(x, \alpha)$ are null.

Later, we shall use the term "limit point" for limit (turning) point (in agreement with the terminology suggested by Keller et al. [1.83], "bifurcation point" for double (bifurcation) point and "bifurcation-limit point" for a singular turning (double) point, because we believe that these terms better describe what happens at these points.

As can be seen in Fig. 1.1, the total number of solutions varies with the change of the parameter α. Hence, the limit points play an important role in determining the number of solutions to a specific problem for a given value of α. If more than one solution exists, we say that for a given value of α, multiple solutions exist (multiplicity of solutions). An overall picture of dependence of stationary solutions X on the parameter α will be called the solution diagram. Figure 1.1 corresponds to sections of the solution diagram. A chosen characteristic value of a solution (value of one solution component, a conveniently defined norm of a solution, the value of a solution in a defined spatial location in the case of DPS, etc.) is usually plotted in the solution diagram. If branches of such a solution diagram cross each other, it does not necessarily mean that the crossing point corresponds to a branch point. A crossing point may occur for some choices of plotted functions but not for others (e.g., for other components of the solution). Our task is to obtain a complete solution diagram, i.e., a solution diagram containing all branches of solutions with all branch points. A numerical technique enabling us to obtain one branch of solutions (or more branches of solutions mutually connected at branch points) is called the continuation technique. In principle, this technique enables us to start from a known solution and continuously compute solutions along a chosen solution branch. Ideally, the continuation technique would enable us to compute a total solution diagram without man-machine communication. However, this ideal is rarely realized.

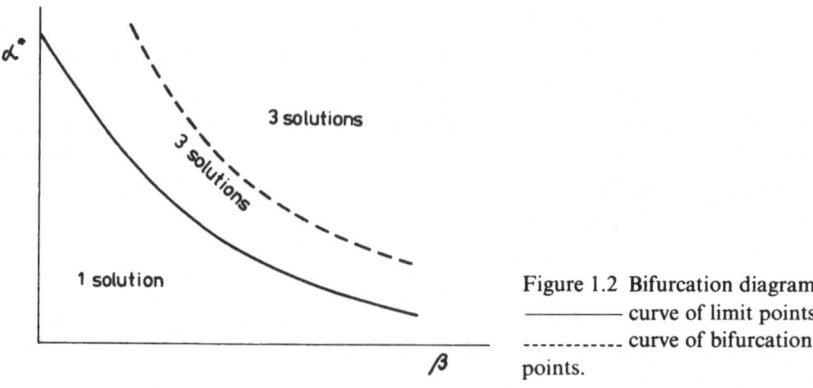

Figure 1.2 Bifurcation diagram.
————— curve of limit points,
------------ curve of bifurcation
points.

Let us consider the case where $A \in R^2$, i.e., $A = (\alpha, \beta)$, $\alpha, \beta \in R^1$. We assume that we are able to locate limit and bifurcation points (X^*, α^*) of a given problem for each fixed β. Then the dependences (branches) $\alpha^*(\beta)$ can be plotted in the parametric plane "$\alpha - \beta$," and we obtain the "bifurcation diagram." In some articles, the solution diagram described above is called a "bifurcation diagram." Here we use the term "bifurcation diagram," to refer to the plot of the dependence on another parameter of α values at bifurcation points. If we cross the curves of bifurcation-limit points in the bifurcation diagram, the number of solutions changes by two (for example, from three solutions to one solution or from one solution to three solutions when the direction of variation of the parameter β is reversed), c.f., Fig. 1.2. Thus, the curves of limit points divide the parametric plane "$\alpha - \beta$" into subregions with a constant number of solutions. The curves of bifurcation points are not useful for determining the number of solutions. They can be used to determine changes in stability on a given branch of solutions, which will be discussed later on. The bifurcation diagram is shown schematically in Fig. 1.2.

Particular attention must be paid to formation of new branches in the solution diagram "$X - \alpha$." One case is illustrated in Fig. 1.3, where the

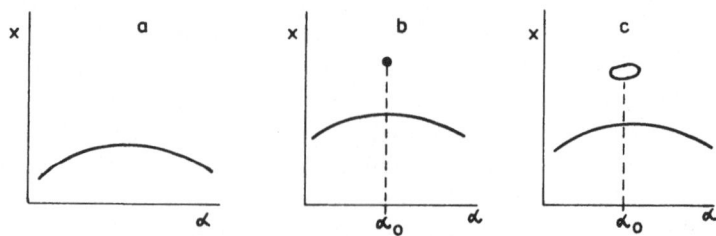

Figure 1.3 Birth of isolas at isola center in the solution diagram. (a) $\beta = \beta_0 - \varepsilon$; (b) $\beta = \beta_0$; (c) $\beta = \beta_0 + \varepsilon$.

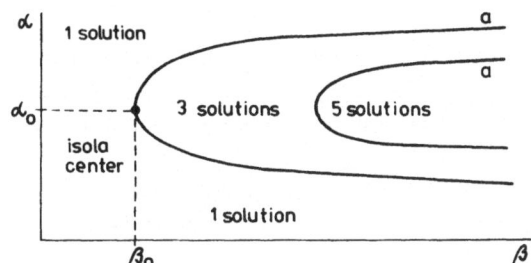

Figure 1.4 Birth of isola at isola center in bifurcation diagram a – curve of limit points.

appearance scheme of isola at the so-called isola center is shown. The point of formation of isola center corresponds to a limit point in the bifurcation diagram, cf., Fig. 1.4.

We shall discuss methods of direct location of isola centers in Section 2.4.3.

1.3.2 Stability of solutions

Let us now return to the dynamical system (1.65). We shall consider a fixed value of α. The time course of $\mathbf{X}$ (trajectory), $\mathbf{X}(t)$, will depend on an initial state of the system, i.e., on the initial condition $\mathbf{X}(0) = \mathbf{X}^0$ at $t = 0$. Let us discuss the asymptotic behavior of a solution, i.e., the behavior of the trajectory for $t \to \infty$. In this case, the solution $\mathbf{X}(t)$ will approach an attractor. We shall consider four types of attractors: an attracting steady-state solution, an attracting limit cycle, quasiperiodic attractor and a "chaotic attractor." The steady-state solution $\overline{\mathbf{X}}$ is attracting when it is locally stable, i.e., if $\varepsilon > 0$ exists such that

$$\|\mathbf{X}(t) - \overline{\mathbf{X}}\| \to 0, \quad t \to \infty, \tag{1.71}$$

whenever $\|\mathbf{X}(0) - \overline{\mathbf{X}}\| < \varepsilon$. Linear stability analysis can be often used to determine whether steady-state solutions are attracting. If the system (1.65) is in the form of a set of nonlinear ordinary differential equations

$$\frac{dx_i}{dt} = f_i(x_1, x_2, \ldots, x_n, \alpha), \quad i = 1, 2, \ldots, n \tag{1.72}$$

and $\overline{\mathbf{X}}$ is a steady-state solution, $\overline{\mathbf{X}}$ is linearly attracting if all eigenvalues of the Jacobian matrix $\mathbf{J}$,

$$\mathbf{J} = \left\{ \frac{\partial f_i(\overline{\mathbf{X}})}{\partial x_j} \right\}, \tag{1.73}$$

have strictly negative real parts. If any eigenvalue of $\mathbf{J}$ lies to the right of the imaginary axis, $\overline{\mathbf{X}}$ cannot be attracting. If some of the eigenvalues of $\mathbf{J}$ lie on the imaginary axis, and if the remaining ones lie in the left half-plane, $\overline{\mathbf{X}}$ may or may not be attracting. We need not compute all eigenvalues—it usually suffices to determine coefficients of the characteristic polynomial. Methods of stability analysis in DPS, i.e., the methods of analysis of the corresponding eigenvalue problem, are more complicated—in principle, an infinite number of eigenvalues may exist. The stability of a particular steady-state solution is usually determined on the basis of results of original dynamical problem (1.65) simulation.

Linearly attracting steady-state solutions cannot disappear by making small changes in the dynamical system (1.65)—the system is structurally stable.

Bifurcation theory is used, for example, to analyze what happens when a part of the spectrum of the Jacobian matrix J of (1.72) moves into the right half-plane (the steady-state solution is no longer attracting). We shall now discuss the relationship between the loss of steady-state solution stability and the branch points in the case of $x \in R^1$. The Jacobian matrix J contains one element only; therefore, it has only one eigenvalue. Thus, a change of stability may occur only in the case where, with the change of parameter α, this eigenvalue passes through the origin of the complex plane, i.e., it equals zero. The element $\partial f / \partial x$ of the Jacobian matrix is therefore equal to zero as well. Thus, this condition is the same as conditions for limit, bifurcation, and bifurcation-limit points. The stability of steady-state solutions may change at these points only. All possibilities of change of stability at these branch points are sketched in Fig. 1.5. To complete the picture, the cases which can occur only in higher-dimensional situations are also shown in Fig. 1.5 (cf., cases m, n, o, p, and q). The conclusions about bifurcations and changes of stability of bifurcating

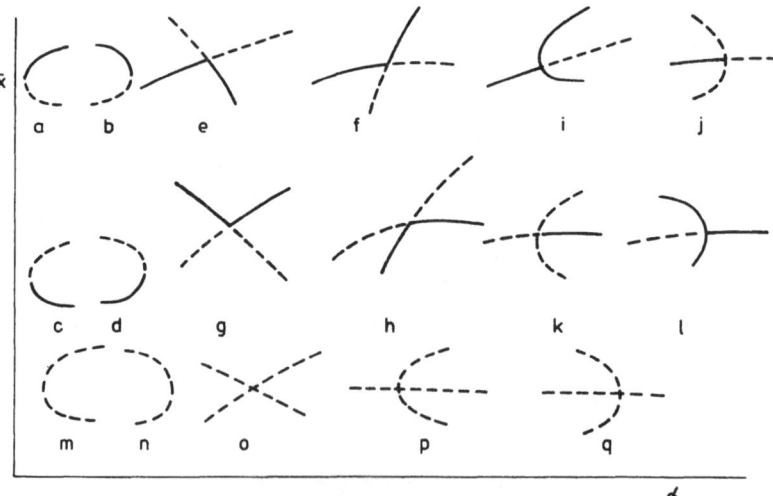

Figure 1.5 Stability of solutions in the neighborhood of branch points, one-dimensional case. ——stable solution; - - - -unstable solution; a, b, c, d: limit (reg. turning) points; e, f, g, h: bifurcation (double) points; i, j, k, l: bifurcation-limit (singular turning) points; m, n, o, p, q: additional possible cases when the dimension of **x** is greater than one.

solutions discussed above strictly apply only to situations where bifurcation occurs from a simple real eigenvalue of **J**. We shall call this situation a real bifurcation.

Another traditional terminology exists. It arose during studies of bifurcation points on a known branch of stable solutions (usually trivial steady-state solutions were considered). With increasing value of the parameter α, the real eigenvalue crosses from the left complex half-plane to the one on the right. The originally stable steady-state solution becomes unstable. Branching of new solutions may then occur, in principle, in three ways schematically shown in solution diagrams in Fig. 1.6. These three situations are called supercritical, subcritical, and transcritical bifurcation. We can easily find analogous situations in Fig. 1.5. Transcritical bifurcation always occurs at the bifurcation point, supercritical and subcritical bifurcations appear at the bifurcation-limit point.

Let us discuss higher-dimensional situations, i.e., the situations where **X** is a vector. Now we have to consider not only the passage of the real eigenvalue through zero, but also the loss of stability when a pair of complex-conjugate eigenvalues crosses the imaginary axis. If a pair of complex-conjugate (nonreal) eigenvalues crosses the imaginary axis, a normal Hopf bifurcation can occur. This type of bifurcation we shall also call complex bifurcation.

In complex bifurcation, a limit cycle appears, i.e., the solution forms a closed curve in phase space $(x_1, x_2, \ldots, x_n)$. If the limit cycle is stable, it represents a second type of attractor—an asymptotically orbitally stable periodic solution. The points of real and complex bifurcation are often plotted to show their dependence on another parameter β in a complete bifurcation diagram. When norms of steady-state and periodic solutions plotted in a

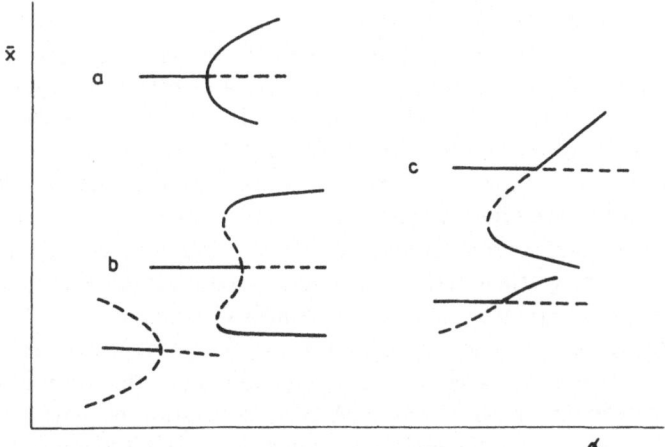

Figure 1.6 Stability of steady-state solutions arising through supercritical (a), subcritical (b), and transcritical (c) bifurcation; —— stable, ----- unstable.

solution diagram are chosen conveniently, we can plot both solutions together. In this case, also, we can obtain supercritical bifurcation or subcritical bifurcation from the considered steady-state-solutions branch of a solution diagram. Transcritical bifurcation is not possible at a complex bifurcation point.

A periodic solution $X(t)$ of system (1.65) with period T satisfies

$$X(t + T) = X(t). \tag{1.74}$$

In the space B, this solution forms a closed orbit (in the case of LPS, it is a curve, i.e., limit cycle) consisting of the points $X(t)$, $t \in [0, T)$. A periodic solution $X(t)$ is asymptotically orbitally stable if, with increasing time, all trajectories starting in its neighborhood end up on the trajectory (see Appendix C for more exact formulation of asymptotic orbital stability).

As the eigenvalues of the Jacobian matrix J of (1.72) determine stability of a steady-state solution, Floquet multipliers can be used to determine stability of a periodic solution. Floquet multipliers are eigenvalues of a certain matrix (monodromy matrix) which is obtained by integrating a linearized system (i.e., variational system) along an orbit of the periodic solution. If all Floquet multipliers are inside the unit circle, the studied periodic solution is locally orbitally stable.

As in the case of steady-state solutions, we can construct and study the dependence of periodic solutions on the parameter α and plot branches of periodic solutions in the solution diagram. The amplitude or period of the solution can be chosen, for example, as the characteristic to be plotted.

When a simple periodic solution (limit cycle) ceases to be attracting, one of the possibilities is that an attracting two-dimensional torus bifurcates. Trajectories move on a two-dimensional torus. Such a quasi-periodic solution will have two characteristic frequencies. In a similar manner, an m-dimensional torus with m characteristic frequencies may arise after $m - 1$ bifurcations. A similar mechanism was suggested by Landau in his theoretical picture of transition to turbulence in hydrodynamics (with $m \to \infty$). In addition to attracting steady-state solutions, limit cycles, and attracting quasi-periodic solutions, we also meet chaotic attractors. In autonomous systems, they appear when the dimension of the system is not less than 3. They are called chaotic attractors because they are not simple or smooth surfaces like spheres or tori, but rather they have a complicated infinitely many-layered structure. The trajectories are often denoted as chaotic (sometimes turbulent). The Poincaré mapping (return mapping) is often used for analyzing the geometry of trajectories in the phase space. (cf., Appendix C for a more exact definition). A point p in a given cross section is mapped into a point q, where the solution curve through p next intersects the cross section. Thus the number of dimensions of the problem is reduced by 1. A chaotic solution is characterized by a sensitive dependence of the solution trajectory on initial conditions. Two trajectories, starting close together on a chaotic attractor, separate exponentially. This exponential rate of separation can be expressed in terms of charac-

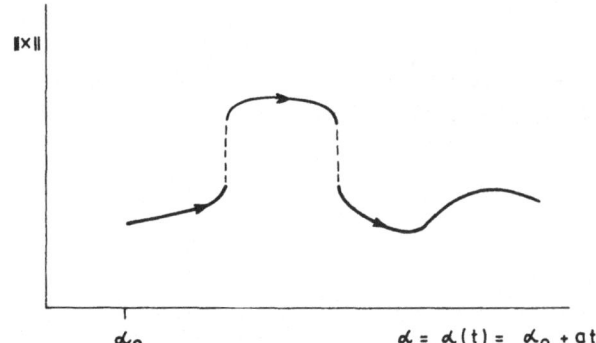

Figure 1.7 Example of an
evolution diagram; linear
variation of α over time.

teristic exponents λ (Liapunov exponents). A positive characteristic exponent confirms the existence of a chaotic attractor. Thus, chaotic attractors introduce probabilistic considerations into the study of deterministic systems (1.65). Time averages are evaluated from trajectories and are analyzed in the form of power spectra.

1.3.3 Evolution systems

A dynamical system (1.65) in which the parameters vary in time is one type of system of evolution. We are interested in the sequence of states (steady states, time periodic states, chaotic states) which occur when one or several parameters vary. Such variation may be deterministic (slow, fast, etc.) or stochastic. If the course of proper behavior norms of the system states is followed continuously, we obtain an evolution diagram. Here a proper norm of solution is plotted as a function of the parameter $\alpha(t)$ which varies in time. Let us note that only stable solutions are followed in the evolution diagram. An example of an evolution diagram is shown schematically in Fig. 1.7. In a neighborhood of a branch point, a rather abrupt change of state may occur.

We have described bifurcations on LPS. Similar phenomena are obtained for DPS, when bifurcation occurs from an isolated unique eigenvalue which crosses the imaginary axis and when, by application of standard functional analysis methods, the DPS can be reduced to a problem of finite dimensions, cf., e.g., Henry [1.84].

1.4 Examples

In this section, we shall provide examples of physical, chemical, and biological systems which will be used in Chapters 2–4 to illustrate described numerical methods. Problems involving nonlinear ordinary differential equations (LPS) will be presented first.

EXAMPLE 1: Dynamical behavior of a continuous stirred tank reactor with a product recycle (first-order exothermic chemical reaction occurs in the reactor) can be described by the system of two coupled nonlinear ordinary differential equations [1.68, 1.71, 1.85, 1.86]:

$$\frac{dy}{dt} = -\Lambda y + \text{Da}(1 - y) \exp \frac{\theta}{1 + \theta/\gamma}, \tag{E1.1}$$

$$\frac{d\theta}{dt} = -\Lambda\theta + \text{Da}B(1 - y) \exp \frac{\theta}{1 + \theta/\gamma} - \beta(\theta - \theta_c). \tag{E1.2}$$

Examples of values for these parameters are $\gamma \in (10, \infty)$, $\Lambda \in (0, 1]$, $\text{Da} \in [0, 0.5]$, $B \in [0, 60]$, $\beta \in [0, 3]$, $\theta_c \in [-5, 2]$. The variable $y \in [0, 1]$ describes reactant conversion; the variable $\theta \in [0, B]$ corresponds to dimensionless temperature. A similar problem with two coupled consecutive reactions, using three coupled nonlinear ordinary differential equations, has also been studied [1.87]. In engineering literature, particularly that concerned with chemical reactors, this is a standard problem for studying the multiplicity and stability of steady-state solutions, existence of periodic solutions, etc.

EXAMPLE 2: A simplified kinetic model of the set of chemical reactions occurring during oxidation of malonic acid by bromate with a ceric/cerous ion catalyst was proposed by Noyes et al. [1.88, 1.89]. The rate of change of concentrations of certain reaction components can be described by the following three coupled ordinary differential equations:

$$\varepsilon \frac{dx}{dt} = \mu y - xy + x(1 - x) - \varepsilon k_0 x,$$

$$\varepsilon' \frac{dy}{dt} = -\mu y - xy + fz + \varepsilon' k_0(y^0 - y),$$

$$\frac{dz}{dt} = x - z - k_0 z. \tag{E2.1}$$

Here x, y, and z are dimensionless concentrations (positive), and μ, ε, k_0, ε', f, and y_0 are positive parameters.

This set of equations is one of the simplest models of the Belousov–Zhabotinsky reaction which is generally used to study dissipative structures in chemical systems. Multiple steady states, periodic, multiple periodic, and chaotic solutions were observed both in lumped- and distributed-parameter systems using this reaction.

EXAMPLE 3: Dynamical behavior of a cascade of two continuous stirred tank reactors with recycle (first-order exothermic chemical reaction occurs in the reactors) can be described by the system of four ordinary differential equations [1.90]:

$$\frac{dy_1}{dt} = (1 - \Lambda)y_2 - y_1 + \mathrm{Da}_1(1 - y_1)\exp\frac{\theta_1}{1 + \theta_1/\gamma}, \tag{E3.1}$$

$$\frac{d\theta_1}{dt} = (1 - \Lambda)\theta_2 - \theta_1 + \mathrm{Da}_1 B(1 - y_1)\exp\frac{\theta_1}{1 + \theta_1/\gamma} - \beta_1(\theta_1 - \theta_{c1}),$$

$$\tag{E3.2}$$

$$\frac{\mathrm{Da}_2}{\mathrm{Da}_1}\frac{dy_2}{dt} = y_1 - y_2 + \mathrm{Da}_2(1 - y_2)\exp\frac{\theta_2}{1 + \theta_2/\gamma}, \tag{E3.3}$$

$$\frac{\mathrm{Da}_2}{\mathrm{Da}_1}\frac{d\theta_2}{dt} = \theta_1 - \theta_2 + \mathrm{Da}_2 B(1 - y_2)\exp\frac{\theta_2}{1 + \theta_2/\gamma} - \beta_2(\theta_2 - \theta_{c2}).$$

$$\tag{E3.4}$$

Here y is reactant conversion, θ is dimensionless temperature, and $\Lambda \in (0, 1]$, $\gamma \in [10, \infty)$, $B \in [0, 60]$, $\mathrm{Da} \in [0, 0.5]$, $\theta_c \in [-5, 2]$, and $\beta \in [0, 3]$ are physical parameters. Indices 1 and 2 correspond to the first and second reactor, respectively (cf., Fig. 1.8).

EXAMPLE 4: The dynamics of concentrations in a network of n mutually connected well-stirred reaction cells, cf., Fig. 1.9, can be studied by solving the set of nm ordinary differential equations

$$\frac{dc_i^k}{dt} = R_i(c_1^k, \ldots, c_m^k) + H_k(c_i^0 - c_i^k) + \sum_{j=1}^{n} D_i^{jk}(c_i^j - c_i^k),$$

$$k = 1, \ldots, n; \quad i = 1, \ldots, m \tag{E4.1}$$

Here the index i denotes the reaction components, k corresponds to the particular reaction cell (hence, c_i^k denotes the concentration of the ith reaction component in the kth cell), and $R_i(c_1^k, \ldots, c_m^k)$ is the reaction source term. H_k are transport coefficients indicating the rate of mass exchange between the kth cell and the environment (where the concentration of the ith component

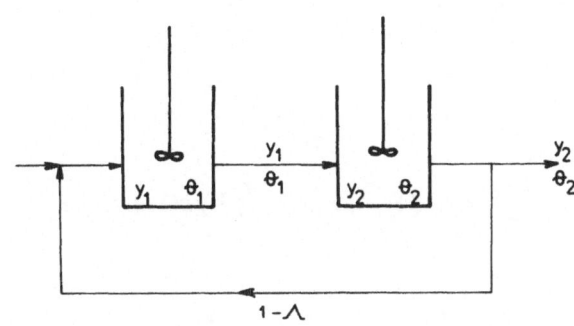

Figure 1.8 The cascade of two continuous stirred tank reactors with recycle.

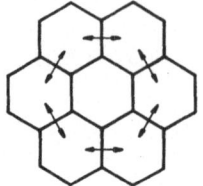 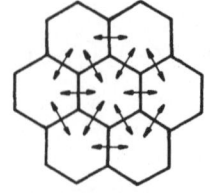

Figure 1.9 Several characteristic configurations of mutually connected reaction cells.

is c_i^0), and D_i^{jk} is the transport coefficient for the mutual mass exchange of the ith component between the jth and kth cell (e.g., we can assume that these coefficients are constant and symmetric, $D_i^{jk} = D_i^{kj}$). The above model is used in the description of morphogen pattern formation in cell assembleges (Turing structures).

A simple example is the model proposed by Prigogine et al. [1.91], called a Brusselator (see Example 6). Here $m = 2$ and

$$R_1(c_1, c_2) = A - (B + 1)c_1 + c_1^2 c_2, \tag{E4.2}$$

$$R_2(c_1, c_2) = Bc_1 - c_1^2 c_2. \tag{E4.3}$$

Different configurations of coupled cells, describing both stationary and dynamical behavior, were studied by Schreiber [1.92, 1,93].

EXAMPLE 5: Graef and Andrews [1.94] described dynamical behavior of an anaerobic digester by the following set of six differential equations:

$$\frac{dx_1}{dt} = \left(\mu - \frac{F}{V}\right)x_1 - K_T x_6 + \frac{F}{V}\xi_1, \tag{E5.1}$$

$$\frac{dx_2}{dt} = -\frac{\mu x_1}{S_{XS}} - \frac{F}{V}(x_2 - \xi_2), \tag{E5.2}$$

$$\frac{dx_3}{dt} = \frac{F}{V}(\xi_3 - x_3), \tag{E5.3}$$

$$\frac{dx_4}{dt} = \mu\left(S_{CX} - \frac{1}{S_{XS}}\right)x_1 - \frac{F}{V}(x_4 - \xi_4) + K_L a(KPx_5 - x_4), \tag{E5.4}$$

$$\frac{dx_5}{dt} = -\phi S_{MX}\mu x_1 x_5 - \phi K_L a(1 + x_5)(KPx_5 - x_4), \tag{E5.5}$$

$$\frac{dx_6}{dt} = \frac{F}{V}(\xi_6 - x_6), \tag{E5.6}$$

$$\mu = \frac{\hat{\mu}}{(1 + K_S/C_{HS} + C_{HS}/K_i)}, \quad C_{HS} = \frac{K_1 x_4 x_2}{K_a(x_3 - x_2)}, \tag{E5.7}$$

where

x_1 = concentration of microorganisms,
x_2 = concentration of the substrate,
x_3 = net cation concentration,
x_4 = concentration of the dissolved carbon dioxide,
x_5 = P_c/P (P_c = partial pressure of carbon dioxide in the gas phase),
x_6 = concentration of toxic components,
ξ_i = value of x_i in the inlet stream into the digester,
ξ_1 = 0.001 mol/l,
ξ_2 = 0.09 mol/l,
ξ_3 = 0.1 mol/l,
ξ_4 = 0.009 mol/l,
ξ_6 = 0 mol/l,
V = volume of the reactor (10 l),
F = flow rate (1 l/day),
K_T = toxicity rate constant (2 day^{-1}),
S_{XS} = microorganisms yield (0.02 gmol/gmol),
S_{CX} = carbon dioxide yield (47 gmol/gmol),
S_{MX} = methane yield (47 gmol/gmol),
$\phi = S_P V/V_G$, S_P = conversion factor (25 l/gmol), V_G = volume of the gas phase in the digester (2 l),
P = total pressure in the digester (0.946 × 10^5 Pa),
$K_L a$ = overall mass transport coefficient in the gas phase (30 day^{-1}),
K = Henry law constant (0.432 × 10^{-2} gmol/l Pa),
μ = growth rate function,
$\hat{\mu}$ = maximal specific growth rate (0.4 day^{-1}),
K_S = saturation constant (0.303 × 10^{-4} gmol/l),
K_i = inhibition constant (0.667 × 10^{-3} gmol/l),
C_{HS} = concentration of unionized volatile acids,
K_a = dissociation constant (0.316 × 10^{-4} gmol/l),
K_1 = bicarbonate ionization constant (10^{-6} gmol/l).

The values in parentheses illustrate typical values used in computations.

Anaerobic digestion is a natural process of decomposition of organic waste. However, it is also being used increasingly for producing methane from agricultural, industrial, and community organic waste materials. Multiple steady states and oscillations which are predicted by the model of Example 5 have been observed in experiments.

Now we shall present several examples concerned with distributed parameter systems.

EXAMPLE 6: *Reaction-diffusion system*. Reaction-diffusion systems have already been discussed in Section 1.2. One of the most famous is the "Brusselator" (cf., Nicolis and Prigogine [1.95]).

A sequence of reactions taking place under open-system conditions is considered in the following form:

$$A \underset{k_{-1}}{\overset{k_1}{\rightleftharpoons}} X$$

$$B + X \underset{k_{-2}}{\overset{k_2}{\rightleftharpoons}} Y + D$$

$$2X + Y \underset{k_{-3}}{\overset{k_3}{\rightleftharpoons}} 3X$$

$$X \underset{k_{-4}}{\overset{k_4}{\rightleftharpoons}} E$$

We can set $k_{-4} = k_{-2} = k_{-1} = k_{-3} = 0$. The course of concentrations in the above reaction system can be described by a set of equations:

$$\frac{d\tilde{X}}{d\tilde{t}} = k_1\tilde{A} - (k_2\tilde{B} + k_4)\tilde{X} + k_3\tilde{X}^2\tilde{Y},$$

$$\frac{d\tilde{Y}}{d\tilde{t}} = k_2\tilde{B}\tilde{X} - k_3\tilde{X}^2\tilde{Y}.$$

The following scaled variables can be introduced [1.96]:

$$t = k_4\tilde{t}, \quad x = \left(\frac{k_3}{k_4}\right)^{1/2}\tilde{X}, \quad y = \left(\frac{k_3}{k_4}\right)^{1/2}\tilde{Y}, \quad A = \left(\frac{k_1^2 k_3}{k_4^3}\right)^{1/2}A,$$

$B = (k_2/k_4)\tilde{B}$. By substituting into the above rate equations, we obtain the source terms for the components x and y:

$$f(x, y) = A - (B + 1)x + x^2y, \tag{E6.1}$$

$$g(x, y) = Bx - x^2y. \tag{E6.2}$$

The parameters A and B are non-negative and usually $A \in (0, 50)$, $B \in (0, 50)$. Reaction-diffusion systems, where one-dimensional diffusion of components x and y with constant diffusion coefficients D_x and D_y takes place in the direction of the spatial coordinate z, can be described by a set of two parabolic partial differential equations:

$$\frac{\partial x}{\partial t} = \frac{D_x}{L^2}\frac{\partial^2 x}{\partial z^2} + f(x, y), \tag{E6.3}$$

$$\frac{\partial y}{\partial t} = \frac{D_y}{L^2}\frac{\partial^2 y}{\partial z^2} + g(x, y). \tag{E6.4}$$

Here L is the system dimension. Boundary conditions of Dirichlet, Neumann, and Robin types are used in the studies of dissipative structures—both symmetric and asymmetric spatially non-monotonic steady-state and periodic solutions have often been discussed in the literature. The trivial solution of (E6.3), (E6.4) is obtained as $\bar{x} = A$, $\bar{y} = B/A$.

EXAMPLE 7: Gierer and Meinhardt's [1.97] model for gradient pattern formation in Hydra is based on a reaction-diffusion system. In this case, we have source terms in Eqs. (E6.3) and (E6.4) in the form

$$f(x, y) = \rho_0 \rho + c\rho \frac{x^2}{y} - \mu x, \tag{E7.1}$$

$$g(x, y) = c'\rho' x^2 - vy. \tag{E7.2}$$

Here x and y are concentrations of morphogens and c, μ, c', ρ_0, and v are positive parameters. ρ and ρ' describe densities for activators and inhibitors, respectively. The trivial solution of Eqs. (E7.1) and (E7.2) can be written in the form

$$\bar{x} = \frac{\rho_0 \rho}{\mu} + \frac{c\rho v}{c'\rho'\mu},$$

$$\bar{y} = c'\rho' \frac{\bar{x}^2}{v}.$$

EXAMPLE 8: *Reaction-diffusion system with multiplicities in the trivial solution—SH model.* The model describing the results of observations of oscillatory changes of concentrations of S-H and S-S groups in metabolism of low molecular thiols and proteins was developed by Selkov [1.56]. If we denote the dimensionless concentrations of low molecular thiols in reduced and disulphide form (S-H and S-S groups) as x and y, respectively, we obtain the source terms in mass balances (E6.3), (E6.4) in the form

$$f(x, y) = \alpha \frac{v_0 + x^\gamma}{1 + x^\gamma} - x(1 + y), \tag{E8.1}$$

$$g(x, y) = (\beta + y) - \delta y. \tag{E8.2}$$

Here α, β, γ, δ, and v_0 are kinetic constants, where α, δ, $v_0 > 0$, $\beta > 1$; $\gamma \geq 1$ is a positive integer. The trivial solution satisfies the two nonlinear algebraic equations $f(\bar{x}, \bar{y}) = g(\bar{x}, \bar{y}) = 0$. We immediately obtain

$$\bar{y} = \frac{\beta \bar{x}}{\delta - \bar{x}}, \tag{E8.3}$$

whence

$$(\beta - 1)\bar{x}^{\gamma+2} + (\alpha + \delta)\bar{x}^{\gamma+1} - \alpha\delta\bar{x}^\gamma - (\beta - 1)\bar{x}^2 + (\alpha v_0 + \delta)\bar{x}$$
$$- \alpha\delta v_0 = 0. \tag{E8.4}$$

For $\gamma = 3$, (E8.4) is a fifth-order polynomial in $\bar{x}$. There are either one or three positive (physically acceptable) roots of (E8.4) [1.64].

EXAMPLE 9: Axial dispersion in a tubular non-adiabatic reactor with the first-order exothermic reaction can be characterized by two partial differential equations [1.77, 1.81]:

$$\frac{\partial y}{\partial t} = \frac{1}{Pe_y} \frac{\partial^2 y}{\partial z^2} - \frac{\partial y}{\partial z} + Da(1 - y) \exp \frac{\theta}{1 + \theta/\gamma},$$ (E9.1)

$$\frac{\partial \theta}{\partial t} = \frac{1}{Pe_\theta} \frac{\partial^2 \theta}{\partial z^2} - \frac{\partial \theta}{\partial z} + B \, Da(1 - y) \exp \frac{\theta}{1 + \theta/\gamma} - \beta(\theta - \theta_c)$$ (E9.2)

with boundary conditions

$$z = 0: Pe_y \, y = \frac{\partial y}{\partial z}, \quad Pe_\theta \theta = \frac{\partial \theta}{\partial z};$$

$$z = 1: \quad \frac{\partial y}{\partial z} = \frac{\partial \theta}{\partial z} = 0.$$ (E9.3)

Here y is dimensionless conversion, $y \in [0, 1)$; θ is the dimensionless temperature, Da is the Damköhler number, γ is dimensionless activation energy, Pe_y is the Péclet number for axial mass transport, Pe_θ is the Péclet number for axial heat transport, B is the dimensionless parameter of heat evolution, and θ_c is the dimensionless cooling temperature. All parameters are positive. θ_c can also be negative. Both multiple-steady-state solutions and periodic behavior of temperature and concentration have been reported in the literature.

EXAMPLE 10: Flow of viscous fluid between two rotating disks (in steady state) can be described by a set of three ordinary differential equations [1.98, 1.99, 1.100]:

$$F'' = \sqrt{Re} \, HF' + Re(F^2 - G^2 + k),$$ (E10.1)

$$G'' = 2Re \, FG + \sqrt{Re} \, G'H,$$ (E10.2)

$$H' = -2\sqrt{Re} \, F$$ (E10.3)

with boundary conditions

$$H(0) = F(0) = H(1) = F(1) = 0,$$

$$G(0) = 1, \quad G(1) = S.$$ (E10.4)

Altogether we have six boundary conditions for the fifth-order system. The problem seems to be overdetermined. However, the parameter k is to be determined, and the problem is therefore correctly formulated. Re denotes the Reynolds number (Re > 0) and S the ratio of the velocity of the upper disk to the velocity of the lower disk, $S \in [-1, 1]$.

EXAMPLE 11: Non-isothermal diffusion and reaction within a porous catalytic particle where a first-order exothermic reaction takes place can be described by two parabolic partial differential equations [1.76, 1.101, 1.102]:

$$Lw \frac{\partial y}{\partial t} = \frac{\partial^2 y}{\partial z^2} + \frac{a}{z} \frac{\partial y}{\partial z} - \phi^2 y \exp \frac{\theta}{1 + \theta/\gamma},$$ (E11.1)

$$\frac{\partial \theta}{\partial t} = \frac{\partial^2 \theta}{\partial z^2} + \frac{a}{z} \frac{\partial \theta}{\partial z} + \phi^2 \gamma \beta y \exp \frac{\theta}{1 + \theta/\gamma}, \tag{E11.2}$$

with boundary conditions

$$z = 0: \frac{\partial y}{\partial z} = 0, \quad \frac{\partial \theta}{\partial z} = 0, \tag{E11.3}$$

$$z = 1: y = 1, \quad \theta = 0. \tag{E11.4}$$

Here y is concentration, $y \in (0, 1]$, θ is temperature, $\theta \in (0, \gamma\beta)$, $\gamma > 0$ is the activation energy, ϕ is the Thiele modulus, β is the parameter of heat evolution, and Lw is the Lewis number (all variables and parameters are dimensionless). The parameter $a = 0$, 1, or 2 is used for plane, cylindrical, or spherical symmetry, respectively. The above system of equations may also yield multiple-steady-state and periodic solutions in certain ranges of parameters.

CHAPTER 2

Multiplicity and Stability in Lumped-Parameter Systems (LPS)

Differentiation of dynamic systems into lumped-parameter systems (LPS) and distributed-parameter systems (DPS) was discussed in Section 1.3. The phase space of LPS is a finite-dimensional space. Let us consider n-dimensional Euclidean space. The LPS' can usually be written in the form

$$\frac{dx_i}{dt} = f_i(x_1, x_2, \ldots, x_n, \alpha_1, \ldots, \alpha_m), \quad i = 1, 2, \ldots, n, \tag{2.1}$$

or in vector form

$$\frac{d\mathbf{x}}{dt} = \mathbf{f}(\mathbf{x}, \boldsymbol{\alpha}), \tag{2.2}$$

where

$$\begin{aligned}
\mathbf{x} &= (x_1, x_2, \ldots, x_n)^T, \\
\mathbf{f} &= (f_1, f_2, \ldots, f_n)^T, \\
\boldsymbol{\alpha} &= (\alpha_1, \alpha_2, \ldots, \alpha_m).
\end{aligned} \tag{2.3}$$

Here x_i denote state variables, t is time, and $\boldsymbol{\alpha}$ is generally a vector of system parameters. However, we shall often consider only a scalar parameter α ($m = 1$). The system is autonomous, i.e., the time variable does not occur in the right-hand sides (RHS) explicitly. We shall usually assume that RHS' are continuous and continuously differentiable functions of their arguments. Five examples of LPS were presented in Section 1.4 (Examples 1–5).

2.1 Steady-State Solutions

The steady-state (stationary) solution $\mathbf{x}$ of system (2.2) satisfies the set of nonlinear equations

$$f_i(x_1, x_2, \ldots, x_n, \alpha) = 0, \quad i = 1, 2, \ldots, n. \tag{2.4}$$

How can we obtain a stationary solution, i.e., the solution of equations (2.4) for the chosen fixed α? In principle, we can use any method for solving a system of nonlinear equations, see, e.g., [2.1]. If the functions f_i can be easily differentiated analytically, Newton's method is usually used. This produces a sequence of approximations $\{x^k\}$ described by the relations:

$$x^{k+1} = x^k + \lambda \Delta x^k, \quad k = 0, 1, \ldots, \tag{2.5}$$

where, for a given k, Δx^k is obtained as the solution of the system of linear algebraic equations

$$J(x^k, \alpha)\Delta x^k = -f(x^k). \tag{2.6}$$

Here $J = \{\partial f_i / \partial x_j\}$ is the Jacobian matrix of system (2.4).

The iteration parameter $\lambda \in (0, 1]$ is usually chosen so that

$$\|f(x^{k+1})\| < \|f(x^k)\|. \tag{2.7}$$

If relation (2.7) is satisfied for $\lambda = 1$, then we let $\lambda = 1$, otherwise the value of λ is halved until (2.7) holds. To start the iteration, we have to choose an initial estimate of x, x^0.

If analytical differentiation of functions f_i is not feasible (or practically impossible), any one of the methods which do not require evaluating derivatives is used, or the partial derivatives are approximated by differences, e.g.,

$$\frac{\partial f_i(x)}{\partial x_j} \cong \frac{f_i(x + he_j) - f_i(x)}{h}, \tag{2.8}$$

where h is a small step and e_j is a unit vector with unity at the jth position. To obtain one iteration step in the Newton method, we have to calculate the vector f $n + 1$ times. This computational effort roughly corresponds to that required in the generalized secant method.

How can we obtain all stationary solutions of the given set of Eqs. (2.4)? The task may be particularly demanding in situations where we have poor preliminary estimates of the solution and the chosen iteration method diverges. In such situations, we can use randomly generated initial estimates of solution x^0; for every initial estimate, either our iteration algorithm works or it will diverge. The solutions thus obtained are subsequently summarized into a "memory of solutions." If a sufficiently high number of initial estimates x^0 is chosen, the probability of solving the problem is high. The number of random initial estimates necessary can sometimes be rather high, e.g., in the particular nonlinear system of equations connected with Example 5 of Chapter 1, it was necessary to use more than 20 initial estimates before a single stationary solution was obtained.

There may be many stationary solutions in some cases. For instance, consider the linear array of reaction cells (cf., Fig. 1.9) in Example 4 of Chapter 1. Here the relation $D_i^{jk} = D_i$ is valid for $|j - k| = 1$, otherwise $D_i^{jk} = 0$. Let us consider an absence of mass exchange with the environment ($H_k = 0$) and choose the following values of parameters A and B in Eqs. (E4.2) and (E4.3);

$A = 2$, $B = 6$. The values of the diffusion coefficients D_1, D_2 are equal to 0.4 and 4, respectively. The resulting stationary solutions are given in Table 2.1 [1.92]. We can see from this table that for $n = 2, 3, 4$ the number of solutions obtained is 3, 7, and 15, respectively.

Table 2.1 Steady-state solutions for Example 4, Chapter 1 and linear array of reaction cells, $A = 2$, $B = 6$, $D_1 = 0.4$, $D_2 = 4$.

	Steady-State Solution Number	Reactor 1	2	3	4	Stability
Two reaction cells ($n = 2$)	0	2.00 3.00	2.00 ← c_1 3.00 ← c_2			unstable
	1[a]	0.57 2.61	3.43 1.97			stable
Three reaction cells ($n = 3$)	0	2.00 3.00	2.00 3.00	2.00 3.00		unstable
	1[a]	4.96 1.40	0.68 2.57	0.36 3.01		stable
	2[a]	2.81 2.24	2.66 2.46	0.53 3.04		unstable
	3	2.64 2.47	0.71 2.83	2.64 2.47		unstable
	4	0.66 2.29	4.69 1.55	0.66 2.29		stable
Four reaction cells ($n = 4$)	0	2.00 3.00	2.00 3.00	2.00 3.00	2.00 3.00	unstable
	1[a]	0.36 4.02	0.39 3.61	0.87 2.75	6.38 1.10	stable
	2[a]	0.37 3.70	0.54 3.27	2.82 2.27	4.27 1.56	unstable
	3[a]	0.54 3.10	2.73 2.52	0.81 2.53	3.92 1.74	unstable
	4a	3.43 1.97	0.57 2.61	0.57 2.61	3.43 1.97	unstable
	4b	0.57 2.61	3.43 1.97	3.43 1.97	0.57 2.61	unstable
	5[a]	0.51 3.38	2.25 2.84	2.46 2.50	2.77 2.28	unstable
	6[a]	2.29 2.80	0.68 3.03	2.50 2.59	2.53 2.46	unstable
	7[a]	0.36 2.91	0.78 2.46	6.10 1.21	0.76 2.05	stable

[a] Each solution denoted by the superscript a represents two solutions. One is presented in the table and the second is a mirror image of the solution presented.

2.2 Dependence of Steady-State Solutions on a Parameter—Solution Diagram

The stationary solution $\mathbf{x}(\alpha)$ depends on α and satisfies

$$\mathbf{f}[\mathbf{x}(\alpha), \alpha] = \mathbf{0}, \tag{2.9}$$

for any α in the interval. The dependence $\mathbf{x}(\alpha)$ can consist of various branches, but it is unique on each branch. The branches begin at branch points (cf., Section 1.3). Here we shall study steady-state solutions of (2.1). The dynamics of steady-state solutions and periodic solutions will be studied in Sections 2.5–2.8. Here we shall call branch points "real bifurcations" (cf., Section 1.3), and we shall classify them either as limit points or bifurcation points (cf., Section 1.3). By changing the value of α, the number of solutions will change at a limit point by two, but in most situations the number of solutions does not change when passing through a bifurcation point.

A large number of numerical procedures can be used for finding the dependence $\mathbf{x}(\alpha)$. The simplest is the sequential one. The interval of α values is divided by the nodes $\alpha_0, \alpha_1, \ldots, \alpha_p$ into sections. The solution $\mathbf{x}(\alpha_j)$ is used as the initial estimate of a solution for $\alpha = \alpha_{j+1}$ (obtained, e.g., by the Newton method). If the points $\{\alpha_j\}$ are sufficiently close, one or two iterations usually suffice to obtain the solution for each α_j.

That is how we can obtain solutions on one branch; we can not cross a limit point, and when passing through a bifurcation point, the behavior is more or less random.

The second method for computing the dependence $\mathbf{x}(\alpha)$ is that of differentiation with respect to the parameter. When differentiating (2.9) with respect to α, we can derive the set of differential equations for $\mathbf{x}(\alpha)$:

$$\frac{d\mathbf{x}}{d\alpha} = -\mathbf{J}^{-1}(\mathbf{x}, \alpha)\frac{\partial \mathbf{f}}{\partial \alpha}, \quad \mathbf{x}(\alpha_0) = \mathbf{x}_0. \tag{2.10a}$$

We have denoted $\partial \mathbf{f}/\partial \alpha = (\partial f_1/\partial \alpha, \ldots, \partial f_n/\partial \alpha)^T$. For $\alpha = \alpha_0$, we shall assume that

$$\mathbf{f}(\mathbf{x}_0, \alpha_0) = \mathbf{0}. \tag{2.11}$$

Equation (2.10a) can be regarded as a system of linear algebraic equations for unknowns $d\mathbf{x}/d\alpha$:

$$\mathbf{J}(\mathbf{x}, \alpha)\frac{d\mathbf{x}}{d\alpha} = -\frac{\partial \mathbf{f}}{\partial \alpha}, \quad \mathbf{x}(\alpha_0) = \mathbf{x}_0. \tag{2.10b}$$

If the matrix $\mathbf{J}(\mathbf{x}(\alpha), \alpha)$ is regular on the interval $[\alpha_0, \alpha_1]$, the dependence $\mathbf{x}(\alpha)$ obtained by integrating (2.10) satisfies the following relation:

$$\mathbf{f}(\mathbf{x}(\alpha), \alpha)) = \mathbf{0}, \quad \alpha \in [\alpha_0, \alpha_1]. \tag{2.12}$$

The matrix J is singular at a branch point (see below). Thus the continuation procedure fails here and must be modified in order to work at branch points (particularly at limit points). The method must be useful for generating the complete dependence $\mathbf{x}(\alpha)$, which forms a continuous smooth curve in $(n + 1)$-dimensional space $(x_1, \ldots, x_n, \alpha)$. We shall take the arc length of the solution curve as a parameter of the method, i.e., we shall continue the solution along the above-mentioned curve in the $(n + 1)$-dimensional space. Upon differentiating (2.4) with respect to z (the length of arc), we obtain

$$\frac{df_i}{dz} = \sum_{j=1}^{n} \frac{\partial f_i}{\partial x_j} \frac{dx_j}{dz} + \frac{\partial f_i}{\partial \alpha} \frac{d\alpha}{dz} = 0, \quad i = 1, 2, \ldots, n. \tag{2.13}$$

The additional equation,

$$\left(\frac{dx_1}{dz}\right)^2 + \cdots + \left(\frac{dx_n}{dz}\right)^2 + \left(\frac{d\alpha}{dz}\right)^2 = 1, \tag{2.14}$$

determines z as the length of arc on the above-mentioned curve of solutions. The initial condition for (2.13) can be rewritten in the form

$$z = 0: \mathbf{x} = \mathbf{x}_0, \quad \alpha = \alpha_0. \tag{2.15}$$

For each z, Eqs. (2.13) form a system of n linear algebraic equations in $n + 1$ unknowns dx_i/dz, $i = 1, \ldots, n$ and $d\alpha/dz$. Let us assume that the matrix

$$J_k = \begin{bmatrix} \dfrac{\partial f_1}{\partial x_1}, \ldots, \dfrac{\partial f_1}{\partial x_{k-1}}, \dfrac{\partial f_1}{\partial x_{k+1}}, \ldots, \dfrac{\partial f_1}{\partial x_{n+1}} \\[2ex] \dfrac{\partial f_2}{\partial x_1}, \ldots \\[2ex] \cdots\cdots\cdots \\[2ex] \dfrac{\partial f_n}{\partial x_1}, \ldots, \dfrac{\partial f_n}{\partial x_{k-1}}, \dfrac{\partial f_n}{\partial x_{k+1}}, \ldots, \dfrac{\partial f_n}{\partial x_{n+1}} \end{bmatrix} \tag{2.16}$$

is regular (for certain values z and k, $1 \le k \le n + 1$). For clarity we let

$$x_{n+1} = \alpha. \tag{2.17}$$

Equations (2.13) can be solved for the unknowns

$$dx_1/dz, \ldots, dx_{k-1}/dz, dx_{k+1}/dz, \ldots, dx_{n+1}/dz$$

in the form

$$\frac{dx_i}{dz} = \beta_i \frac{dx_k}{dz}, \quad i = 1, 2, \ldots, k - 1, k + 1, \ldots, n + 1. \tag{2.18}$$

This solution, i.e., evaluating the coefficients β_i, can be carried out by means of Gauss elimination applied to matrix (2.16). If the Gauss elimination with pivoting is used for the original rectangular $n \times (n + 1)$ matrix

$\boldsymbol{J} = \{\partial f_i/\partial x_j\}$, $i = 1, \ldots, n$; $j = 1, \ldots, n + 1$, the column index k and $(-\beta_i)$ in the column k are produced.

If we introduce (2.18) into condition (2.14), we obtain

$$\left(\frac{dx_k}{dz}\right)^2 = \left(1 + \sum_{\substack{i=1 \\ i \neq k}}^{n+1} \beta_i^2\right)^{-1}. \tag{2.19}$$

The sign of the derivative dx_k/dz is given by the orientation of the parameter z along the curve, i.e., by the direction of movement along the curve. The other derivatives are computed from (2.18). The system of differential equations (2.18), (2.19) can be solved by any numerical technique for the integration of initial-value problems. However, we have to keep in mind that the evaluation of right-hand sides of these differential equations requires the solution of a system of linear algebraic equations and thus can be laborious and time consuming. Hence, it is advantageous to use numerical methods which require relatively few evaluations of the right-hand sides for the required overall accuracy. The explicit Adams–Bashforth methods are convenient from the above-mentioned point of view [2.2, 2.3].

The errors of approximation accumulate during integration, and the computed solution deviates from the correct solution with increasing z. In this case, it is necessary to improve accuracy of the obtained solution by Newton's method applied to (2.4), considering as variables $(x_1, \ldots, x_{k-1}, x_{k+1}, \ldots, x_{n+1})$ and a fixed value of x_k. Since we always have good initial estimates for Newton's method, one or two iterations suffice for the required accuracy. On the basis of the above procedures, we can construct a continuation algorithm, where numerical integration of Eqs. (2.18) and (2.19) plays the role of predictor and Newton's method the role of corrector. A simplified block diagram of such an algorithm is shown in Fig. 2.1.

A general computer routine based on the above algorithm [2.4] is given in Appendix A.

When the choice of control parameters in the routine is made properly, the algorithm generates a continuous curve of solutions in $(n + 1)$-dimensional space $(x_1, \ldots, x_n, \alpha)$; it passes through limit points without difficulties, and at bifurcation points it continues on the next branch more or less at random (though it usually retains the direction of the original branch). Results obtained by this procedure are illustrated in Figs. 2.2–2.5.

This algorithm can produce a continuous curve (consisting of branches of solutions) in $(n + 1)$-dimensional space. For any other curve (branch of solutions), we need to obtain an initial estimate of the vector $(x_1, \ldots, x_n, \alpha)$ again in order to begin the continuation procedure. This holds particularly for the isolated curves $\mathbf{x}(\alpha)$, isolas (cf. Sections 1.3 and 2.4.3). The isolas can be obtained only if at least one point lying on them is known. To obtain a full portrait of stationary solutions depending on the parameter α, we must again use the generator of random initial estimates (but now in $(n + 1)$-dimensional space

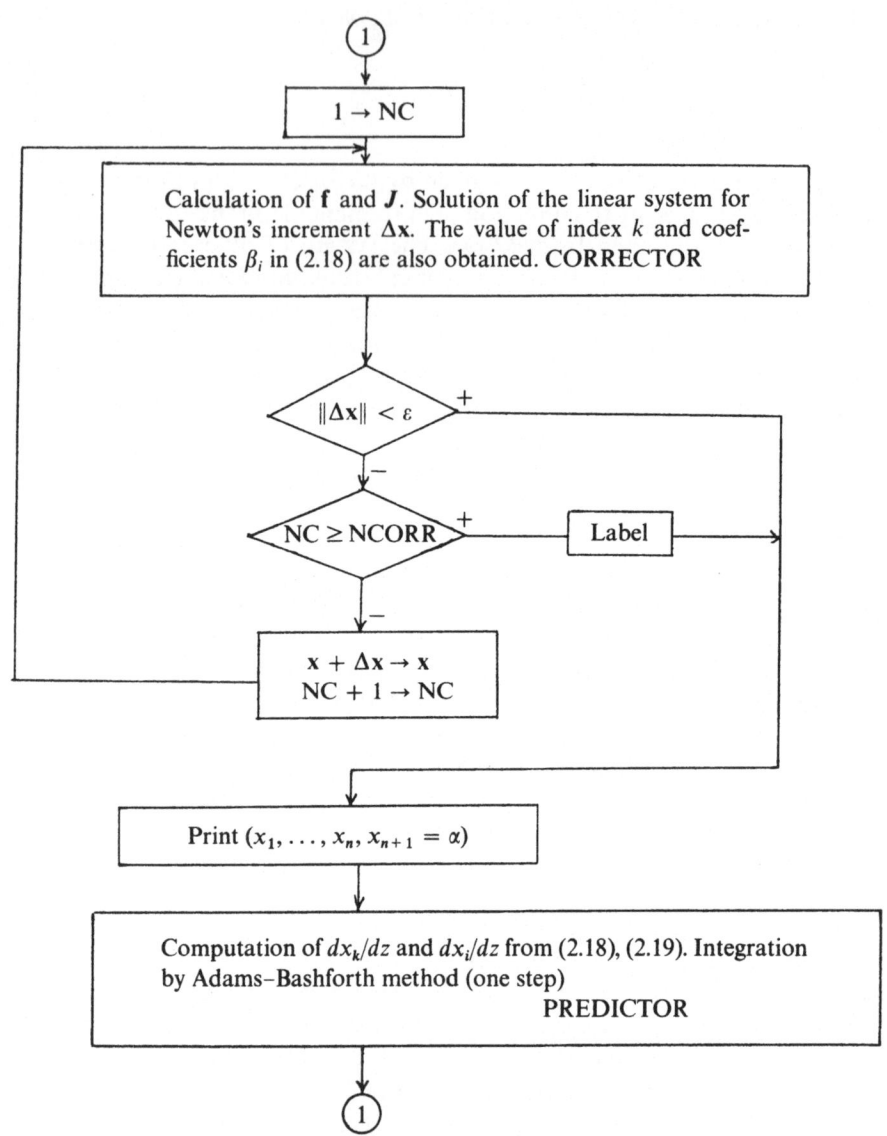

Figure 2.1 Simplified block schema of continuation.

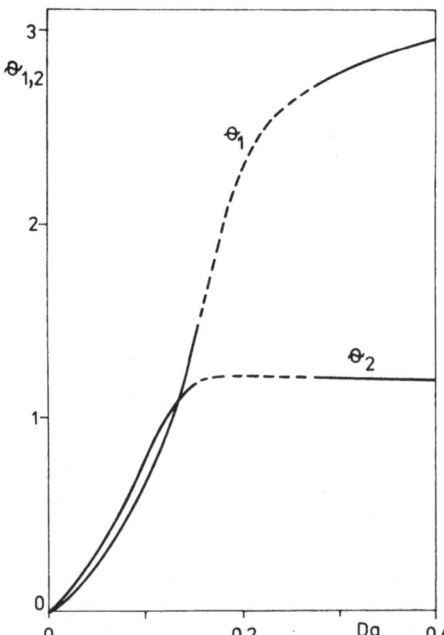

Figure 2.2 Dependence of the steady-state solution on the parameter $Da = Da_1 = Da_2$, Example 3, Chapter 1, $\gamma = 1000$, $B = 12$, $\theta_{c1} = \theta_{c2} = 0$, $\beta_1 = \beta_2 = 2$, $\Lambda = 0.8$. —— stable, ---- unstable.

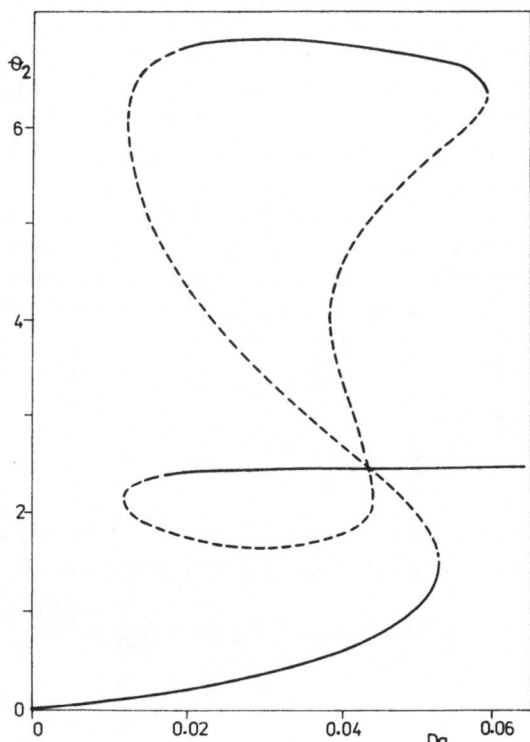

Figure 2.3 Dependence of the steady-state solution on the parameter $Da = Da_1 = Da_2$, Example 3, Chapter 1, $\gamma = 1000$, $B = 22$, $\theta_{c1} = \theta_{c2} = 0$, $\beta_1 = \beta_2 = 2$, $\Lambda = 1$. —— stable, ---- unstable.

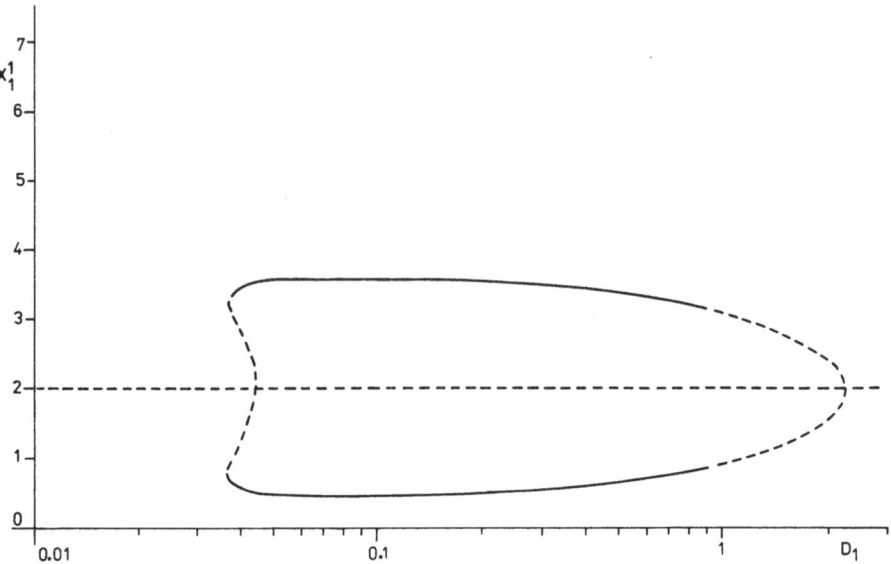

Figure 2.4 Dependence of the steady-state solution on the parameter D_1, $D_2 = 10D_1$, Example 4, Chapter 1, $A = 2$, $B = 6$. —— stable, ---- unstable.

$x_1, \ldots, x_n, \alpha$) and test whether we have obtained, by means of Newton's method, a known or new solution of the problem. If we have a new solution we can supplement the already known family of solutions by means of the continuation algorithm and repeat the whole procedure.

Very often we can make use of some special properties of the stationary equations. For example, if the system is nonlinear only in x_k, we can employ an inverse procedure. We choose x_k, and from the stationary equations we calculate $x_1, \ldots, x_{k-1}, x_{k+1}, \ldots, x_n, \alpha$. If the choice of x_k in the expected interval is sufficiently dense, we obtain the dependence $\mathbf{x}(\alpha)$. Let us illustrate it with Examples 1 and 5 of Chapter 1 (though details of the procedure vary with the particular problem and the ingenuity of the user).

Upon combining (E1.1) and (E1.2) in the stationary state, we obtain an invariant

$$y = \frac{\theta}{B} + \frac{\beta(\theta - \theta_c)}{B\Lambda} \tag{2.20}$$

After substituting (E1.1) into the RHS, we obtain one nonlinear algebraic equation for the required stationary solution

$$f(\theta, \mathrm{Da}, \Lambda, \gamma, B, \beta, \theta_c) = -\Lambda\frac{\theta}{B} - \frac{\beta(\theta - \theta_c)}{B}$$
$$+ \mathrm{Da}\left[1 - \frac{\theta}{B} - \frac{\beta(\theta - \theta_c)}{B\Lambda}\right]\exp\left[\frac{\theta}{1 + \theta/\gamma}\right] = 0. \tag{2.21}$$

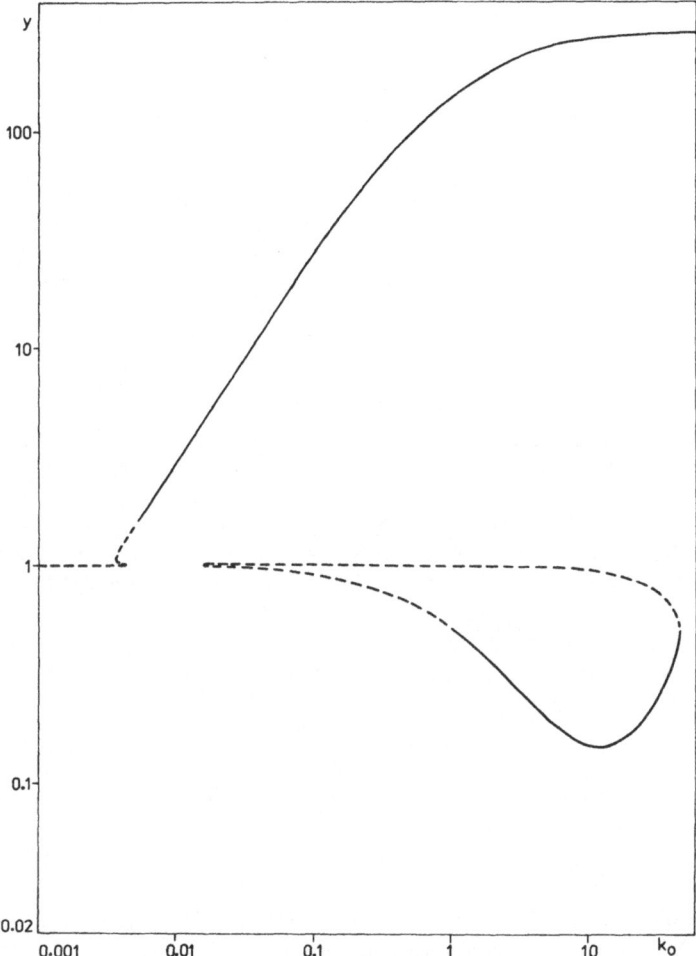

Figure 2.5 Dependence of the steady-state solution on the parameter k_0, Example 2, Chapter 1, $f = 1$, $\mu = 8.4 \times 10^{-6}$, $y_0 = 280$. —— stable, ---- unstable.

In this equation, θ occurs in nonlinear terms, the parameters Da, β, θ_c, and B appear linearly (after multiplication by $B\Lambda$) and the parameter Λ occurs quadratically. From (2.21), we can express, e.g., Da, in an explicit way:

$$\text{Da} = g(\theta, \Lambda, \gamma, B, \beta, \theta_c), \tag{2.22}$$

and similarly we can obtain Λ as a solution of quadratic equation (from the two resulting values only the values of $\Lambda \in (0, 1]$ are considered).

Thus, for a properly dense net of values of θ (for fixed values of parameters Da, γ, B, β, θ_c), we can obtain corresponding values of Λ, and construct the dependence $y(\Lambda)$ and $\theta(\Lambda)$ using (2.20). An example is shown in Fig. 2.6.

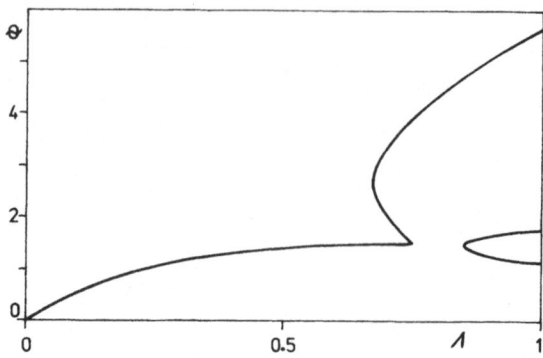

Figure 2.6 Dependence of the dimensionless temperature θ in the continuous stirred tank reactor on the parameter Λ, Example 1, Chapter 1, $\gamma = 20$, $B = 20$, $\beta = 1.975$, $\theta_c = 0$, $Da = 0.07$.

The above procedure is evident. Let us now present an analogous approach for Example 5. The algorithm may have the following structure:

0. Stationary values $x_3 = \xi_3$ and $x_6 = \xi_6$ (see (E5.3) and (E5.4)) are calculated. The values of all parameters are fixed with the exception of that for which we shall generate the dependence (see point 5).
1. The value of x_2 is chosen.
2. From stationary forms of (E5.1) and (E5.2), x_1 and μ are computed.
3. The values thus obtained are inserted into (E5.4) and (E5.5). From (E5.4), we obtain

$$x_4 = b_1 + b_2 x_5, \tag{2.23}$$

and after introducing it into (E5.5), the following quadratic equation results:

$$a_1 x_5^2 + a_2 x_5 + a_3 = 0. \tag{2.24}$$

4. Positive roots x_5 of Eq. (2.24) are computed, x_4 is determined from (2.23), and x_1 and μ are obtained as in step 2.
5. The relation (E5.7) for μ is used to calculate one of the unknown parameters which appear linearly — $\hat{\mu}$, K_s, K_i, K_a, or K_1.

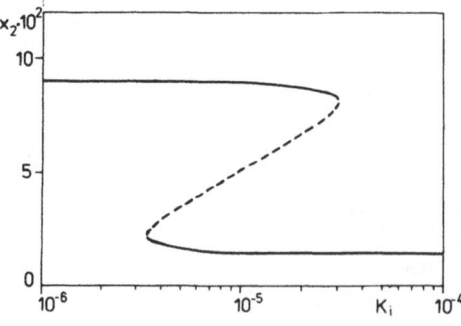

Figure 2.7 Dependence of the concentration of substrate on the parameter K_i, Example 5, Chapter 1. —— stable, ----- unstable.

6. Thus we are able to map x_2 (and all other variables in a stationary state) into values of the given parameter. The whole procedure is repeated starting with step 1, until a sufficiently dense set of x_2 values is calculated. An example of this is shown in Fig. 2.7.

2.3 Stability of Steady-State Solutions

Stationary state $\bar{x}$ of the differential equation (2.1) is called locally stable if

$$\lim_{t \to \infty} \|x(t) - \bar{x}\| = 0, \tag{2.25}$$

for every $x(0)$ chosen in a sufficiently small neighborhood of $\bar{x}$, i.e., for $x(0)$ such that

$$\|x(0) - \bar{x}\| < \delta, \tag{2.26}$$

where δ is a conveniently chosen small number.

Consider a system of linear differential equations with constant coefficients

$$\frac{d\mathbf{x}}{dt} = A\mathbf{x}, \tag{2.27}$$

where $\mathbf{x} = (x_1, \ldots, x_n)^T$, $A = \{a_{ij}\}$. Let the only stationary solution of (2.27) be $\bar{x} = \mathbf{0}$. If the eigenvalues of matrix A are known, we can determine the stability of this unique stationary solution. If, for every eigenvalue λ_i,

$$\text{Re}\{\lambda_i\} < 0, \quad i = 1, \ldots, n, \tag{2.28}$$

the zero solution is stable and all trajectories approach it for $t \to \infty$. If, on the contrary, at least one eigenvalue has a positive real part, the solution is unstable since there are trajectories that approach ∞ for $t \to \infty$. To determine whether all eigenvalues have negative real parts, we examine the coefficients a_i of the characteristic polynomial of matrix A:

$$P(\lambda) = (-1)^n \det(A - \lambda I) = \lambda^n + a_1 \lambda^{n-1} + \cdots + a_{n-1}\lambda + a_n. \tag{2.29}$$

Several methods exist for evaluating the coefficients of the polynomial (2.29). Let us discuss the two best known: the Krylov method and the interpolation method. Others can be found in any textbook on numerical methods in linear algebra, e.g., [2.5].

The Krylov method uses the Hamilton–Cayley identity,

$$P(\mathbf{A}) = A^n + a_1 A^{n-1} + \cdots + a_{n-1}A + a_n I = \mathbf{0}. \tag{2.30}$$

If (2.30) is multiplied by a suitable vector $\mathbf{h}^0$, we obtain

$$a_1 \mathbf{h}^{n-1} + a_2 \mathbf{h}^{n-2} + \cdots + a_{n-1} \mathbf{h}^1 + a_n \mathbf{h}^0 = -\mathbf{h}^n, \tag{2.31}$$

where the vectors

$$\mathbf{h}^k = A^k \mathbf{h}^0$$

can be obtained recursively,

$$\mathbf{h}^{k+1} = A \mathbf{h}^k, \quad k = 0, 1, \ldots, n-1. \tag{2.32}$$

Equation (2.31) thus forms a system of n linear algebraic equations with n unknown coefficients $a_1, \ldots, a_n$, which can be determined, e.g., by Gauss elimination (see [2.5]).

The interpolation method is based on the fact that a unique nth-order polynomial passing through $n + 1$ points can be constructed. Hence, we choose $\lambda_0, \lambda_1, \ldots, \lambda_n$ and compute

$$P(\lambda_i) = (-1)^n \det(A - \lambda_i I). \tag{2.33}$$

If we construct the Lagrange interpolation polynomial passing through points $\lambda_i, P(\lambda_i), i = 0, 1, \ldots, n$, we find $P(\lambda)$ and the coefficients $a_1, \ldots, a_n$.

Without calculating eigenvalues, the Routh–Hurwitz criterion is usually used to determine, from the known coefficients of the polynomial, whether all eigenvalues have negative real parts. Let us set $D_1 = a_1$ and

$$D_k = \det \begin{bmatrix} a_1 & a_3 & a_5 & \cdots & a_{2k-1} \\ 1 & a_2 & a_4 & \cdots & a_{2k-2} \\ 0 & a_1 & a_3 & \cdots & a_{2k-3} \\ 0 & 1 & a_2 & \cdots & a_{2k-4} \\ \vdots & & & & \\ 0 & 0 & & \cdots & a_k \end{bmatrix}, \tag{2.34}$$

for $k = 2, 3, \ldots, n$, while a_j equals zero for $j > n$. If $D_k > 0, k = 1, 2, \ldots, n$, all eigenvalues have negative real parts.

Evaluating all eigenvalues, e.g., by the Lin–Bairstow method [2.5, 2.6], is generally a more laborious procedure. Accumulation of rounding errors can cause computational difficulties in high-order systems.

Let us now deal with the nonlinear system (2.2). We shall follow the behavior of the solutions in the neighborhood of a stationary solution $\bar{\mathbf{x}}$ for a given α. Let us introduce a vector of deviations of solutions from $\bar{\mathbf{x}}$,

$$\boldsymbol{\delta} = \mathbf{x} - \bar{\mathbf{x}}. \tag{2.35}$$

Equation (2.2) can then be rewritten approximately in the form ($' = d/dt$),

$$x_i' = \bar{x}_i' + \delta_i' = f_i(\mathbf{x}, \alpha) = f_i(\bar{\mathbf{x}} + \boldsymbol{\delta}, \alpha) \doteq f_i(\bar{\mathbf{x}}, \alpha)$$

$$+ \sum_{j=1}^{n} \frac{\partial f_i(\bar{\mathbf{x}}, \alpha)}{\partial x_j} \delta_j, \quad i = 1, 2, \ldots, n, \tag{2.36}$$

where the functions f_i are expanded into a Taylor series up to linear terms. Since $\bar{x}_i' = f_i(\bar{\mathbf{x}}, \alpha) = 0$, we obtain from (2.36):

$$\delta_i' = \sum_{j=1}^{n} a_{ij} \delta_j, \tag{2.37}$$

where $a_{ij} = \partial f_i(\bar{\mathbf{x}}, \alpha)/\partial x_j$ are constants.

Let us apply the above conclusions about stability of linear systems to (2.37). Hence, for $\boldsymbol{\delta} \to \mathbf{0}$ (and thus $\mathbf{x} \to \bar{\mathbf{x}}$), as $t \to \infty$, all eigenvalues of matrix $A = \{a_{ij}\}$ must have negative real parts. If $\mathbf{x}(t) \to \bar{\mathbf{x}}$ for initial conditions taken in a sufficiently small neighborhood of $\bar{\mathbf{x}}$, the stationary solution $\bar{\mathbf{x}}$ is called locally stable. This method of evaluating stability of a stationary solution is the well-known first Liapunov method. If some of the eigenvalues of the linearized RHS have zero real parts, we cannot determine stability of the stationary state as described above, and we must resort to investigating the full nonlinear system (2.2) or its higher-order approximations (see, e.g., [1.3, 1.4] and Appendix C). Stability of stationary states as the parameter varies was determined in Examples by the first Liapunov method, cf., Figs. 2.2–2.7 and Table 2.1. Analysis of the stability of time-delay systems is more complicated. We, generally, obtain a transcendental rather than polynomial equation for the "eigenvalues". Moreover, we usually lack preliminary information about the location of roots in the complex plane.

2.4 Branch Points—Real Bifurcation

To do a parameter study, we often face the problem of locating all solution branches in the solution diagram. We can reduce computations considerably if we can first evaluate all branch points and then use the continuation algorithm to compute solution branches.

Let us define, in more detail, some of the concepts we shall use. The unique and continuous (in most cases smooth as well) dependence of solutions $\mathbf{x}(\alpha)$ of Eq. (2.4) with respect to α is called a branch of solutions. Specifically, we have in mind a "geometric picture" of the branch of solutions. This term will be used in this sense. Let us assume that the branch is continued in the parameter α as far as possible. We shall omit the case where for an isolated value $\alpha = \alpha_0$, a solution of (2.4) exists (this occurs for linear systems and eigenvalue problems). The branch either exists in the interval $\alpha \in (-\infty, \infty)$ or it has boundary points beyond which it does not exist, i.e., it exists in the interval $\alpha \in [\alpha_l, \alpha_r]$. Let us assume that we know the branch of solutions $\bar{\mathbf{x}}(\alpha)$ in the neighborhood of $\alpha = \alpha_0$ and that α_0 is not the boundary point of the branch. The point $(\mathbf{x}_0, \alpha_0) \in R^{n+1}$, $\mathbf{x}_0 = \bar{\mathbf{x}}(\alpha_0)$ will be called a regular point of the branch when every solution of (2.4) in the neighborhood of the point (the neighborhood in R^{n+1} is considered) belongs to the branch. If the Jacobian

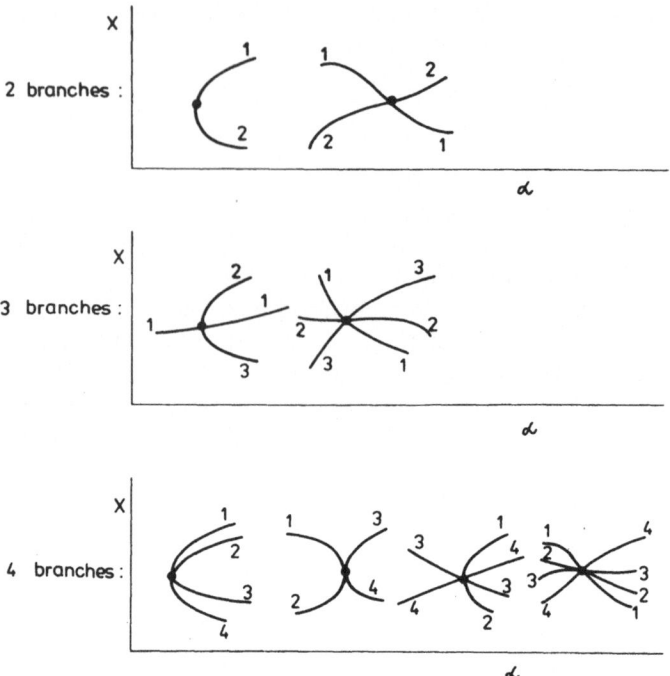

Figure 2.8 Branch points with 2, 3, and 4 branches in the solution diagram.

matrix $J = \{\partial f_i/\partial x_j\}$, evaluated at the point $(\mathbf{x}_0, \alpha_0)$, is regular, the point is also regular. This follows from the implicit function theorem. Conversely, if, in every neighborhood of this point, a solution $(\mathbf{x}, \alpha)$ of (2.4) exists such that $\|\mathbf{x}(\alpha) - \bar{\mathbf{x}}(\alpha)\| > 0$, the point $(\mathbf{x}_0, \alpha_0)$ is called the branch point [2.7]. This definition of the branch point also applies where $(\mathbf{x}_0, \alpha_0)$ is a boundary point of the branch $\bar{\mathbf{x}}(\alpha)$. If the point $(\mathbf{x}_0, \alpha_0)$ is the left (or right) boundary point of the branch $\bar{\mathbf{x}}(\alpha)$, and if the solution $(\mathbf{x}, \alpha)$ in the neighborhood of the branch point of (2.4), for which $\|\mathbf{x}(\alpha) - \bar{\mathbf{x}}(\alpha)\| > 0$, exists for $\alpha > \alpha_0$ only (or for $\alpha < \alpha_0$), the point $(\mathbf{x}_0, \alpha_0)$ is called the limit point (turning point).[1] A branch point which is not a boundary point will be called a bifurcation point. The bifurcation point and the limit point may coincide, c.f., Fig. 1.3, cases 3 and 4. A complex situation arises when a higher number of branches join at the branch point. Examples of 2, 3, and 4 branches joining at the branch point are shown in Fig. 2.8.

In many problems there exists a trivial solution (this solution does not depend on the value of the parameter in question, c.f., Example 4, Chapter 1;

---

[1] The terms limit points and bifurcation points are used here in the sense suggested by Bauer *et al.* [1.83]. Thus, at a simple limit point only two different branches of solutions join and end.

the solution $c_1^k = A$, $c_2^k = B/A$, $k = 1, 2, \ldots, n$, is independent of the values of D_i^{jk}). The branching (or bifurcation) from the trivial solution is called primary bifurcation [2.8]. The determination of such points is relatively easy here (compare below). All other bifurcations (i.e. those occurring from other than trivial branches) are called secondary bifurcations. To determine such a bifurcation point we can use a trial-and-error technique involving the calculation of solutions along a branch [1.83]. A second possibility involves a straightforward iteration technique to determine branch points [2.9, 2.10, 2.11].

These methods will be described in Section 2.4.1. We shall show how to evaluate individual branch directions at bifurcation points in Section 2.4.2. It is important to know these directions in order to start the continuation algorithm in a neighborhood of the bifurcation point (the simple limit point does not cause any problems, see Section 2.2). The appearance of isolated solutions will be studied in Section 2.4.3.

2.4.1 Evaluation of limit and bifurcation points

We have used the term "bifurcation point" in the foregoing discussions to mean the branching of solutions. When we deal with branching of steady-state solutions, the term "real bifurcation" is often used because the branching is related to how the zero (real) eigenvalue of the Jacobian matrix J crosses the imaginary axis in the complex plane. We say "complex bifurcation" occurs when a pair of complex-conjugate eigenvalues crosses the imaginary axis. We shall study this type of bifurcation in Section 2.5.

We shall discuss necessary conditions and computational techniques by using a system of nonlinear algebraic equations

$$f_i(x_1, x_2, \ldots, x_n, \alpha) = 0, \quad i = 1, 2, \ldots, n. \tag{2.4}$$

The necessary condition which determines the branch points follows from the implicit function theorem, namely:

$$f_{n+1}(x_1, x_2, \ldots, x_n, \alpha) = \det J(x_1, \ldots, x_n, \alpha) = 0. \tag{2.38}$$

Here J is the Jacobian matrix with the elements

$$g_{ij} = \frac{\partial f_i(x_1, \ldots, x_n, \alpha)}{\partial x_j}, \quad i, j = 1, 2, \ldots, n. \tag{2.39}$$

Eigenvalues λ of the Jacobian matrix J satisfy the equation

$$\det(J(x_1, \ldots, x_n, \alpha) - \lambda I) = 0. \tag{2.40}$$

Equality (2.38) is therefore equivalent to the statement: zero is an eigenvalue of J. This is another definition of point of real bifurcation. At a bifurcation point, the stability of solutions in the transient sense can change. If, as parameter α varies, the eigenvalue moves along the real axis and crosses from the left complex half-plane into the right one or vice versa, then the stability of solution can change.

Let us set

$$f_{n+2}(x_1, x_2, \ldots, x_n, \alpha) = \det \bar{J}(x_1, \ldots, x_n, \alpha) = 0, \qquad (2.41)$$

where $\bar{J} = \{\bar{g}_{ij}\}$, $\bar{g}_{ij} = g_{ij}$ for $j = 1, 2, \ldots, n - 1$, $\bar{g}_{in} = \partial f_i/\partial \alpha$. Thus the matrix $\bar{J}$ is obtained from J by substituting the column $\{\partial f_i/\partial \alpha\}$ for the last column. Obviously, any column other than the last could be used with the same results. The following inequality is valid in most practical problems for a limit point $(\mathbf{x}^*, \alpha^*)$:

$$f_{n+2}(\mathbf{x}^*, \alpha^*) \neq 0. \qquad (2.42)$$

Such a limit point will be called regular.[2] This inequality is derived from the fact that a unique dependence $\alpha(x_n)$ exists in the neighborhood of this point and from the implicit function theorem.

On the other hand, for a bifurcation point $(\mathbf{x}^{**}, \alpha^{**})$, it follows from the implicit function theorem that

$$f_{n+2}(\mathbf{x}^{**}, \alpha^{**}) = 0 \qquad (2.43)$$

because the dependence of α on x_n is not unique in the neighborhood of $(\mathbf{x}^{**}, \alpha^{**})$. Therefore, we can often distinguish between limit and bifurcation points by using Eqs. (2.42) and (2.43).

Let us return to the problem of evaluating a primary bifurcation point, i.e., the point of bifurcation from trivial solution $\bar{\mathbf{x}}$, $\mathbf{x}(\alpha) = \bar{\mathbf{x}}$ for α from a sufficiently large interval of α. Equations (2.4) are then satisfied by inserting $\bar{\mathbf{x}}$ regardless the value of α. We now need only satisfy (2.38), i.e., the relation

$$\varphi(\alpha) = \det \mathbf{J}(\bar{x}_1, \ldots, \bar{x}_n, \alpha) = 0. \qquad (2.44)$$

Equation (2.44) is the transcendental equation for evaluating α^{**} and determines the bifurcation point $(\mathbf{x}^{**}, \alpha^{**}) = (\bar{\mathbf{x}}, \alpha^{**})$. α can often be derived from (2.44) in an explicit way, without iteration procedures. Otherwise, Newton's method for a single equation

$$\alpha_{k+1} = \alpha_k - \frac{\varphi(\alpha_k)}{\varphi'(\alpha_k)},$$

for example, can be used. In this equation, α_0 is an initial estimate and φ' is obtained either by analytical or numerical differentiation.

The original equations (2.4) and the necessary condition (2.38) form a set of $n + 1$ equations in $n + 1$ unknowns $x_1^*, \ldots, x_n^*, \alpha^*$, corresponding to coordinates of the branch points (both limit and bifurcation points) in question.

---

[2] We can illustrate the difference between a regular and irregular limit point by the following simple example. Let $n = 1$ and (x^*, α^*) be a regular limit point of $f(x, \alpha) = 0$, i.e., $\partial f/\partial \alpha \neq 0$ at this point. Let us consider the equation $\bar{f} = [f(x, \alpha)]^2 = 0$, which has the same limit point (x^*, α^*) on the same solution curve. However, $\partial \bar{f}/\partial \alpha = 0$ at this point; (x^*, α^*) is therefore irregular for $\bar{f}$.

To solve this system, we can use Newton's method:

$$F'(X^k)\Delta X^k = -F(X^k) \tag{2.45}$$

and

$$X^{k+1} = X^k + v\Delta X^k. \tag{2.46}$$

Here $X = (x_1, x_2, \ldots, x_n, \alpha)^T$, $F = (f_1, \ldots, f_{n+1})^T$, and F' is the Jacobian matrix of the $(n+1) \times (n+1)$ system

$$F(X) = 0. \tag{2.47}$$

For $i = 1, 2, \ldots, n+1$, the elements of $F' = \{\gamma_{ij}\}$ are equal to

$$\gamma_{ij} = \frac{\partial f_i}{\partial x_j}, \quad j = 1, 2, \ldots, n, \quad \gamma_{i,n+1} = \frac{\partial f_i}{\partial \alpha}. \tag{2.48}$$

In order to ensure that the iteration process converges, we test for the condition

$$\|F(X^{k+1})\| < \|F(X^k)\|.$$

We begin each iteration with $v = 1$. If the inequality is not satisfied, the value of v is reduced (e.g., halved).

The expression for γ_{ij}, $i \neq n+1$ can be easily developed. For $i = n+1$ we can use the following procedure. From the definition of the determinant, we obtain

$$\frac{\partial \det J}{\partial g_{ij}} = (-1)^{i+j} G_{ij}, \tag{2.49}$$

where G_{ij} is a minor, i.e., a determinant of the $(n-1) \times (n-1)$ matrix which is formed from J by eliminating row i and column j. Partial derivatives of (2.38) are

$$\gamma_{n+1,m} = \frac{\partial f_{n+1}}{\partial x_m} = \sum_{i,j=1}^{n} (-1)^{i+j} G_{ij}(X) \frac{\partial^2 f_i}{\partial x_j \partial x_m}, \quad m = 1, 2, \ldots, n, \tag{2.50}$$

and

$$\gamma_{n+1,n+1} = \frac{\partial f_{n+1}}{\partial \alpha} = \sum_{i,j=1}^{n} (-1)^{i+j} G_{ij}(X) \frac{\partial^2 f_i}{\partial x_j \partial \alpha}. \tag{2.51}$$

Note that we must evaluate second derivatives $\partial^2 f_i/\partial x_j \partial x_m$ and $\partial^2 f_i/\partial x_j \partial \alpha$ to obtain the matrix F'. If analytical differentiation is difficult, we can use difference formulas to approximate these derivatives.

A limit point for which $F'(X)$ is regular will be called strongly regular. For such a point, Newton's method guarantees quadratic rate of convergence, cf., e.g., [2.1]. The strongly regular limit point is regular, too, as can easily be seen by expanding $\det F'$ with respect to the last row. On the other hand, a regular limit point need not be strongly regular: cf., the equation $x^4 - \alpha = 0$.

Let us consider points (x, α) satisfying Eqs. (2.4), (2.38), and (2.41), i.e., points which satisfy the necessary conditions for the existence of a bifurcation point. We have a system of $n + 2$ equations in $n + 1$ unknowns $x_1, x_2, \ldots, x_n$, α. Let us denote the $(n + 2) \times (n + 1)$ Jacobian matrix of (2.4), (2.38), and (2.41) by F' and let $F = (f_1, f_2, \ldots, f_{n+2})^T$. We can solve this overdetermined system by the Gauss–Newton method [2.1]:

$$[F'^T(X^k)F'(X^k)]\Delta X^k = -F'^T(X^k)F(X^k),\tag{2.52}$$

$$X^{k+1} = X^k + \Delta X^k.$$

This iteration is similar to Newton's and enables us to find a stationary (e.g., minimum) point of the functional $\varphi = F^T F$; φ has a zero minimum at the bifurcation point X^{**}. If $F'^T(X^{**}) \cdot F'(X^{**})$ is regular, (2.52) converges quadratically.

The most probable situations are summarized in Table 2.2. Counter examples can easily be found; however, the conclusions presented in Table 2.2 should be correct for most practical problems.

Table 2.2 Probable situations.

| | Type of convergence with | |
	(2.45), (2.46)	(2.52)
Limit point	Quadratic	None
Bifurcation point	Linear	Quadratic

Let us consider a simple algebraic example

$$f_1 = (x^2 - \alpha)[(x - 1)^2 + \alpha - 4](1 - k + k\alpha) = 0.\tag{2.53}$$

Solutions of (2.53) are sketched in Fig. 2.9. $k = 0$ or 1 is a real parameter. Specific values of f_2 and f_3 and of their derivatives are obvious. The number of iterations needed to satisfy the relation

$$\max(|x^k - x^*|, |\alpha^k - \alpha^*|) \leq \varepsilon\tag{2.54}$$

is summarized in Table 2.3 for various initial approximations. It can be inferred from Table 2.3 that the iteration process (2.45), (2.46) is quadratically convergent when approaching limit points, but only linearly (if at all) convergent when reaching bifurcation points. The iteration process (2.52) converges quadratically to bifurcation points. Point 5 in Fig. 2.9 is the double bifurcation point for $k = 1$. The behavior of both iteration processes is shown in the second part of Table 2.3. (2.52) is then only linearly convergent to the double bifurcation point 5.

Limit points can be easily evaluated where only a single state variable appears as nonlinear in the system and where an inverse procedure for finding

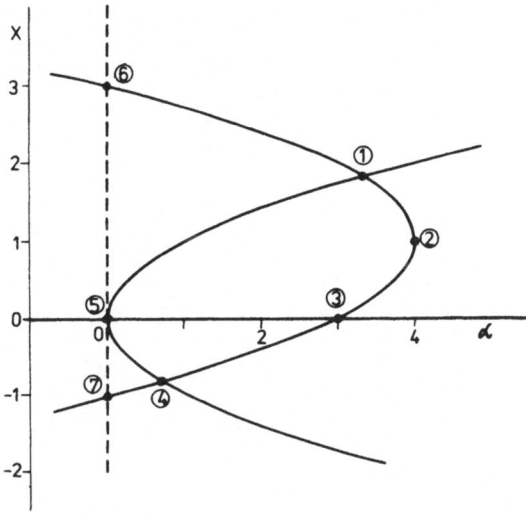

Point	x	α	$k = 0$	$k = 1$
1	1.8229	3.3229	BP	BP
2	1	0	LP	LP
3	0	3	BP	BP
4	−0.8229	0.6771	BP	BP
5	0	0	BP	double BP
6	3	0	–	BP
7	−1	0	–	BP

BP or LP denotes a bifurcation or limit point, respectively.

Figure 2.9 Solutions to (2.53).

the parametric dependence can be used. Consider Example 1 of Chapter 1 and the resulting decomposition (2.22). The limit point located on the curve $\theta(\text{Da})$ is characterized by relations $d\,\text{Da}/d\theta = 0$ or

$$\frac{\partial g(\theta, \Lambda, \gamma, B, \beta, \theta_c)}{\partial \theta} = 0. \tag{2.55}$$

If we solve for θ^* (for fixed values of Λ, γ, B, β, and θ_c), the value of Da at the limit point can be found from relation (2.22):

$$\text{Da}^* = g(\theta^*, \Lambda, \gamma, B, \beta, \theta_c). \tag{2.56}$$

An application of this procedure will be discussed in Section 2.6.

Table 2.3   Results for example (2.53); the number of iterations necessary to satisfy (2.54).

		Iteration Formula (2.45), (2.46)				Iteration Formula (2.52)			
	Initial	ε			Converged to the	ε			Converged to the
k	Approximation (x, α)	10^{-2}	10^{-4}	10^{-6}	point[a]	10^{-2}	10^{-4}	10^{-6}	point[a]
0	0.01, 0.01	0	11	20	5	0	2	2	5
	1.1, 4.1	2	2	3	2	9	19	27	(1.02, 3.07)[b]
	1.5, 3.0	8	15	27	1	3	4	4	1
	0,1, 3.1	6	13	19	3	2	2	3	3
	− 1.0, 0.5	7	14	21	4	2	4	5	4
	1.0, 1.0	7	15	25	5	4	5	5	5
	2.2, 2.0	8	14	21	1	3	4	4	1
	10, 10	19	26	>31	1	12	12	13	1
	− 10, − 10	19	26	>31	4	13	14	16	4
1	0.01, 0.01	0	>23		singul.	0	7	14	5
	1.0, 1.0	16	28	>31	5	16	28	>31	5
	− 1.1, − 0.1				singul.	2	4	5	7
	3.1, − 0.1				singul.	2	3	4	6

[a] For numbering of points, cf., Fig. 2.9.

[b] Stationary point of φ.

2.4.2 Direction of branches at a bifurcation point

We shall discuss a simple method for finding the directions of branches at a bifurcation point [2.12]. This information can be used to determine starting points for the continuation algorithm (cf., Section 2.2). Let us assume that we know the bifurcation point (x^{**}, α^{**}). All calculations will be made for the values $(x, \alpha) = (x^{**}, \alpha^{**})$. Let $\tilde{J}$ denote the expanded $n \times (n + 1)$ Jacobian matrix of (2.4) with elements

$$f'_{ij} = \frac{\partial f_i}{\partial x_j}, \quad f'_{i\alpha} = \frac{\partial f_i}{\partial \alpha};$$

hence, $\tilde{J}$ can be written as

$$\tilde{J} = \begin{pmatrix} f'_{11}, f'_{12}, \ldots, f'_{1n}, f'_{1\alpha} \\ f'_{21}, \ldots \\ \vdots \\ f'_{n1}, f'_{n2}, \ldots, f'_{nn}, f'_{n\alpha} \end{pmatrix}. \tag{2.57}$$

$\tilde{J}$ has rank less than n at the bifurcation point (this follows from conditions (2.38) and (2.41)). To derive the algorithm, we assume that the rank of $\tilde{J}$ is $n - 1$ (lower ranks will be considered later).

Let us assume that the submatrix $\boldsymbol{J}_1$ of the matrix $\tilde{\boldsymbol{J}}$,

$$\boldsymbol{J}_1 = \begin{pmatrix} f'_{12}, f'_{13}, \cdots, & f'_{1n} \\ f'_{22} \cdots & \\ \vdots & \\ f'_{n-1,2}, f'_{n-1,3}, \cdots, f'_{n-1,n} \end{pmatrix}, \tag{2.58}$$

is regular (if not, we must rearrange the equations). It follows from the implicit function theorem that in the neighborhood of the bifurcation point, there exist unique dependences

$$x_i = \varphi_i(x_1, \alpha), \quad i = 2, 3, \ldots, n, \tag{2.59}$$

with derivatives denoted as

$$\varphi'_{i1} = \frac{\partial \varphi_i}{\partial x_1}, \quad \varphi'_{i\alpha} = \frac{\partial \varphi_i}{\partial \alpha}, \quad i = 2, 3, \ldots, n, \tag{2.60}$$

for which the linear equations

$$\sum_{j=2}^{n} f'_{ij} \varphi'_{j1} = -f'_{i1}, \quad i = 1, 2, \ldots, n-1 \tag{2.61}$$

and

$$\sum_{j=2}^{n} f'_{ij} \varphi'_{j\alpha} = -f'_{i\alpha}, \quad i = 1, 2, \ldots, n-1 \tag{2.62}$$

hold.

Consider the second derivatives to be

$$\varphi'_{i11} = \frac{\partial^2 \varphi_i}{\partial x_1^2}, \quad \varphi'_{i1\alpha} = \frac{\partial^2 \varphi_i}{\partial x_1 \partial \alpha}, \quad \varphi'_{i\alpha\alpha} = \frac{\partial^2 \varphi_i}{\partial \alpha^2}. \tag{2.63}$$

Upon differentiating (2.61) with respect to x_1, we obtain

$$\sum_{j=2}^{n} f'_{ij} \varphi'_{j11} = - \sum_{j=2}^{n} \left[2f'_{ij1} \varphi'_{j1} + \sum_{k=2}^{n} (f'_{ijk} \varphi'_{k1}) \varphi'_{j1} \right] - f'_{i11},$$
$$i = 1, 2, \ldots, n-1. \tag{2.64}$$

In like manner, differentiating (2.61) with respect to α, we obtain

$$\sum_{j=2}^{n} f'_{ij} \varphi'_{j1\alpha} = - \sum_{j=2}^{n} \left[f'_{ij\alpha} \varphi'_{j1} + f'_{ij1} \varphi'_{j\alpha} + \sum_{k=2}^{n} (f'_{ijk} \varphi'_{k\alpha}) \varphi'_{j1} \right] - f'_{i1\alpha}$$
$$i = 1, 2, \ldots, n-1. \tag{2.65}$$

Finally, differentiating (2.62) with respect to α, we get

$$\sum_{j=2}^{n} f'_{ij} \varphi'_{j\alpha\alpha} = - \sum_{j=2}^{n} \left[2f'_{ij\alpha} \varphi'_{j\alpha} + \sum_{k=2}^{n} (f'_{ijk} \varphi'_{k\alpha}) \varphi'_{j\alpha} \right] - f'_{i\alpha\alpha},$$
$$i = 1, 2, \ldots, n-1. \tag{2.66}$$

Here we set

$$f'_{ijk} = \frac{\partial^2 f_i}{\partial x_j \partial x_k}, \quad f'_{ij\alpha} = \frac{\partial^2 f_i}{\partial x_j \partial \alpha}.$$

Note that the derivatives (2.63) are generated by one application of Gauss elimination for three vectors of the right-hand sides. The systems have the same matrix J_1 (we have assumed that J_1 is regular). Derivatives φ'_{i1}, $\varphi'_{i\alpha}$, computed earlier, are inserted into the right-hand sides.

Let us substitute the existing (but unknown) dependences (2.59) in the last equation of (2.4), i.e.,

$$F(x_1, \alpha) = f_n(x_1, \varphi_2(x_1, \alpha), \dots, \varphi_n(x_1, \alpha), \alpha) = 0. \tag{2.67}$$

This procedure of reducing the original n equations to a subsystem of $n - 1$ equations together with equation (2.67) corresponds to the Liapunov–Schmidt method for finite-dimensional problems. The next method of analysing Eq. (2.67) at the branch point is analogous to that described by Iooss and Joseph [1.82].

Evidently,

$$F(x_1, \alpha) = 0,$$

since $\varphi_i(x_1, \alpha) = x_i$ by (2.59). We compute partial derivatives of F in (2.67) with respect to x_1 and α at the bifurcation point (x_1^{**}, α^{**}):

$$\frac{\partial F}{\partial x_1} = f'_{n1} + \sum_{j=2}^{n} f'_{nj} \varphi'_{j1},$$

$$\frac{\partial F}{\partial \alpha} = f'_{n\alpha} + \sum_{j=2}^{n} f'_{nj} \varphi'_{j\alpha}. \tag{2.68}$$

It can be proved (e.g., by means of Cramer's rule for solving (2.61) and (2.62)) that, assuming (2.38) and (2.41),

$$\frac{\partial F}{\partial x_1} = 0, \quad \frac{\partial F}{\partial \alpha} = 0, \tag{2.69}$$

at the point (x_1^{**}, α^{**}). Relations (2.69) are the same as relations (2.38) and (2.41), except for a nonzero multiple. We shall now derive relations for the second derivatives of the function F:

$$A = \frac{\partial^2 F}{\partial x_1^2} = f'_{n11} + \sum_{j=2}^{n} \left[2f'_{n1j} \varphi'_{j1} + f'_{nj} \varphi'_{j11} + \sum_{k=2}^{n} (f'_{nkj} \varphi'_{k1}) \varphi'_{j1} \right]$$

$$B = \frac{\partial^2 F}{\partial x_1 \partial \alpha} = f'_{n1\alpha} + \sum_{j=2}^{n} \left[f'_{n1j} \varphi'_{j\alpha} + f'_{nj\alpha} \varphi'_{j1} + f'_{nj} \varphi'_{j1\alpha} \right.$$

$$\left. + \sum_{k=2}^{n} (f'_{njk} \varphi'_{k\alpha}) \varphi'_{j1} \right]$$

$$C = \frac{\partial^2 F}{\partial \alpha^2} = f'_{n\alpha\alpha} + \sum_{j=2}^{n} \left[2f'_{nj\alpha} \varphi'_{j\alpha} + f'_{nj} \varphi'_{j\alpha\alpha} + \sum_{k=2}^{n} (f'_{njk} \varphi'_{k\alpha}) \varphi'_{j\alpha} \right]. \tag{2.70}$$

Behavior of F in the neighborhood of the point (x_1^{**}, α^{**}) can be described approximately as (using Taylor's expansion)

$$2F(x_1, \alpha) \approx A(x_1 - x_1^{**})^2 + 2B(x_1 - x_1^{**})(\alpha - \alpha^{**}) + C(\alpha - \alpha^{**})^2 = 0. \tag{2.71}$$

Dividing (2.71) by $(x_1 - x_1^{**})^2$ or $(\alpha - \alpha^{**})^2$, and taking the limits $\alpha \to \alpha^{**}$, $x_1 \to x_1^{**}$, respectively, produces equations:

$$C\left(\frac{d\alpha}{dx_1}\right)^2 + 2B\frac{d\alpha}{dx_1} + A = 0, \tag{2.72}$$

or

$$A\left(\frac{dx_1}{d\alpha}\right)^2 + 2B\frac{dx_1}{d\alpha} + C = 0. \tag{2.73}$$

We shall now consider consequences of Eqs. (2.72) and (2.73).

Case 1 ($A \neq 0$): From (2.73), we obtain

$$\left.\frac{dx_1}{d\alpha}\right|_{1,2} = \frac{-B \pm \sqrt{B^2 - AC}}{A}, \tag{2.74}$$

assuming $D = B^2 - AC \geq 0$. The case $D > 0$ is depicted in Fig. 2.10(a); two existing branches intersect each other. The case $D = 0$ is shown in Fig. 2.10(b), where the branches have a common point of contact (and the same slope at the bifurcation point).

Case 2 ($A = 0, C \neq 0$): From (2.72),

$$\left.\frac{d\alpha}{dx_1}\right|_{1,2} = \frac{-B \pm \sqrt{B^2 - AC}}{C} = \begin{cases} 0 \\ -\dfrac{2B}{C} \end{cases}. \tag{2.75}$$

A solution diagram for this case is given in Fig. 2.10(c).

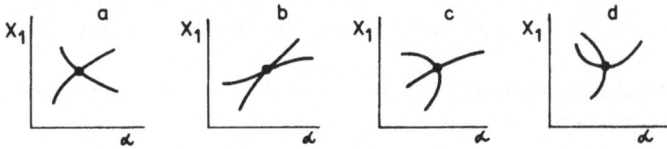

Figure 2.10 Types of bifurcation points in the solution diagram.

Case 3 ($A = C = 0$): Upon solving Eqs. (2.72) and (2.73), we obtain the derivatives

$$\frac{d\alpha}{dx_1} = 0, \quad \frac{dx_1}{d\alpha} = 0,$$ (2.76)

respectively.

This case is shown in Fig. 2.10(d).

To determine branch directions with respect to the remaining variables, we shall simply use the relations

$$dx_i = \varphi'_{i1} \, dx_1 + \varphi'_{i\alpha} \, d\alpha, \quad i = 2, \ldots, n,$$ (2.77)

where the derivatives φ'_{i1} and $\varphi'_{i\alpha}$ have already been computed. Depending on whether we deal with Cases 1, 2, or 3, we obtain either

$$\frac{dx_i}{dx_1} = \varphi'_{i1} + \varphi'_{i\alpha} \frac{d\alpha}{dx_1},$$ (2.78)

or

$$\frac{dx_i}{d\alpha} = \varphi'_{i1} \frac{dx_1}{d\alpha} + \varphi'_{i\alpha}.$$ (2.79)

2.4.2.1 Selecting starting points for the continuation algorithm

We shall describe the construction of starting points for the use of the continuation algorithm DERPAR (see Appendix A). To begin the algorithm, we have to choose:

1. An initial estimate of a point on the branch of solutions, i.e., $x_1, x_2, \ldots, x_n$, α.
2. A direction of continuation; the variables increase or decrease along the branch according to the values of N_i, $i = 1, 2, \ldots, n + 1$. For $N_i = 1$, the value of x_i increases along the branch, and for $N_i = -1$ it decreases. N_{n+1} corresponds to the parameter α. (N_i correspond to variables NDIR(i) in Appendix A.)
3. A variable which is kept fixed in the course of the initial Newton method. The variable will be indexed by k, $k \in [1, n + 1]$. This selection corresponds to the choice of preferences in the algorithm in Appendix A. The preferences are of the type PREF(k) $\doteq$ 0, PREF(i) $\neq$ 0, $i \neq k$. The preferences can be changed to usual values after the initial run of Newton's method (zero value of PREF(k) might hinder the continuation process).

Let us now determine the starting points. We shall introduce two small numerical constants

$$h_1 > 0, \quad h_\alpha > 0.$$ (2.80)

Table 2.4 Starting points for the continuation algorithm. For each direction, two starting points are used, $P = \pm 1$.

Direction	Starting Point	N_i	k
1	$\alpha = \alpha^{**} + Ph_\alpha$ $x_1 = x_1^{**} + Ph_\alpha S_1$ $x_i = x_i^{**} + P[\varphi'_{i1}hS_1 + \varphi'_{i\alpha}h_\alpha]$ $i = 2,\ldots,n$	$N_{n+1} = P$ $N_1 = P\,\text{sign}(S_1)$ $N_i = P\,\text{sign}(\varphi'_{i1}S_1 + \varphi'_{i\alpha})$ $i = 2,\ldots,n$	$n+1$
2	$x_1 = x_1^{**} + Ph_1$ $\alpha = \alpha^{**} + Ph_1 S_2$ $x_i = x_i^{**} + P[\varphi'_{i1}h_1 + \varphi'_{i\alpha}h_1 S_2]$ $i = 2,\ldots,n$	$N_1 = P$ $N_{n+1} = P\,\text{sign}(S_2)$ $N_i = P\,\text{sign}(\varphi'_{i1} + \varphi'_{i\alpha}S_2)$ $i = 2,\ldots,n$	1

At least one of the two directions computed can be characterized by a finite value of derivative $dx_1/d\alpha$ (we take the smaller value if both are finite).

$$\text{Direction 1}: \frac{dx_1}{d\alpha} = S_1. \tag{2.81}$$

Similarly, the second direction can be characterized by a finite value of the derivative $d\alpha/dx_1$.

$$\text{Direction 2}: \frac{d\alpha}{dx_1} = S_2. \tag{2.82}$$

All four starting points for continuation are given in Table 2.4. The direction of continuation and index k for the fixed variable are also given in Table 2.4. The point $(x_1,\ldots,x_n,\alpha)$ results from the initial Newton iteration. We can test whether the point actually belongs to the branch in question. For example, for direction 1 (ε small),

$$\left| \frac{x_i - x_i^{**}}{\alpha - \alpha^{**}} - \varphi'_{i1}S_1 - \varphi'_{i\alpha} \right| < \varepsilon, \quad i = 2,\ldots,n \tag{2.83a}$$

and

$$\left| \frac{x_1 - x_1^{**}}{\alpha - \alpha^{**}} - S_1 \right| < \varepsilon. \tag{2.83b}$$

Similar relations hold for the second direction. In this way, we can check all four points.

Remark 1. We could compute second derivatives $d^2x_1/d\alpha^2$ and $d^2\alpha/dx_1^2$, respectively, to obtain better approximations of branches in the neighborhood of the bifurcation point. However, we would have to compute third derivatives of functions f_i which might prove time consuming. On the other hand, we could determine convexity or concavity of functions in the case of a zero first derivative (perpendicular bifurcations).

Remark 2. When applying the algorithm, we have to evaluate first and second derivatives of the functions f_i. It appears to be effective to compute f'_{ij} from analytical formulas, and f'_{ijk} from the values of f'_{ij} by means of difference formulas. This means we must compute the matrix J $(n + 2)$ times altogether, (we can save some work by using symmetry of the second derivatives, i.e., the relations $f'_{ijk} = f'_{ikj}$).

2.4.2.2 Illustrative examples

We demonstrate the method on a simple example for $n = 2$. We shall choose

$$f_1 = x_1 - 2x_2 - \alpha - x_1 x_2 - 3x_2 \alpha + x_1^2 - 2x_2^2 - \alpha^2 = 0,$$
$$f_2 = x_2^2 - x_2^2 x_1^2 + \alpha^2 x_1^2 - \alpha^2 = 0. \tag{2.84}$$

An example of convergence of the Gauss–Newton method to one bifurcation point (Section 2.4.1) is given in Table 2.5. For this bifurcation point, the values of derivatives and resulting S_1, S_2 are presented in Table 2.6, together with proposed starting points for $h_1 = 0.01$, $h_\alpha = 0.01$. The solution dependences on parameter α, $x_1(\alpha)$, $x_2(\alpha)$ in the neighborhood of the bifurcation point are shown in Fig. 2.11.

We consider the problem of two interconnected reaction cells with mutual mass exchange, Example 4, Chapter 1, to illustrate how we compute directions of branching at the bifurcation point. Here the system of four equations takes the form (E4.1). We set $\alpha = D_1 = D_1^{12} = D_1^{21}$. The vector of variables is $x = (c_1^1, c_2^1, c_1^2, c_2^2)$. The other parameters are fixed at $D_2 = D_2^{12} = D_2^{21} = D_1/\rho$, $H_1 = H_2 = 0$. We shall use reaction mechanism (E4.2, 3) with the values of parameters A and B equal to 2 and 6, respectively. In this case, there exist primary bifurcation points on a homogeneous solution curve $c_1^1 = c_1^2 = A$, $c_2^1 = c_2^2 = B/A$ (cf., Fig. 2.4). The values $\alpha^{**} = D_1^{**}$ at the bifurcation points are $D_1^{**} = 0.0443$, $D_1^{**} = 2.256$.

Table 2.5 Convergence of the Gauss–Newton method to a bifurcation point of system (2.84).

Iteration	x_1	x_2	α
0	1.000	1.000	1.000
1	0.992	0.497	0.502
2	0.996	0.348	0.349
3	1.000	0.333	0.333
4	1.000	0.333	0.333

Table 2.6 Results for the bifurcation point $(1; 0.33333; 0.33333)$, Eq. (2.84).

$$J = \begin{bmatrix} 2.667 & -5.333 & -2.667 \\ 0 & 0 & 0 \end{bmatrix}$$

$$\varphi'_{21} = 0.5, \quad \varphi'_{2\alpha} = -0.5$$

	$j = 1$	$j = 2$	$j = \alpha$
f'_{11j}	2	-1	0
f'_{12j}	-1	-4	-3
$f'_{1\alpha j}$	0	-3	-2
f'_{21j}	0	-1.333	1.333
f'_{22j}	-1.333	0	0
$f'_{2\alpha j}$	1.333	0	0

$$A = -1.333, \quad B = 2.000, \quad C = 0.000$$

$$S_1 = 0 = \frac{dx_1}{d\alpha}, \frac{dx_2}{d\alpha} = -0.5$$

$$S_2 = 0.333 = \frac{d\alpha}{dx_1}, \frac{dx_2}{dx_1} = 0.333$$

Starting points $(h_1 = 0.01, h_\alpha = 0.01)$

x_1	x_2	α	N_1	N_2	N_3	k
1.000	0.328	0.343	0	-1	1	3
1.000	0.338	0.323	0	1	-1	3
1.010	0.336	0.336	1	1	1	1
0.990	0.330	0.330	-1	-1	-1	1

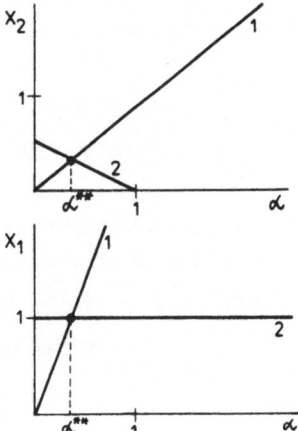

Figure 2.11 Solution diagram of (2.84) in the neighborhood of a bifurcation point.

Table 2.7 Directions of branching at the bifurcation point of Example 4, Chapter 1. $A = 2$, $B = 6$, $\rho = 0.1$, $\mathbf{x}^{**} = (2, 3, 2, 3)$, $\alpha^{**} = D_1^{**} = 0.04433$.

i	1	2	3	4	$\dfrac{d\alpha}{d\cdot}$
Direction 1: $\dfrac{dx_i}{d\alpha}$	0	0	0	0	(1)
Direction 2: $\dfrac{dx_i}{dx_1}$	(1)	-1.2278	-1.0000	1.2278	0

The results are summarized in Table 2.7, and they agree with those shown in Fig. 2.4.

2.4.2.3 Bifurcation points with higher degeneration

Let us consider the case where the matrix $\tilde{J}$ in (2.57) has rank $n - 2$ (if the rank were lower, the derivation would be similar, but less clear). Let us assume that a submatrix J_2 of $\tilde{J}$,

$$J_2 = \begin{pmatrix} f'_{13}, f'_{14}, \ldots, & f'_{1n} \\ f'_{23}, \ldots & \\ \vdots & \\ f'_{n-2,3}, f'_{n-2,4}, \ldots, f'_{n-2,n} \end{pmatrix}, \tag{2.85}$$

is regular. As in (2.59), the functions

$$x_i = \varphi_i(x_1, x_2, \alpha), \quad i = 3, 4, \ldots, n, \tag{2.86}$$

exist. For the derivatives φ'_{i1}, φ'_{i2}, $\varphi'_{i\alpha}$ we can derive a system of equations analogous to (2.61), (2.62) which can be solved for three right-hand sides simultaneously. To determine the second derivatives φ'_{i11}, φ'_{i12}, $\varphi'_{i1\alpha}$, φ'_{i22}, $\varphi'_{i2\alpha}$, $\varphi'_{i\alpha\alpha}$, we shall use equations similar to (2.64)–(2.66). Now we shall substitute (2.86) into the last two equations of (2.4) and obtain two equations in three unknowns:

$$F_1(x_1, x_2, \alpha) = f_{n-1}(x_1, x_2, \varphi_3(x_1, x_2, \alpha), \ldots, \varphi_n(x_1, x_2, \alpha), \alpha) = 0,$$
$$F_2(x_1, x_2, \alpha) = f_n(x_1, x_2, \varphi_3(x_1, x_2, \alpha), \ldots, \varphi_n(x_1, x_2, \alpha), \alpha) = 0. \tag{2.87}$$

When we compute the partial derivatives of the functions F_1 and F_2, we get a relation analogous to (2.71), i.e., the system of two equations

$$F_i(x_1, x_2, \alpha) \approx A_{i1}(x_1 - x_1^{**})^2 + A_{i2}(x_1 - x_1^{**})(x_2 - x_2^{**})$$
$$+ A_{i3}(x_1 - x_1^{**})(\alpha - \alpha^{**}) + A_{i4}(x_2 - x_2^{**})^2$$
$$+ A_{i5}(x_2 - x_2^{**})(\alpha - \alpha^{**}) + A_{i6}(\alpha - \alpha^{**})^2 = 0, \quad i = 1, 2. \tag{2.88}$$

Let us assume that $A_{i1} \neq 0$, $A_{i4} \neq 0$ and divide (2.88) by the factor $(\alpha - \alpha^{**})^2$. For $\alpha \to \alpha^{**}$, we have a system of two "quadratic equations" in two unknowns $dx_1/d\alpha$, $dx_2/d\alpha$:

$$A_{i1}\left(\frac{dx_1}{d\alpha}\right)^2 + A_{i2}\frac{dx_1}{d\alpha}\frac{dx_2}{d\alpha} + A_{i3}\frac{dx_1}{d\alpha} + A_{i4}\left(\frac{dx_2}{d\alpha}\right)^2$$

$$+ A_{i5}\frac{dx_2}{d\alpha} + A_{i6} = 0, \quad i = 1, 2. \tag{2.89}$$

Upon eliminating one unknown by means of resultant theory e.g., [2.13], we obtain a fourth-order equation for the unknown $dx_2/d\alpha$:

$$P\left(\frac{dx_2}{d\alpha}\right) = \det R\left(\frac{dx_2}{d\alpha}\right) = a_0\left(\frac{dx_2}{d\alpha}\right)^4 + a_1\left(\frac{dx_2}{d\alpha}\right)^3 + a_2\left(\frac{dx_2}{d\alpha}\right)^2$$

$$+ a_3\frac{dx_2}{d\alpha} + a_4 = 0, \tag{2.90}$$

the matrix $\mathbf{R}$ being defined as

$$\mathbf{R}(t) = \begin{pmatrix} A_{11} & A_{12}t + A_{13} & A_{14}t^2 + A_{15}t + A_{16} & 0 \\ 0 & A_{11} & A_{12}t + A_{13} & A_{14}t^2 + A_{15}t + A_{16} \\ A_{21} & A_{22}t + A_{23} & A_{24}t^2 + A_{25}t + A_{26} & 0 \\ 0 & A_{21} & A_{22}t + A_{23} & A_{24}t^2 + A_{25}t + A_{26} \end{pmatrix}$$

$$\tag{2.91}$$

Solutions of Eq. (2.90) yield the branch directions at the bifurcation point. We can obtain either four or two directions. If the degeneracy is higher (rank of $\tilde{J}$ is lower), the process becomes complicated.

2.4.3 Occurrence of isolas, isola formation

We shall now consider how new isolated families of solutions (isolas) arise when a parameter varies [2.14, 2.15]. We shall introduce a new parameter, β, into the set of stationary equations, obtaining

$$f_i(x_1, \ldots, x_n, \alpha, \beta) = 0, \quad i = 1, 2, \ldots, n. \tag{2.92}$$

In Fig. 2.12, for $n = 1$ we show what happens when isolas appear. They arise at the point P for $\alpha > \alpha_{is}$, and for $\alpha < \alpha_{is}$ they cease to exist. Let us derive necessary conditions for P. In the neighborhood of this point,

$$\alpha = \varphi(x_k, \beta) \tag{2.93}$$

is unique and has an extremum at P. Here we select component x_k.

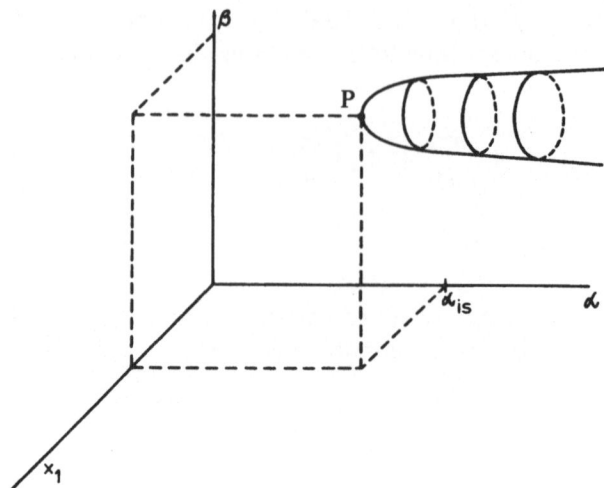

Figure 2.12 Occurrence of isolas, shown schematically.

When we choose values for x_k and β we can, by (2.92), calculate $x_1, \ldots, x_{k-1}$, $x_{k+1}, \ldots, x_n$, α and thus define the dependence φ. However, this is unnecessary for the computational algorithm. Since φ has an extremum at the point P,

$$\frac{\partial \varphi}{\partial x_k} = 0, \quad \frac{\partial \varphi}{\partial \beta} = 0. \tag{2.94}$$

Let us denote

$$J_k = \begin{vmatrix} \dfrac{\partial f_1}{\partial x_1}, \ldots, \dfrac{\partial f_1}{\partial x_{k-1}}, \dfrac{\partial f_1}{\partial \alpha}, \dfrac{\partial f_1}{\partial x_{k+1}}, \ldots, \dfrac{\partial f_1}{\partial x_n} \\ \dfrac{\partial f_2}{\partial x_1}, \\ \vdots \\ \dfrac{\partial f_n}{\partial x_1}, \ldots, \dfrac{\partial f_n}{\partial x_{k-1}}, \dfrac{\partial f_n}{\partial \alpha}, \dfrac{\partial f_n}{\partial x_{k+1}}, \ldots, \dfrac{\partial f_n}{\partial x_n} \end{vmatrix}, \quad \frac{\partial \mathbf{f}}{\partial x_k} = \begin{vmatrix} \dfrac{\partial f_1}{\partial x_k} \\ \dfrac{\partial f_2}{\partial x_k} \\ \vdots \\ \dfrac{\partial f_n}{\partial x_k} \end{vmatrix}, \tag{2.95}$$

and

$$\mathbf{r} = (r_1, \ldots, r_n)^T = \left(\frac{dx_1}{dx_k}, \ldots, \frac{dx_{k-1}}{dx_k}, \frac{d\alpha}{dx_k}, \frac{dx_{k+1}}{dx_k}, \ldots, \frac{dx_n}{dx_k} \right)^T. \tag{2.96}$$

Differentiating (2.92) with respect to x_k (here α is a function of x_k), we obtain a system of linear equations

$$J_k \cdot \mathbf{r} = -\frac{\partial \mathbf{f}}{\partial x_k}. \tag{2.97}$$

The kth component satisfies the first condition (2.94):

$$r_k(x_1, x_2, \ldots, x_n, \alpha, \beta) = 0. \tag{2.98}$$

Similarly, for realization of the second equation (2.94), we set

$$\frac{\partial \mathbf{f}}{\partial \beta} = \left(\frac{\partial f_1}{\partial \beta}, \frac{\partial f_2}{\partial \beta}, \ldots, \frac{\partial f_n}{\partial \beta}\right)^T,$$

and

$$\mathbf{s} = (s_1, \ldots, s_n)^T = \left(\frac{dx_1}{d\beta}, \ldots, \frac{dx_{k-1}}{d\beta}, \frac{d\alpha}{d\beta}, \frac{dx_{k+1}}{d\beta}, \ldots, \frac{dx_n}{d\beta}\right)^T.$$

Again we have a system of linear equations for fixed x_k:

$$\mathbf{J}_k \cdot \mathbf{s} = -\frac{\partial \mathbf{f}}{\partial \beta}. \tag{2.99}$$

Table 2.8 Course of the Newton iteration to determine the point of occurrence of isolas. Example 2, Chapter 1, $f = 1$, $\mu = 8.4 \times 10^{-6}$.

Iteration	x	y	z	y^0	k_0
0	0.200	1.000	0.100	0.300	1.000
1	0.251	0.797	0.125	2022	1.018
2	0.251	0.748	0.125	3496	1.010
3	0.249	0.750	0.125	3508	0.997
4	0.249	0.750	0.125	3508	0.997
0	0.300	2.000	0.500	0.500	1.000
1	0.226	1.093	0.117	3580	0.986
2	0.281	0.640	0.133	3576	1.125
3	0.257	0.736	0.127	3527	1.021
4	0.249	0.750	0.125	3509	0.997
5	0.249	0.750	0.125	3508	0.997
6	0.249	0.750	0.125	3508	0.997
0	2.000	2.000	2.000	2.000	2.000
1	0.767	3.584	1.772	54260	0.726
2	0.443	3.650	1.143	43880	0.509
3	0.276	3.437	0.698	29715	0.389
4	0.186	3.053	0.417	18715	0.323
5	0.148	2.431	0.250	11154	0.321
6	0.193	1.117	0.171	6039	0.551
7	0.221	0.732	0.116	3180	0.790
8	0.246	0.758	0.127	3554	0.953
9	0.249	0.750	0.125	3509	0.993
10	0.249	0.750	0.125	3508	0.997

We use the kth component to formulate the second condition (2.94):

$$s_k(x_1, x_2, \ldots, x_n, \alpha, \beta) = 0. \tag{2.100}$$

Note that (2.97) and (2.99) have the same matrices and thus can be solved simultaneously for two different right-hand sides with one use of Gaussian elimination.

Altogether we have obtained $n + 2$ nonlinear equations (2.92), (2.98), and (2.100) for $n + 2$ unknowns $x_1, \ldots, x_n, \alpha, \beta$. We can compute the residuals (left-hand sides) if we choose the values of $x_1, \ldots, x_n, \alpha, \beta$. We now have sufficient information to solve the system of nonlinear equations without evaluating derivatives, or to apply Newton's method with the Jacobian matrix evaluated by means of differences. It would be difficult to perform an analytical evaluation of the Jacobian matrix.

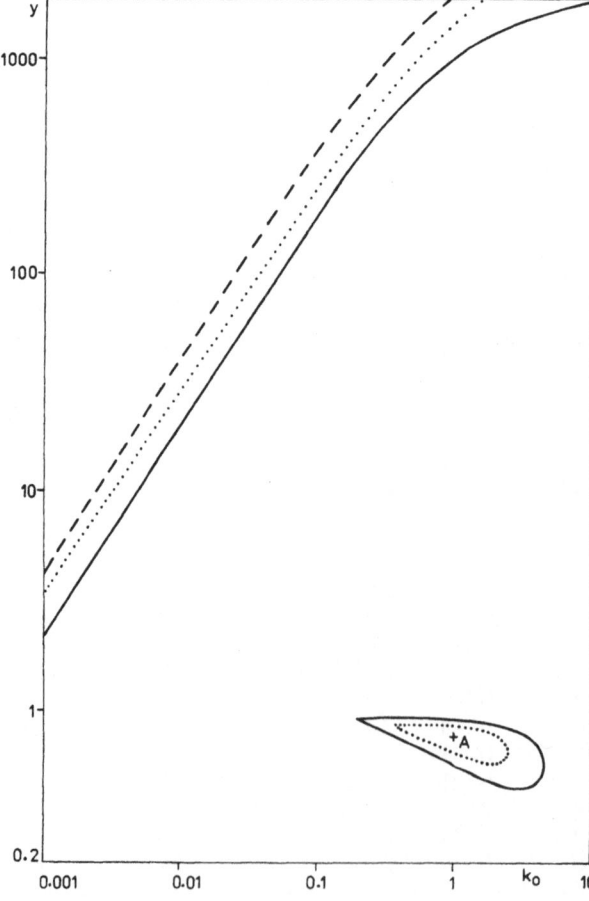

Figure 2.13 Dependence of the steady-state solution on the parameter k_0, Example 2, Chapter 1, $f = 1$, $\mu = 8.4 \times 10^{-6}$,
——— $y^0 = 2000$,
···· $y^0 = 3000$,
– – – $y^0 = 4000$; the point A corresponds to the resulting point of occurrence of isolas, cf., Table 2.8.

We have two ways to determine whether isolas exist for $\alpha > \alpha_{is}$ or for $\alpha < \alpha_{is}$, where α_{is} as well as β_{is} and $x_1^{is}, \ldots, x_n^{is}$ result when solving this system of $n + 2$ nonlinear equations.

First we choose (by trial and error) the point $x_1^{is}, \ldots, x_n^{is}$ for $\beta = \beta_{is}$ and $\alpha = \alpha_{is} \pm \varepsilon$ and solve (2.92). This yields the direction of isola formation. Another possibility is to calculate the values of the second derivatives $\partial^2 \varphi / \partial x_k^2$, $\partial^2 \varphi / \partial \beta^2$, $\partial^2 \varphi / \partial x_k \partial \beta$ and to use Sylvester's conditions to differentiate between maxima and minima of φ. Let us illustrate this procedure on Example 2, Chapter 1. Table 2.8 enumerates the results obtained by Newton's method. Isolas depicting dependence of steady state solutions on the parameter k_0 (corresponds to β) are shown in Fig. 2.13. Here the isolas exist for $y^0 < 3508$ (y^0 corresponds to α), cf., Table 2.8.

2.5 Branch Points—Complex Bifurcations

Let us reconsider the autonomous set of differential equations (2.1) dependent on α. As stated in Section 2.3, the stability of the stationary solution x for a given α is determined by the eigenvalues of linearized right-hand sides, i.e., by the eigenvalues of the Jacobian matrix $J = \{\partial f_i / \partial x_j\}$ evaluated for this stationary solution. The elements of the Jacobian matrix J depend on α continuously (with continuous dependence of the stationary solution $x(\alpha)$ on α). Hence, the eigenvalues of J also depend on α continuously. What happens to the stationary solution when, with α increasing, one or more eigenvalues cross the imaginary axis? The stability of the stationary solution can change. If, for $\alpha < \alpha_0$ all eigenvalues of J were in the complex left half-plane while for $\alpha > \alpha_0$, at least one eigenvalue were located in the right half-plane, the stationary solution will have lost its stability for $\alpha = \alpha_0$, and for $\alpha > \alpha_0$ it will be unstable. This travelling of eigenvalues from one half-plane to another can be realized in one of two qualitatively different ways (cf., Section 1.3). The case where the real eigenvalue crosses the imaginary axis (and hence is equal to zero) has already been discussed in Section 2.4. There the assumptions of the implicit function theorem are not satisfied, and usually new branches of stationary solutions arise (real bifurcation). Here we shall deal with the second type of bifurcation, i.e., where a pair of complex-conjugate eigenvalues cross the imaginary axis ("complex bifurcation"). We shall also assume that no other eigenvalue with a zero real part (in the entire neighborhood of $\alpha = \alpha_0$) occurs among the eigenvalues. The assumptions of the implicit function theorem are thus satisfied and the branch of stationary solutions continues for $\alpha > \alpha_0$. The stationary solution, however, has lost its stability at the point $\alpha = \alpha_0$ (when it was stable for $\alpha < \alpha_0$). Under certain conditions, a branch of periodic solutions (limit cycles) arises adjacent to the branch of stationary solutions.

We shall present the classical Hopf theorem (in a somewhat shortened and, consequently, incomplete version).

Then we shall briefly discuss results (obtained by means of perturbation techniques) concerning the behavior of the bifurcating periodic solution in the neighborhood of the bifurcation point.

2.5.1 The Hopf bifurcation theorem

Let us consider a steady-state solution $\bar{x}$ of Eqs. (2.4) for $\alpha = \alpha^+$ and a branch of stationary solutions $\bar{x}(\alpha)$ in the neighborhood of $\alpha = \alpha^+$. The functions f_i are taken to be sufficiently smooth.

Let us assume that all eigenvalues of the Jacobian matrix J are nonzero and that only two eigenvalues are purely imaginary; let us call them $\lambda(\alpha)$ and $\bar{\lambda}(\alpha)$ in the neighborhood of $\alpha = \alpha^+$. Then

$$\mathrm{Re}\{\lambda(\alpha^+)\} = \mathrm{Re}\{\bar{\lambda}(\alpha^+)\} = 0. \tag{2.101}$$

Let us further assume that the real part of $\lambda'(\alpha^+)$ is nonzero; that is,

$$\mathrm{Re}\{\lambda'(\alpha^+)\} \neq 0. \tag{2.102}$$

We then conclude that a branch of periodic solutions of (2.1) exists for $\alpha > \alpha^+$, $\alpha < \alpha^+$, or $\alpha = \alpha^+$. In greater detail: then there exist functions $x(t, \varepsilon)$, $T(\varepsilon)$, and $\alpha(\varepsilon)$ defined for all t and for a small ε [2.16], such that:

1. $\alpha(\varepsilon)$ and $T(\varepsilon)$ are power series in ε.
2. $T(0) = 2\pi/|\lambda(\alpha^+)|$ and $\alpha(0) = \alpha^+$.
3. $x(t, 0) = \bar{x}$, but $x(t, \varepsilon) \neq \bar{x}(\alpha(\varepsilon))$.
4. $x(t, \varepsilon)$ is a solution of equation

   $$x' = f(x, \alpha(\varepsilon)),$$

   with period $T(\varepsilon)$,
5. $\alpha'(\varepsilon)$ is either non-positive, or non-negative, or identically equal to zero for small ε.

If all eigenvalues of J have negative real parts for $\alpha < \alpha^+$, one of the following alternatives holds:

(a) $\alpha(\varepsilon) > \alpha^+$ for all small $\varepsilon \neq 0$ and each periodic solution $x(t, \varepsilon)$ is stable;
(b) $\alpha(\varepsilon) < \alpha^+$ for all small $\varepsilon \neq 0$ and each periodic solution is unstable.

From the theorem, we can see a very close connection between the complex bifurcation point (i.e., the point where two pure imaginary eigenvalues of the Jacobian matrix exist) and the appearance of periodic solutions (oscillations). We shall denote the points of complex bifurcation as (x^+, α^+).

Several papers dealing with the Hopf bifurcation have appeared recently. A good review can be found in [2.16]. However, there is almost no literature dealing with a non-analytic (numerical) determination of such bifurcations.

Numerical determination of the Hopf bifurcation point involves two tasks:

1. To determine the point $(\mathbf{x}^+, \alpha^+)$,
2. To determine the character of branching of the periodic solution branch, i.e., direction of branching and stability of the solution. We must also obtain the data required for beginning the algorithm for continuation of periodic solutions.

Hassard [2.17] has used a trial-and-error technique to determine the point of complex (Hopf) bifurcation. This method requires evaluating eigenvalues of the Jacobian matrix. Direct iteration techniques for determinating the point $(\mathbf{x}^+, \alpha^+)$ will be given in Sections 2.5.2 and 2.5.3.

Hassard [2.17] has derived the Hopf bifurcation formulas, i.e., the expansion of a branch of periodic solutions in the neighborhood of the point $(\mathbf{x}^+, \alpha^+)$ into a power series in ε. The governing equations are transformed into normal forms, see Appendix C.

Eigenvalues are assumed to be in the form

$$\lambda_{1,2}(\alpha) = \eta(\alpha) \pm i\omega(\alpha). \tag{2.103}$$

The Jacobian matrix J can be transformed into the Jordan canonical form at the point $(\mathbf{x}^+, \alpha^+)$ (transformation requires that all eigenvalues and eigenvectors of J be known). At the point $(\mathbf{x}^+, \alpha^+)$:

$$\eta(\alpha^+) = 0, \quad \omega(\alpha^+) = \omega_0, \quad \eta'(\alpha^+) \neq 0. \tag{2.104}$$

The values of α, of the period T, and of the exponent β of the bifurcating periodic solution are expressed in the form of a power series:

$$\alpha(\varepsilon) = \alpha^+ + \sum_{i=1}^{\infty} \alpha_i \varepsilon^i, \tag{2.105}$$

$$T(\varepsilon) = T_0 + \frac{2\pi}{\omega(\alpha^+)} \sum_{i=1}^{\infty} T_i \varepsilon^i, \tag{2.106}$$

$$\beta(\varepsilon) = \sum_{i=1}^{\infty} \beta_i \varepsilon^i. \tag{2.107}$$

Here T_0 is defined as $T_0 = 2\pi/\omega(\alpha^+)$, and the exponent β determines the stability of the periodic solution (cf., Appendix C and Section 2.8). For $\beta(\varepsilon) < 0$, the bifurcating periodic solution is asymptotically orbitally stable. If we determine the values of coefficients α_i, T_i, and β_i, we shall find the asymptotic periodic solution of (2.1) in the neighborhood of the point $(\mathbf{x}^+, \alpha^+)$:

$$\mathbf{x}(t, \alpha) = \bar{\mathbf{x}}(\alpha) + \varepsilon \, \mathrm{Re}\{e^{2\pi it/T} \mathbf{v}_1\} + O(\varepsilon^2), \tag{2.108}$$

where $\mathbf{v}_1$ is the eigenvector of J corresponding to the eigenvalue $\lambda_1(\alpha^+)$. The value of ε is approximately determined from $\alpha(\varepsilon)$. To find the coefficients α_i, T_i, and β_i, we can use the central manifold theorem and Poincaré normal forms. If relations (2.104) are satisfied, then $\alpha_1 = \alpha_3 = 0$, $\beta_1 = \beta_3 = 0$, and $T_1 = T_3 = 0$. The sign of the coefficient α_2 determines the direction of bifurcation, and the sign of β_2 determines the stability.

2.5.2 Direct decomposition technique for location of the complex bifurcation point

At a Hopf bifurcation point $(x_1^+, \ldots, x_n^+, \alpha^+)$, the Jacobian matrix J $(x_1, \ldots, x_n, \alpha)$ has a pair of complex-conjugate pure imaginary eigenvalues, i.e.,

$$\text{Re}\{\lambda_{1,2}\} = 0. \tag{2.109}$$

Let the characteristic polynomial of J be

$$P(\lambda) = \lambda^n + a_1 \lambda^{n-1} + a_2 \lambda^{n-2} + \cdots + a_{n-1}\lambda + a_n. \tag{2.110}$$

The necessary condition for the occurrence of the Hopf bifurcation point is expressed by (2.109), i.e., the polynomial (2.110) has two roots

$$\lambda_{1,2} = \pm i \sqrt{\omega^+}, \tag{2.111}$$

where $\omega^+ > 0$. The polynomial $P(\lambda)$ can thus be decomposed into

$$P(\lambda) = (\lambda^2 + \omega^+)P_{n-2}(\lambda), \tag{2.112}$$

where $P_{n-2}(\lambda)$ is a polynomial of degree $n - 2$. For an estimate of ω, we can write Eq. (2.112) in a form similar to that in the Lin–Bairstow method, e.g., [2.6]:

$$P(\lambda) = (\lambda^2 + \omega)(\lambda^{n-2} + p_1 \lambda^{n-3} + \cdots + p_{n-3}\lambda + p_{n-2}) + A\lambda + B. \tag{2.113}$$

Coefficients $p_1, \ldots, p_{n-2}$, A, and B can be calculated recursively:

$$p_{-1} = 0; \quad p_0 = 1; \quad p_k = a_k - \omega p_{k-2}, \quad k = 1, 2, \ldots, n - 2,$$

$$A = a_{n-1} - \omega p_{n-3}; \quad B = a_n - \omega p_{n-2}. \tag{2.114}$$

Here the values of p_k, A, and B depend on $\mathbf{x}$, α, and ω. For $\omega = \omega^+$, we can write [2.18]

$$f_{n+1}(x_1, x_2, \ldots, x_n, \alpha, \omega) = A(x_1, x_2, \ldots, x_n, \alpha, \omega) = 0,$$

$$f_{n+2}(x_1, x_2, \ldots, x_n, \alpha, \omega) = B(x_1, x_2, \ldots, x_n, \alpha, \omega) = 0. \tag{2.115}$$

As a result we have $n + 2$ equations (2.4), (2.115) in $n + 2$ unknowns x_1, $x_2, \ldots, x_n$, α, ω. The solution of these equations gives the Hopf bifurcation point $(x_1^+, x_2^+, \ldots, x_n^+, \alpha^+, \omega^+)$. The Newton iteration process can be used to solve this problem, partial derivatives $\partial f_i/\partial x_j$ and $\partial f_i/\partial \alpha$, $i, j = 1, \ldots, n$, can be evaluated analytically, and the remaining two rows of the Jacobian matrix for Newton's method can be calculated either numerically or analytically.

Let us note that the information on stability of the steady-state solution $\mathbf{x}(\alpha)$ in the vicinity of α^+ is contained in coefficients $p_1, \ldots, p_{n-2}$. The Routh–Hurwitz criterion can be used to test stability, cf., (2.34).

There is another way to determine α and ω at the Hopf bifurcation point. If $\mathbf{J}$ has an eigenvalue λ, $\mathbf{J}^2$ has an eigenvalue λ^2. This follows from $\mathbf{Jx} = \lambda\mathbf{x} \Rightarrow \mathbf{J}^2\mathbf{x} = \lambda\mathbf{Jx} = \lambda^2\mathbf{x}$. Then for the eigenvalues $\lambda_{1,2}$, from (2.111), we have $(\omega^+ > 0)$

$$\lambda_{1,2}^2 = -\omega^+. \tag{2.116}$$

Therefore, $\mathbf{J}^2$ has a double negative eigenvalue $-\omega^+$. Let us denote the polynomial of $\mathbf{J}^2$ as

$$Q(\omega) = \omega^n + q_1\omega^{n-1} + \cdots + q_{n-1}\omega + q_n. \tag{2.117}$$

Obviously, at the Hopf bifurcation point, relations (2.118) must be satisfied:

$$Q(-\omega^+) = 0 \quad \text{and} \quad \frac{dQ(-\omega^+)}{d\omega} = 0. \tag{2.118}$$

Equations (2.115) can thus be replaced by equations

$$f_{n+1}(x_1, x_2, \ldots, x_n, \alpha, \omega) = Q(\omega) = 0,$$

$$f_{n+2}(x_1, x_2, \ldots, x_n, \alpha, \omega) = \frac{dQ(\omega)}{d\omega} = n\omega^{n-1} + (n-1)q_1\omega^{n-2}$$

$$+ \cdots + q_{n-1} = 0. \tag{2.119}$$

We then use Newton's method in much the same way as before. If the iteration leads to a point $(x_1^+, \ldots, x_n^+, \alpha^+, -\omega^+)$, where $\omega^+ < 0$, this is not a point of complex bifurcation because the two eigenvalues found, λ_1 and λ_2, are then real and of opposite sign, $\lambda_{1,2} = \pm\sqrt{-\omega^+}$.

Let us examine the behavior of each method in Example 3, Chapter 1. We designate Da_1 as α (Da_1 is the Damköhler number in the first reactor) and then define the vector $\mathbf{x} = (x_1, \theta_1, x_2, \theta_2)^T$. Several results obtained by Newton's method applied to Eqs. (2.4) and (2.115) are presented in Table 2.9.

For $Da_2 = 0.1$, we find two different complex bifurcation points. The coefficients p_1 and p_2 in (2.113) indicate that the stability of steady-state solutions changes. Detailed numerical investigation reveals that for $Da_1 \in (0; 0.1540) \cup (0.2775; \infty)$, the steady-state solutions are stable and unique. On the other hand, a stable periodic solution exists in the region $Da_1 \in (0.1540; 0.2775)$.

Some results for Eqs. (2.4) and (2.119) are presented in Table 2.10. Note that convergence is slower here than in the results presented in Table 2.9. Therefore, the former method is preferable since no matrix multiplication is necessary and the rate of convergence is higher.

Table 2.9 Results for Example 3, Chapter 1, Newton's method for (2.4) and (2.115).
($\gamma = 1000, B = 12, \beta_1 = \beta_2 = 2, \theta_{c1} = \theta_{c2} = 0, \Lambda = 0.8, Da_2 = 0.1$.)

Iteration	x_1	θ_1	x_2	θ_2	Da_1	ω	$\sqrt{\sum_{i=1}^{6} f_i^2}$	
0	0.9000	3.0000	0.9000	3.0000	0.4000	20.0000	8.7	$E + 2$
1	0.9184	3.0557	0.8068	0.4138	0.4317	16.7030	9.4	$E + 2$
2	0.8958	2.9268	0.9011	0.9676	0.3525	12.7501	3.5	$E + 2$
3	0.8723	2.8402	0.8986	1.0483	0.3099	10.1811	1.1	$E + 2$
4	0.8546	2.7793	0.8869	1.0552	0.2881	8.6348	3.0	$E + 1$
5	0.8458	2.7492	0.8801	1.0538	0.2793	7.9527	4.7	$E - 0$
6	0.8438	2.7426	0.8786	1.0533	0.2775	7.8152	1.8	$E - 1$
7	0.8437	2.7424	0.8785	1.0533	0.2775	7.8098	2.8	$E - 4$
0	0.6000	2.0000	0.7000	1.0000	0.2000	1.0000	2.5	$E + 1$
1	0.5576	1.7730	0.6516	0.9565	0.1692	1.1902	6.2	$E - 0$
2	0.5031	1.5912	0.6042	0.9342	0.1570	0.6706	1.6	$E - 0$
3	0.4805	1.5156	0.5849	0.9229	0.1541	0.8066	4.1	$E - 2$
4	0.4801	1.5143	0.5845	0.9224	0.1540	0.8111	1.5	$E - 5$
0	0.5000	1.0000	0.7000	1.9000	0.1000	0.5000	1.1	$E - 0$[a]
1	0.2880	0.7285	0.6841	1.8581	0.0960	0.4507	1.4	$E - 1$[a]
2	0.2712	0.6648	0.6795	1.8549	0.0956	0.4497	1.8	$E - 3$[a]
3	0.2712	0.6651	0.6794	1.8543	0.0956	0.4491	1.6	$E - 6$[a]

[a] $Da_2 = 0.2$.

Table 2.10 Results for Example 3, Chapter 1, Newton's method for (2.4) and (2.119).
($\gamma = 1000, B = 12, \beta_1 = \beta_2 = 2, \theta_{c1} = \theta_{c2} = 0, \Lambda = 0.8, Da_2 = 0.1$.)

Iteration	x_1	θ_1	x_2	θ_2	Da_1	$-\omega$	$\sqrt{\sum_{i=1}^{6} f_i^2}$	
0	0.4000	1.4000	0.5000	0.9000	0.1200	-0.7000	6.5	$E - 0$
1	0.4945	1.5579	0.6042	0.9487	0.1539	-0.9563	1.4	$E - 1$
2	0.5276	1.6744	0.6238	0.9432	0.1605	-0.8277	5.8	$E - 0$
3	0.5064	1.6025	0.6065	0.9348	0.1575	-0.8221	1.5	$E - 0$
4	0.4938	1.5605	0.5960	0.9290	0.1558	-0.8183	4.2	$E - 1$
5	0.4872	1.5380	0.5904	0.9252	0.1550	-0.8151	1.2	$E - 1$
6	0.4837	1.5263	0.5875	0.9242	0.1545	-0.8132	3.1	$E - 2$
7	0.4819	1.5204	0.5860	0.9233	0.1543	-0.8122	8.0	$E - 3$
8	0.4810	1.5174	0.5853	0.9229	0.1541	-0.8116	2.1	$E - 3$
9	0.4806	1.5158	0.5849	0.9227	0.1541	-0.8114	5.4	$E - 4$
10	0.4803	1.5151	0.5847	0.9226	0.1540	-0.8112	1.5	$E - 4$
11	0.4802	1.5147	0.5846	0.9225	0.1540	-0.8112	4.3	$E - 5$
12	0.4802	1.5145	0.5846	0.9225	0.1540	-0.8111	1.4	$E - 5$
13	0.4801	1.5144	0.5846	0.9225	0.1540	-0.8111		

Information about $n - 2$ eigenvalues of the Jacobian matrix is contained in coefficients $p_1, \ldots, p_{n-2}$ of the polynomial $P_{n-2}(\lambda)$, cf., (2.112) and (2.113). If Routh–Hurwitz conditions (2.34) are satisfied, these $n - 2$ eigenvalues are located in the complex left half-plane and at the point $(x_1^+, x_2^+, \ldots, x_n^+, \alpha^+)$ periodic solutions bifurcate from the stable stationary solution. The direction of the loss of stationary-solution stability at the point $(\mathbf{x}^+, \alpha^+)$ is determined by the signs of the derivatives

$$\frac{d\,\mathrm{Re}\{\lambda_1\}}{d\alpha} = \frac{d\,\mathrm{Re}\{\lambda_2\}}{d\alpha}. \tag{2.120}$$

For $d\,\mathrm{Re}\{\lambda_1\}/d\alpha > 0$, in case the point $\alpha = \alpha^+$ is crossed in the positive direction, the steady-state solution loses its stability. If the periodic solution bifurcates supercritically (in the direction $\alpha > \alpha^+$), it is stable; if the bifurcation is subcritical (in the direction $\alpha < \alpha^+$), it is unstable. On the other hand, if $d\,\mathrm{Re}\{\lambda_1\}/d\alpha < 0$, the steady-state solution is stable for $\alpha > \alpha^+$ and unstable for $\alpha < \alpha^+$. How can we determine the signs of (2.120)? Let

$$\lambda_1 = r(\cos\varphi + i\sin\varphi). \tag{2.121}$$

Then, at the point of complex bifurcation $(x_1^+, \ldots, x_n^+, \alpha^+)$,

$$r^+ = \sqrt{-\omega^+}, \quad \varphi^+ = \frac{\pi}{2}, \tag{2.122}$$

and the sign of (2.120) is determined by the expression

$$\frac{d\varphi(x_1^+, \ldots, x_n^+, \alpha^+)}{d\alpha} \lessgtr 0 \tag{2.123}$$

By substituting (2.121) into the characteristic polynomial (2.110), we obtain, for real (R) and imaginary (I) parts,

$$P_R(\mathbf{x}, \alpha, r, \varphi) = r^n \cos n\varphi + a_1 r^{n-1} \cos(n-1)\varphi + \cdots + a_n = 0, \tag{2.124}$$

$$P_I(\mathbf{x}, \alpha, r, \varphi) = r^n \sin n\varphi + a_1 r^{n-1} \sin(n-1)\varphi + \cdots + a_{n-1} r \sin\varphi = 0, \tag{2.125}$$

where $a_1, \ldots, a_n$ depend on $\mathbf{x}$ and α.

Equations (2.124) and (2.125) form a system which determines r and φ; we must solve (2.123). By applying the implicit function theorem, we obtain the derivatives $dr/d\alpha$ and $d\varphi/d\alpha$ as the solution of the linear equations

$$\begin{pmatrix} \dfrac{\partial P_R}{\partial r}, & \dfrac{\partial P_R}{\partial \varphi} \\[2mm] \dfrac{\partial P_I}{\partial r}, & \dfrac{\partial P_I}{\partial \varphi} \end{pmatrix} \begin{pmatrix} \dfrac{dr}{d\alpha} \\[2mm] \dfrac{d\varphi}{d\alpha} \end{pmatrix} = \begin{pmatrix} -\dfrac{\partial P_R}{\partial \alpha} \\[2mm] -\dfrac{\partial P_I}{\partial \alpha} \end{pmatrix}, \tag{2.126}$$

where the matrix of the system and the right-hand side are evaluated at the point of complex bifurcation $(x_1^+, \ldots, x_n^+, \alpha^+)$ and r, φ are taken as r^+, φ^+, respectively. The derivatives $\partial P_R/\partial r$, $\partial P_R/\partial \varphi$, $\partial P_I/\partial r$, and $\partial P_I/\partial \varphi$ can be evaluated by differentiating (2.124) and (2.125) with respect to r and φ.

For the derivative $\partial P_R/\partial \alpha$, we can form the following relation:

$$\frac{\partial P_R}{\partial \alpha} = \left[\sum_{j=1}^{n} \frac{\partial a_1}{\partial x_j} \frac{dx_j}{d\alpha} + \frac{\partial a_1}{\partial \alpha} \right] r^{n-1} \cos(n-1)\varphi + \cdots + \left[\sum_{j=1}^{n} \frac{\partial a_n}{\partial x_j} \frac{dx_j}{d\alpha} + \frac{\partial a_n}{\partial \alpha} \right].$$

(2.127)

A similar expression is obtained for $\partial P_I/\partial \alpha$ by differentiating (2.125) with respect to α. $dx_j/d\alpha$ can be found by solving the system of linear equations

$$J(x_1, \ldots, x_n, \alpha) \frac{dx}{d\alpha} = -\frac{\partial f}{\partial \alpha},$$

(2.128)

since J does not have a zero eigenvalue at this point and hence is regular. The values of derivatives $\partial a_i/\partial x_j$ and $\partial a_i/\partial \alpha$ can be obtained, for example, by means of difference formulas or by analytical differentiation (for small n), e.g., in the form

$$\frac{\partial a_1}{\partial x_j} = \left(-\frac{\partial^2 f_1}{\partial x_1\, \partial x_j} - \frac{\partial^2 f_2}{\partial x_2\, \partial x_j} - \cdots - \frac{\partial^2 f_n}{\partial x_n\, \partial x_j} \right) \cdot (-1)^n.$$

2.5.3 Direct iteration techniques

The method of locating the point of complex bifurcation discussed in the previous section required repeated evaluation of coefficients of the characteristic polynomial (2.110). Here we shall present two iterative algorithms where this is unnecessary [2.19]. Similar algorithm has been suggested in [2.20]. For an eigenvalue λ of the Jacobian matrix J,

$$J\mathbf{u} = \lambda\mathbf{u},$$

(2.129)

where $\mathbf{u}$ is an eigenvector. We shall consider complex λ, $\lambda = r + is$, and therefore the vector $\mathbf{u}$ will also be complex, i.e., $\mathbf{u} = \mathbf{v} + i\mathbf{w}$, where $\mathbf{v}$ and $\mathbf{w}$ are real vectors. Equation (2.129) can then be rewritten for the case of complex bifurcation (i.e., for $r = 0$)

$$J\mathbf{v} + s\mathbf{w} = 0,$$

(2.130)

$$J\mathbf{w} - s\mathbf{v} = 0.$$

(2.131)

Let us consider Eqs. (2.130) and (2.131) in the form

$$f_i(x_1, \ldots, x_n, \alpha, s, v_1, \ldots, v_n, w_1, \ldots, w_n) = 0, \quad i = n+1, n+2, \ldots, 3n.$$

Every complex multiple of the eigenvector $\mathbf{u} = \mathbf{v} + i\mathbf{w}$ is also an eigenvector of J belonging to the eigenvalue $\lambda = is$. In practice, this means that we can choose two components of the vectors $\mathbf{v}$ and $\mathbf{w}$ arbitrarily. This statement is valid "almost always" and for a general eigenvalue ($r \neq 0$) of $\boldsymbol{J}$. We shall not discuss conditions under which individual choices of two such components may be made. Let us arrange the unknowns into a vector

$$\boldsymbol{\psi} = (x_1, x_2, \ldots, x_n, \alpha, s, v_1, \ldots, v_n, w_1, \ldots, w_n). \tag{2.132}$$

In fact we have $3n$ nonlinear equations (2.4), (2.130), and (2.131) in $3n + 2$ unknowns, the components of the vector $\boldsymbol{\psi}$. However, two of the last $2n$ components of $\boldsymbol{\psi}$ can be chosen (fixed) when the set of $3n$ equations (2.4), (2.130), and (2.131) is solved for the remaining $3n$ components of $\boldsymbol{\psi}$ (i.e., the point of complex bifurcation). We use the Newton iteration method to solve this $3n \times 3n$ system. Part of the Jacobian matrix for the Newton method can be evaluated analytically, the rest by means of finite difference approximations for partial derivatives. An occurrence matrix of the solved system (and also of the Jacobian matrix used in the Newton method) is sketched in Fig. 2.14.

The elements of the Jacobian matrix which can easily be obtained analytically are denoted by a circle. As mentioned above, two components of the vectors $\mathbf{v}$ and $\mathbf{w}$ can be chosen and fixed in the course of iteration. Thus, three possibilities can occur in Fig. 2.14.

(a) dim $\tilde{\mathbf{v}} = n - 2$, dim $\tilde{\mathbf{w}} = n$ (two components of vector $\mathbf{v}$ are chosen);
(b) dim $\tilde{\mathbf{v}} = n - 1$, dim $\tilde{\mathbf{w}} = n - 1$ (one component of vector $\mathbf{v}$ and one component of vector $\mathbf{w}$ are chosen);
(c) dim $\tilde{\mathbf{v}} = n$, dim $\tilde{\mathbf{w}} = n - 2$ (two components of vector $\mathbf{w}$ are chosen).

We can therefore formulate a new algorithm:

Algorithm A

1. Choose an initial estimate $(x_1, \ldots, x_n, \alpha, s, v_1, \ldots, v_n, w_1, \ldots, w_n)$ and determine two of the last $2n$ components which remain unchanged over the entire course of iteration (we may not choose zero values).

	$x_1 \ldots x_n$	α	s	$\tilde{\mathbf{v}}$	$\tilde{\mathbf{w}}$
Eqs. (2.4)	$\otimes$	$\otimes$	—	—	—
Eqs. (2.130)	$\times$	$\times$	$\otimes$	$\otimes$	$\otimes$
Eqs. (2.131)	$\times$	$\times$	$\otimes$	$\otimes$	$\otimes$

Figure 2.14 Occurrence matrix for systems (2.4), (2.130), (2.131).

2. Evaluate the Jacobian matrix J.
3. Evaluate $3n$ residuals (2.4), (2.130), and (2.131).
4. Evaluate $3n \times 3n$ Jacobian matrix for Newton's method. (For elements which can be easily determined analytically, see Fig. 2.14.)
5. Compute the next Newton iterate and repeat step 2.

Next we shall derive an algorithm which will be of lower dimension. From Eqs. (2.130) and (2.131), we can easily obtain

$$J^2 v + s^2 v = 0 \tag{2.133}$$

or

$$J^2 w + s^2 w = 0. \tag{2.134}$$

Consider Eqs. (2.133) in the form $f_i(x_1, \ldots, x_n, \alpha, s^2, v_1, \ldots, v_n) = 0$, $i = n + 1, \ldots, 2n$. We have $2n$ equations (2.4) and (2.133) in $2n + 2$ unknowns $x_1, \ldots, x_n, \alpha, s^2, v_1, \ldots, v_n$. Similarly, we can again choose two components of the vector v arbitrarily. They remain fixed during the Newton iteration. An occurrence matrix of (2.4) and (2.133) is sketched in Fig. 2.15. The elements of the Jacobian matrix for Newton's method which can easily be determined analytically are again denoted by a circle. The dimensionality of the problem is lower than in Algorithm A. Let us summarize this algorithm:

Algorithm B

1. Choose an initial estimate $(x_1, \ldots, x_n, \alpha, s, v_1, \ldots, v_n)$ and determine two components of vector v which remain unchanged in the course of iteration (we choose non zero values of these components).
2. Evaluate the Jacobian matrix J and its square J^2.
3. Evaluate $2n$ residuals (2.4) and (2.133).
4. Evaluate a $2n \times 2n$ Jacobian matrix in the Newton method. (For elements which can be easily calculated analytically, see Fig. 2.15.)
5. Compute the next Newton iterate and repeat step 2.

The results of iterating according to Algorithms A and B in Example 3, Chapter 1 are given in Tables 2.11 and 2.12. The rates of convergence of both

	$x_1 \ldots x_n$	α	s^2	$\tilde{v}$
Eqs. (2.4)	$\otimes$	$\otimes$	$-$	$-$
Eqs. (2.133)	$\times$	$\times$	$\otimes$	$\otimes$

Figure 2.15 Occurrence matrix for systems (2.4), (2.133), dim $\tilde{v} = n - 2$.

Table 2.11 Determination of Hopf bifurcation points. Results for Algorithm A. Newton's method, double precision arithmetics, Example 3, Chapter 1. ($\gamma = 1000$, $B = 12$, $\beta_1 = \beta_2 = 2$, $\theta_{c1} = \theta_{c2} = 0$, $\Lambda = 0.8$, $Da_2 = 0.2$.)

Iteration	x_1	θ_1	x_2	θ_2	$\alpha = Da_1$	s	v_1	v_2	v_3	v_4	w_1	w_2	w_3	w_4	$(\sum_{i=1}^{12} f_i^2)^{1/2}$
0	0.5000	1.0000	0.7000	1.9000	0.1000	0.7000	1.0000	1.0000	1.0000	1.0000	1.0000	1.0000	1.0000	1.0000	1.2 E + 1
1	0.3531	0.9915	0.6149	1.0634	0.1410	0.7346	1.0000	1.0000	4.5930	11.725	0.2824	3.4724	3.1896	33.016	3.0 E + 1
2	0.2917	0.7723	0.6178	1.5015	0.1139	0.6250	1.0000	1.0000	3.9275	8.2677	0.4104	3.2657	3.0732	26.870	5.6 E0
3	0.2713	0.6701	0.6696	1.8082	0.0980	0.6619	1.0000	1.0000	3.8409	6.8199	0.4261	3.3781	3.5048	30.756	9.6 E − 1
4	0.2714	0.6658	0.6793	1.8533	0.0957	0.6710	1.0000	1.0000	3.7973	6.4110	0.4341	3.4675	3.5808	32.030	4.0 E − 2
5	0.2712	0.6651	0.6794	1.8543	0.0956	0.6702	1.0000	1.0000	3.7990	6.4239	0.4339	3.4648	3.5813	32.055	9.0 E −6

Table 2.12 Determination of Hopf bifurcation points. Results for Algorithm B. Newton's method, double precision arithmetics, Example 3, Chapter 1. ($\gamma = 1000$, $B = 12$, $\beta_1 = \beta_2 = 2$, $\theta_{c1} = \theta_{c2} = 0$, $\Lambda = 0.8$, $Da_2 = 0.2$.)

Iteration	x_1	θ_1	x_2	θ_2	$\alpha = Da_1$	s^2	v_1	v_2	v_3	v_4	$(\sum_{i=1}^{8} f_i^2)^{1/2}$
0	0.5000	1.0000	0.7000	1.9000	0.1000	0.4900	1.0000	1.0000	1.0000	1.0000	9.3 E0
1	0.2715	0.6620	0.7016	2.0588	0.0847	0.4574	1.0000	1.0000	2.9591	1.9184	4.6 E0
2	0.2687	0.6573	0.6738	1.8255	0.0949	0.4166	1.0000	1.0000	3.8157	6.9071	8.5 E − 2
3	0.2712	0.6647	0.6795	1.8550	0.0955	0.4485	1.0000	1.0000	3.8019	6.4398	1.2 E − 2
4	0.2712	0.6651	0.6794	1.8543	0.0956	0.4491	1.0000	1.0000	3.7990	6.4239	1.0 E − 5

algorithms seem to be similar. Comparison of Algorithms A and B with those presented in Section 2.5.2 (with respect to computer-memory requirements and convergence properties) shows that Algorithms A and B are as effective as the first algorithm of Section 2.5.2; the second algorithm of Section 2.5.2 was less successful [2.19]. Algorithms A and B are preferable because they do not require evaluation of coefficients of the characteristic polynomial (for higher n, rounding errors could cause problems in evaluation).

The values of derivatives (2.120) can also be computed after solving by Algorithms A or B. Equations (2.130) and (2.131) are of the form (before the terms with r are deleted)

$$\mathbf{Jv} - r\mathbf{v} + s\mathbf{w} = \mathbf{0}, \quad \mathbf{Jw} - s\mathbf{v} - r\mathbf{w} = \mathbf{0}.$$

We can consider that, together with (2.4), we have a system of $3n$ nonlinear equations in $3n + 1$ unknowns. Differentiating with respect to α, we obtain a system of linear equations for evaluating a vector of derivatives

$$\left(\frac{dx_1}{d\alpha}, \ldots, \frac{dx_n}{d\alpha}, \frac{dr}{d\alpha}, \frac{ds}{d\alpha}, \frac{dv_1}{d\alpha}, \ldots, \frac{dv_n}{d\alpha}, \frac{dw_1}{d\alpha}, \ldots, \frac{dw_n}{d\alpha} \right).$$

Again, we set two of the last $2n$ components equal to zero. The component $dr/d\alpha$ of this resulting vector then determines the direction of the "loss of stability" of the steady-state solution in question.

2.6 Bifurcation Diagram

Consider this problem with another parameter, i.e., let us write (2.1) as

$$\frac{dx_i}{dt} = f_i(x_1, \ldots, x_n, \alpha, \beta), \quad i = 1, 2, \ldots, n. \tag{2.135}$$

If we find the limit point $(x_1^*, \ldots, x_n^*, \alpha^* = \alpha_0)$, for $\beta = \beta_0$, we can use the continuation technique of Section 2.2 to calculate limit points that vary continuously with the parameter β. We obtain a curve in the "β-α" plane as schematically shown in Fig. 2.16. We shall call it the "limit points curve".

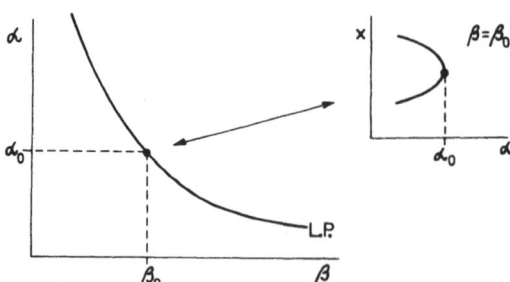

Figure 2.16 Schematic representation of loci of limit points.

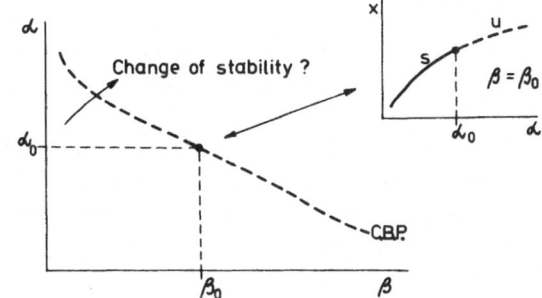

Figure 2.17 Schematic representation of loci of complex bifurcation points. s—stable, u—unstable

When crossing this curve (in the "β-α" parameter plane), the number of stationary solutions of (2.135) changes by two. We remark that the curve, which characterizes bifurcation points (of a non-limit type), does not supply any information about the change in the number of solutions.

In like manner, we can apply continuation techniques where complex bifurcation points depend on β, cf., Fig. 2.17. We shall call this the "complex bifurcation points curve" or "Hopf bifurcation points curve."

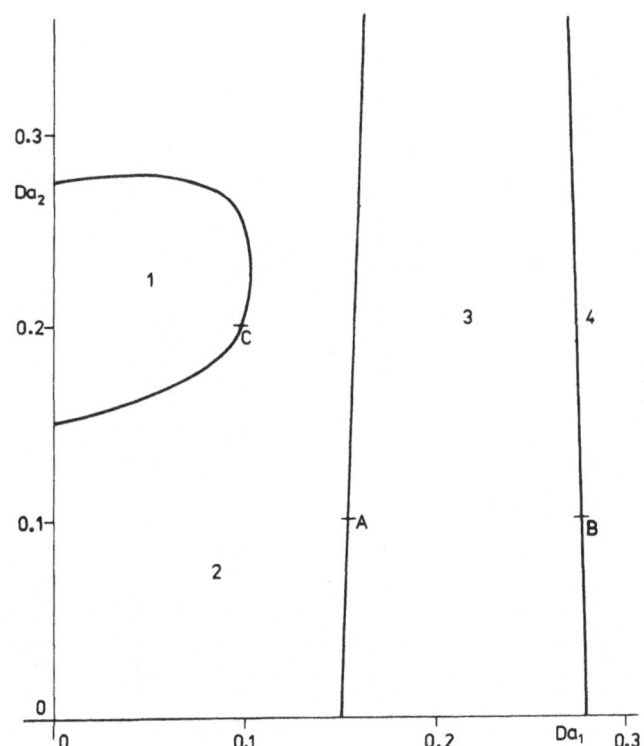

Figure 2.18 Bifurcation diagram for Example 3, Chapter 1, $\gamma = 1000$, $B = 12$, $\beta_1 = \beta_2 = 2$, $\theta_{c1} = \theta_{c2} = 0$, $\Lambda = 0.8$.

If both limit points curves and complex bifurcation points curves are drawn in one figure, a bifurcation diagram results. The above curves will divide the "β-α" parameter plane into regions, where the total number of stationary solutions and the number of stable stationary solutions are constant.

Let us discuss several such diagrams for Example 3, Chapter 1. The diagram of complex bifurcation points for the case where a unique stationary solution of the problem always exists is given in Fig. 2.18, $\alpha = Da_1$, $\beta = Da_2$. The points $A, B,$ and C correspond to results given in Table 2.9. Stationary solutions are unstable in regions 1 and 3, and limit cycles (periodic solutions) exist there. Stationary solutions are stable in regions 2 and 4.

A somewhat more complex situation arises for higher values of B, where multiple solutions already exist. A bifurcation diagram for such a case is given in Fig. 2.19. Each of the 15 regions in the parameter plane is characterized by numbers N-M, where N is the total number of stationary solutions and M is the number of stable stationary solutions. An analogous bifurcation diagram for Example 4, Chapter 1 with $n = 2$, is presented in Fig. 2.20. The reader may compare the bifurcation diagram with the solution diagram in Fig. 2.4, where B equals 6.

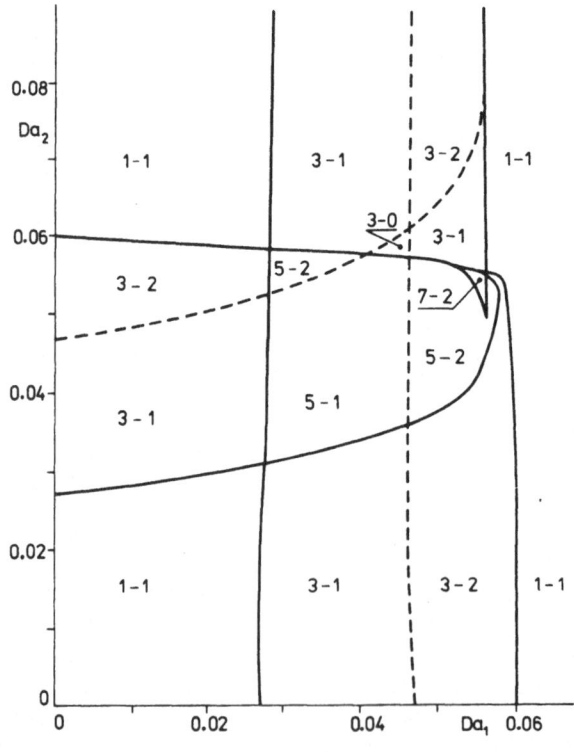

Figure 2.19 Bifurcation diagram for Example 3, Chapter 1, $\gamma = 1000$, $B = 22$, $\beta_1 = \beta_2 = 2$, $\theta_{c1} = \theta_{c2} = 0$, $\Lambda = 0.8$.
——, loci of limit points
– – – –, loci of complex bifurcation points

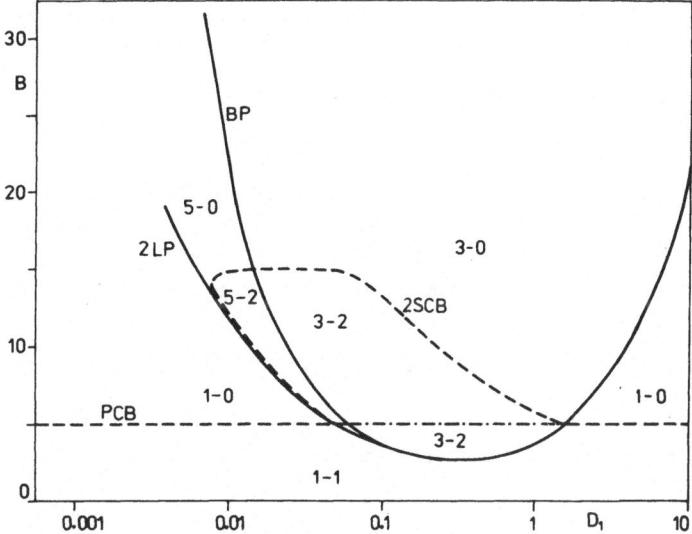

Figure 2.20 Bifurcation diagram for Example 4, Chapter 1, $A = 2$, $D_i = D_i^{12} = D_i^{21}$, $i = 1,2$; $D_1/D_2 = 0.1$. (PCB-primary complex bifurcation, BP-bifurcation-limit point, 2LP-limit point appearing twice due to symmetry, 2SCB-secondary complex bifurcation point appearing twice due to symmetry)

2.7 Transient Behavior of LPS—Numerical Methods

A great number of numerical methods have been developed to solve the system of ordinary differential equations

$$\frac{d\mathbf{x}}{dt} = \mathbf{f}(t, \mathbf{x}), \tag{2.136}$$

with the initial condition $\mathbf{x}(0) = \mathbf{x}^0$. A good review of methods for non-stiff systems can be found in the book by Henrici [2.3]. Let an approximate solution at a point t_j ($j = 0, 1, \ldots, t_0 = 0$, $h_j = t_{j+1} - t_j > 0$) be $\mathbf{x}^j \sim \mathbf{x}(t_j)$. The methods of numerical integration can be classified into single-step and multistep techniques.

By means of the single-step methods, the approximation at the successive points t_{j+1} is constructed solely on the basis of the known solution approximation at the point t_j:

$$\mathbf{x}^{j+1} = \mathbf{x}^j + h_j \boldsymbol{\phi}(t_j; \mathbf{x}^j; h_j), \tag{2.137}$$

where the function $\boldsymbol{\phi}$, which does not depend on t_j for (2.1) (autonomous system), is dependent on the actual form of the right-hand sides $\mathbf{f}$ of differential equations (2.136) to be solved. The function $\boldsymbol{\phi}$ thus characterizes the particular single-step method. (We consider only explicit single-step methods in (2.137), i.e., such methods, where the function ϕ is independent of x^{j+1}.)

2.7.1 Runge–Kutta methods

The most important among the single-step methods are Runge–Kutta methods. Only such methods are practical for dynamic simulation problems, which include an adaptive control of step size, h_j, that depends on the solution and on the maximum admissible approximation error in a single step.

A classical example of such methods is the Merson modification of the fourth-order Runge–Kutta scheme. The function ϕ is expressed recursively in the form

$$\mathbf{k}^1 = h\mathbf{f}(t_j, x^j), \qquad\qquad \mathbf{y}^1 = \mathbf{x}^j + \frac{\mathbf{k}^1}{3},$$

$$\mathbf{k}^2 = h\mathbf{f}\left(t_j + \frac{h_j}{3}, \mathbf{y}^1\right), \qquad \mathbf{y}^2 = \mathbf{x}^j + \frac{\mathbf{k}^1 + \mathbf{k}^2}{6},$$

$$\mathbf{k}^3 = h\mathbf{f}\left(t_j + \frac{h_j}{3}, \mathbf{y}^2\right), \qquad \mathbf{y}^3 = \mathbf{x}^j + \frac{\mathbf{k}^1 + 3\mathbf{k}^3}{8},$$

$$\mathbf{k}^4 = h\mathbf{f}\left(t_j + \frac{h_j}{2}, \mathbf{y}^3\right), \qquad \mathbf{y}^4 = \mathbf{x}^j + 0.5\,\mathbf{k}^1 - 1.5\mathbf{k}^3 + 2\mathbf{k}^4,$$

$$\mathbf{k}^5 = h\mathbf{f}(t_j + h_j, \mathbf{y}^4), \qquad \mathbf{y}^5 = \mathbf{x}^j + \frac{\mathbf{k}^1 + 4\mathbf{k}^4 + \mathbf{k}^5}{6}. \tag{2.138}$$

A good estimate of the approximation error of $\mathbf{y}^5$ is

$$\mathbf{E} = \frac{\mathbf{y}^4 - \mathbf{y}^5}{5}. \tag{2.139}$$

If $\|\mathbf{E}\| < \varepsilon$, where ε is the maximum error in one integration step, $\mathbf{y}^5$ is accepted as a sufficiently accurate approximation of x^{j+1}. If $\|\mathbf{E}\| > \varepsilon$, $\mathbf{y}^5$ is not acceptable and the whole procedure is repeated with a shorter step h_j, usually evaluated from the formula

$$h_j^{\text{new}} = 0.8 h_j^{\text{old}} \left(\frac{\varepsilon}{\|\mathbf{E}\|}\right)^{0.2}. \tag{2.140}$$

We may also define the next interval size h_{j+1} by using the right-hand side of (2.140) when the approximation satisfies $\|\mathbf{E}\| < \varepsilon$ and is acceptable.

The coefficient 0.8 is a safety factor which guarantees that the increased step size will be successful with respect to the chosen tolerance ε. The exponent 0.2 in (2.140) follows from the fact that the local error of the approximation of $\mathbf{y}^5$ (i.e., the error of approximation in a single step) is $O(h_j^5)$.

An entire class of methods of different orders, analogous to the Merson method, was derived by Fehlberg [2.21]. Similarly, as in the Merson method, the number of necessary evaluations of right-hand sides of differential equations (the number of evaluations of **k**'s in a single step) increases in comparison with the minimum number necessary for the method of the given order. For the Merson method, which deals with the fourth-order approximation (hence, a global error of $O(h^4)$), five evaluations of the right-hand sides are necessary in one step, in comparison with the minimal number of four evaluations (compare the standard Runge–Kutta method [2.3]]).

2.7.2 Multistep methods

The difference between single-step and multistep methods lies in the fact that in the multistep algorithms the value of solutions (and the right-hand sides) are used at more than one of the previous node points. For a constant step $h_j = h$, the general linear multistep method can be written in the form

$$\alpha_k \mathbf{x}^{j+1} + \alpha_{k-1} \mathbf{x}^j + \alpha_{k-2} \mathbf{x}^{j-1} + \cdots + \alpha_0 \mathbf{x}^{j+1-k}$$

$$= h[\beta_k \mathbf{f}^{j+1} + \beta_{k-1} \mathbf{f}^j + \cdots + \beta_0 \mathbf{f}^{j+1-k}]. \tag{2.141}$$

Here we have assumed $\alpha_k \neq 0$, $\alpha_0^2 + \beta_0^2 > 0$. The method represented by (2.141) is called the k-step method. It is fully characterized by coefficients α_i and β_i. If $\alpha_k = 1$, $\alpha_{k-1} = -1$, and $\alpha_i = 0$ for $i = 0, 1, \ldots, k - 2$, the methods of the type (2.141) are called the Adams methods.

For $\beta_k = 0$, the method is explicit (of the Adams–Bashforth type) because we need not iterate to calculate $\mathbf{x}^{j+1}$. For $\beta_k \neq 0$, the method is implicit (of the Adams–Moulton type). To obtain $\mathbf{x}^{j+1}$, we have to use an iteration technique. However, a good initial estimate is usually available (e.g., $\mathbf{x}^j$) because the step size h is generally small.

If $\alpha_k = 1$, $\alpha_{k-2} = -1$, and $\alpha_i = 0$ for $i \neq k, k - 2$, we obtain the Nyström and Milne–Simpson methods. Coefficients of these methods as well as coefficients of methods of the Adams type are given by Henrici [2.3]. An algorithm with a non-equidistant step size h is described in [2.22]; however, the algorithm is somewhat complicated.

In comparison with the single-step methods, multistep methods have higher efficiency with respect to the ratio of the right-hand-sides evaluations to the obtained accuracy. On the other hand, it is difficult to begin integration because the initial conditions alone are insufficient. To begin the multistep method, several steps of a single-step method are usually used. The second disadvantage to multistep method is the difficulty in implementing the automatic step-size control. Gear [2.23] has developed a routine based on the Adams–Bashforth–Moulton formulas of a variable order with an automatic step-size control. This program has recently become one of the most often used routines for the solution of non-stiff equations. It involves the programing of only the right-hand sides and the Jacobian matrix of the right-hand sides.

2.7.3 Integration along the solution arc

It often happens that some of the components x_i change rapidly with varying independent variable t (the derivative dx_i/dt is large). This problem can be alleviated to certain extent by using the method of integration along the solution arc. Let us denote the length of the arc as z and differentiate (2.136) formally with respect to z; we thus obtain

$$\frac{dx_i}{dz} = \frac{dx_i}{dt}\frac{dt}{dz}.$$ (2.142)

From the Pythagorean Theorem for the length of the arc z it follows that

$$\frac{dt}{dz} = \pm \left[1 + \left(\frac{dx_1}{dt}\right)^2 + \cdots + \left(\frac{dx_n}{dt}\right)^2\right]^{-1/2};$$ (2.143)

here the signs $+$ or $-$ denote orientation along the arc. When integration of (2.136) is followed in the positive direction, we take the sign $+$ in (2.143). Considering (2.142) and (2.143), we obtain a system of $n + 1$ differential equations from (2.136):

$$\frac{dt}{dz} = f_0(t, \mathbf{x}),$$ (2.144a)

$$\frac{dx_i}{dz} = f_i(t, \mathbf{x}) \cdot f_0(t, \mathbf{x}), \quad i = 1, 2, \ldots, n,$$ (2.144b)

where f_0 is evidently defined by the relation

$$f_0(t, \mathbf{x}) = \{1 + [f_1(t, \mathbf{x})]^2 + \cdots + [f_n(t, \mathbf{x})]^2\}^{-1/2}.$$ (2.144c)

The reader may easily transform the initial conditions himself. Application of the methods of numerical integration is similar; we merely integrate the system of $n + 1$ differential equations instead of the system of n equations. Further,

$$\left|\frac{dx_i}{dz}\right| \le 1, \quad \left|\frac{dt}{dz}\right| \le 1,$$ (2.145)

still holds and this is why high values of derivatives will not occur. Disadvantages of this procedure follow from the fact that we cannot end the integration at a chosen point, $t = t_f$, because t is now a dependent variable, and furthermore, we have to calculate the square root at every evaluation of the right-hand side.

2.7.4 Integration of phase trajectories for autonomous systems

If the system (2.136) is autonomous, as is often the case, i.e., if

$$\frac{d\mathbf{x}}{dt} = \mathbf{f}(\mathbf{x}), \tag{2.146}$$

the number of equations to be integrated can be reduced in situations where we are interested in phase trajectories only and not in the explicit solution dependence on time t. For example, for $f_k(\mathbf{x}) \neq 0$, we can rewrite system (2.146) as

$$\frac{dx_i}{dx_k} = \frac{f_i(\mathbf{x})}{f_k(\mathbf{x})}, \quad i = 1, \ldots, n, \quad i \neq k. \tag{2.147}$$

The index k can be changed during integration when $f_k \to 0$. It is even possible to choose the index k so that $|f_k|$ will be maximal; then the right-hand sides of (2.147) will always be less than unity. We must always consider the orientation of the curve in the n-dimensional space of dependent variables $x_1, \ldots, x_n$.

2.7.5 Numerical methods for stiff systems of ODE

Stiff systems are characterized by such Jacobian matrix of the right-hand sides where the real parts of eigenvalues differ from each other by several orders of magnitude. The definition of a stiff system is given for the stable system, i.e., for the case where, for all eigenvalues λ_i of the Jacobian matrix

$$\mathrm{Re}\{\lambda_i\} < 0.$$

If $\mathrm{Re}\{\lambda_i\}$ differ by several orders of magnitude, the system is called stiff. From this definition it follows that stiffness is a property more or less qualitative, even though there have been attempts to define it quantitatively. The problem of stiffness has arisen in connection with numerical solution of differential equations. The solution of linearized system of differential equations

$$\frac{d\mathbf{x}}{dt} = J\mathbf{x} \tag{2.148}$$

is given in the form

$$\mathbf{x} = \sum_i [\ldots] \exp(\mathrm{Re}\{\lambda_i\}t), \tag{2.149}$$

where, within the square brackets, there are bounded (for nonmultiple eigenvalues λ_i) vector functions of the variable t (constant for real λ_i). Evidently the terms including $\exp \mathrm{Re}\{\lambda_i\}t$, where $\mathrm{Re}\{\lambda_i\}$ is negative and large, will become negligible in comparison with the other terms within a short time period. Thus, numerical approximation of these terms becomes unimportant.

However, if some of the above-given methods of integration are applied, a very short integration step has to be used to guarantee the numerical stability of the method. To ensure stability, most numerical methods require that $h|\lambda_i|$ be bounded, usually roughly by 1 and 10. Hence, we must solve the problem with unnecessarily extreme accuracy, because a less accurate solution is unachievable.

Dahlquist [2.24] has defined so-called A-stability of a numerical method in the following way. Let us consider a scalar equation

$$\frac{dx}{dt} = \lambda x \tag{2.150}$$

for a general complex number λ, Re$\{\lambda\} < 0$. A numerical method of integration that produces a sequence $x^j = x(t_j)$ with a constant integration step h is called A-stable if, in the recurrent relation

$$x^{j+1} = p(h\lambda)x^j, \tag{2.151}$$

the variable p (dependent on the product $h\lambda$) satisfies the relation

$$|p(h\lambda)| < 1 \tag{2.152}$$

for any integration step size h. In principle, this definition expresses the requirement that x^j tend toward 0 for any step size $h > 0$, Re$\{\lambda\} < 0$. There exist generalizations of this definition. For example, the L-stable method is characterized by $|p(h\lambda)|$ tending toward 0 for $h \to \infty$.

There are two aspects of problems connected with the stiff systems: stability and the accuracy. If a method which is not A-stable is used (i.e., the method where the range of values of $h\lambda$ satisfying (2.152) does not cover the entire left complex half-plane), large negative numbers decrease the integration step h so that the integration becomes ineffective. On the contrary, if an A-stable method is used, there are no problems with stability. However, for a certain step size h, the solution component corresponding to the eigenvalue with the largest absolute value is approximated somewhat inaccurately. A good review of the methods for stiff systems can be found in [2.25]. Over the last 10 years, a number of numerical methods have appeared. They are either A-stable or differ from A-stability only in small regions of the left complex half-plane. These methods can be roughly divided into three groups. The widely used modified Gear algorithm (based on multistep methods [2.26]) belongs to the first group and is oriented to stiff systems. So-called semi-implicit Runge–Kutta methods belong to the second group. One of representatives of these methods is the following third-order method for an autonomous system (2.146). Analogously to the Runge–Kutta methods, we proceed with calculations in one step according to relations

$$\begin{aligned}
\mathbf{k}^1 &= h_j[\mathbf{I} - ha_1\mathbf{J}(\mathbf{x}^j)]^{-1}\mathbf{f}(\mathbf{x}^j), \\
\mathbf{k}^2 &= h_j[\mathbf{I} - ha_2\mathbf{J}(\mathbf{x}^j + c_1\mathbf{k}^1]^{-1}\mathbf{f}(\mathbf{x}^j + b_1\mathbf{k}^1), \\
\mathbf{x}^{j+1} &= \mathbf{x}^j + w_1\mathbf{k}^1 + w_2\mathbf{k}^2.
\end{aligned} \tag{2.153}$$

Table 2.13 Coefficients of semi-implicit Runge–Kutta methods (2.153).

Method:	Rosenbrock		Calahan
Order:	2	3	3
a_1	$1 - \sqrt{2}/2$	1.40824829	0.788675134
a_2	$1 - \sqrt{2}/2$	0.59175171	0.788675134
b_1	$(\sqrt{2} - 1)/2$	0.17378667	-1.15470054
c_1	0	0.17378667	0
w_1	0	-0.41315432	0.75
w_2	1	1.41315432	0.25

The coefficients a_1, a_2, b_1, c_1, w_1, and w_2 are given in Table 2.13. All of these methods are A-stable, as can be found by inserting the coefficients into the differential equation (2.150). Let us remark that iteration is not necessary. The Jacobian matrix of the right-hand sides must be evaluated and a system of linear algebraic equations must be solved. Michelsen [2.27] has dealt with an automatic step-size control for the semi-implicit Runge–Kutta methods.

The methods using the second derivatives of a solution belong to the third group. Linniger and Willoughby [2.28] have suggested the following method $(' = d/dt)$:

$$x^{j+1} = x^j + h[\beta_0(x^{j+1})' + \beta_1(x^j)'] + h^2[\gamma_0(x^{j+1})'' + \gamma_1(x^j)''], \qquad (2.154)$$

where x' and x'' are defined as $x' = f(t, x)$ and $x'' = \partial f/\partial t + J \cdot f$. The authors have described two methods: in the first, the coefficients are determined to be

$$\beta_0 = \frac{1+a}{2}, \quad \beta_1 = \frac{1-a}{2}, \quad \gamma_0 = \frac{b+a}{4}, \quad \gamma_1 = \frac{b-a}{4}, \qquad (2.155)$$

and in the second, the coefficients are determined to be

$$\beta_0 = \frac{1+a}{2}, \quad \beta_1 = \frac{1-a}{2}, \quad \gamma_0 = \frac{-(1+3a)}{12}, \quad \gamma_1 = \frac{1-3a}{12}. \qquad (2.156)$$

There are two optional parameters a and b in the method described by (2.155). They can be chosen so that $p(h\lambda)$ will be approximately $\exp(h\lambda)$ for two different values of $h\lambda$. Only one parameter, a, is chosen in the method described by (2.156), according to the following relation:

$$a = \tfrac{1}{3}[q^2 + 6q + 12 - e^q(q^2 - 6q + 12)] \cdot [e^q(q^2 - 2q) + q^2 + 2q]^{-1},$$

where $q = h\lambda$ is the chosen value. Both methods are implicit, which is why we must use an iteration technique in each step.

An explicit (semi-implicit) single-step method for (2.136) has been derived in [2.29]. The method uses the second derivative of the solution, and a system of linear algebraic equations in each step has to be solved:

$$C(\mathbf{x}^{j+1} - \mathbf{x}^j) = \mathbf{d}, \tag{2.157}$$

where $C = \{c_{im}\}$ is an $n \times n$ matrix and $\mathbf{d}$ is an n-dimensional vector. The elements of C and $\mathbf{d}$ are determined by the relations

$$c_{im} = \delta_{im} + \frac{h_j^2}{2} \sum_{s=1}^{n} \frac{\partial f_i^j}{\partial x_s} \frac{\partial f_s^j}{\partial x_m} - h_j \frac{\partial f_i^j}{\partial x_m},$$

$$d_i = h_j f_i^j + \frac{h_j^2}{2} \frac{\partial f_i^j}{\partial t} - \frac{h_j^2}{2} \sum_{s=1}^{n} \frac{\partial f_i^j}{\partial x_s} \left(f_s^j + h_j \frac{\partial f_s^j}{\partial t} \right). \tag{2.158}$$

Here $\mathbf{f}(t_j, \mathbf{x}^j) = \mathbf{f}^j$, and δ_{im} is the Kronecker delta. This L-stable method is of the second order and is useful for smaller systems of differential equations (with increasing n, the necessary work with evaluation of summations in (2.158) also increases). A Fortran routine with an automatic step-size control has been described in [2.30].

In another paper [2.31], an entire class of implicit methods of higher orders (using the second derivates of a solution) is discussed. A Newton-like iteration technique for the solution of the derived systems of nonlinear equations is suggested as well. The methods also require evaluation of the Jacobian matrix of the right-hand sides. Besides the above-mentioned methods, a number of implicit methods exist, e.g., the trapezoidal rule ($\beta = \frac{1}{2}$),

$$\mathbf{x}^{j+1} = \mathbf{x}^j + h_j[(1 - \beta)\mathbf{f}(t_{j+1}, \mathbf{x}^{j+1}) + \beta \mathbf{f}(t_j, \mathbf{x}^j)], \tag{2.159}$$

or the implicit Euler method ($\beta = 0$). The choice of β is discussed in [2.28].

2.7.6 Systems of differential and algebraic equations

We often encounter coupled systems of differential and algebraic equations in the form

$$\frac{d\mathbf{x}}{dt} = \mathbf{f}(t, \mathbf{x}, \mathbf{u}), \tag{2.160}$$

$$\mathbf{0} = \mathbf{g}(t, \mathbf{x}, \mathbf{u}). \tag{2.161}$$

Here $\mathbf{x}$, $\mathbf{u}$, and $\mathbf{g}$ represent $(x_1, \ldots, x_n)^T$, $(u_1, \ldots, u_m)^T$, and $(g_1, \ldots, g_m)^T$, respectively. Such systems arise, for example, when quasi-steady-state assumptions are accepted (as is common, e.g., in chemical kinetics).

In principle, there exist two types of approaches to the solution of the above problem. The first entails integration of the system (2.160), while (2.161) is solved at each invaluating of the right-hand sides $\mathbf{f}$ by a chosen iteration method with respect to $\mathbf{u}$ for fixed t and $\mathbf{x}$. A good initial estimate, $\mathbf{u}(t_j)$, exists for the iteration process at $t = t_{j+1}$. The choice of the initial approximation can be difficult at the beginning of integration for $t = t_0 = 0$, i.e., for the initial conditions

$$t = 0: \mathbf{x}(0) = \mathbf{x}^0, \quad \mathbf{u}(0) = \mathbf{u}^0. \tag{2.162}$$

Evidently

$$g(0, \mathbf{x}^0, \mathbf{u}^0) = \mathbf{0}. \tag{2.163}$$

must hold

The second type of approach consists of the transformation of the system of algebraic equations into a system of differential equations. Let us differentiate (2.161) with respect to t. We obtain a system of linear algebraic equations for $d\mathbf{u}/dt$:

$$\left[\frac{\partial \mathbf{g}}{\partial \mathbf{u}}\right] \cdot \frac{d\mathbf{u}}{dt} = -\left[\frac{\partial \mathbf{g}}{\partial \mathbf{x}}\right] \cdot \mathbf{f} - \frac{\partial \mathbf{g}}{\partial t}, \tag{2.164}$$

where the matrix of the system $\{\partial \mathbf{g}/\partial \mathbf{u}\}$ is the Jacobian matrix of functions $\mathbf{g}$ with respect to variables $\mathbf{u}$. For the system of $n + m$ differential equations (2.160) and (2.164), we have initial conditions (2.162). This approach assumes that $\{\partial \mathbf{g}/\partial \mathbf{u}\}$ is a regular matrix along the trajectory. If the matrix is not regular, branching of solutions $\mathbf{u}(t)$ occurs in the system of equations (2.161). Branching also causes serious problems for the first type of approach. For every concrete problem, we need to set up a proper strategy, for example, on the basis of physical considerations about the character of the solution.

2.7.7 Integration of differential equations with time delay

Differential equations with time delay arise in a number of transport, control, and biological problems [2.32]. Several distinct time delays occurring in a given system can be written in the form

$$\frac{d\mathbf{x}}{dt} = \mathbf{f}(t, \mathbf{x}(t), \mathbf{x}(t - \tau_1), \ldots, \mathbf{x}(t - \tau_k)), \tag{2.165}$$

where τ_i are individual time delays. Let us denote $\tau = \max \tau_i$; for the integration of (2.165) we need initial conditions in the interval $t \in [-\tau, 0]$. Hence, n functions

$$x_i(t) = \phi_i(t), \quad t \in [-\tau, 0] \tag{2.166}$$

must be given for the proper definition of the initial state of the system. During the integration, we must retain the trajectory $\mathbf{x}(t)$ on the interval of length τ and interpolate along this trajectory. When methods with an automatic step-size control are used, the record (i.e., saved table of values) of the trajectory

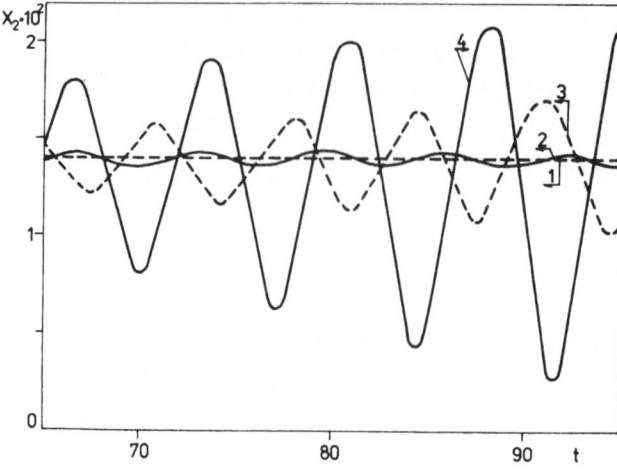

Figure 2.21 Influence of
the time delay τ in
Example 5, Chapter 1,
$1 : \tau = 1.5; 2 : \tau = 1.75;$
$3 : \tau = 1.9; 4 : \tau = 2.$

is not equidistant; the results of each integration step are saved in the form
$t, x_1(t), \ldots, x_n(t)$. The record is revised periodically and the unnecessary
results for lower t are deleted. Either linear or cubic interpolation (from two
or four neighboring points) is usually used. The accuracy thus obtained is
generally sufficient because the integration method with an automatic step-
size control decreases the step size when large gradients occur and the record
of values is denser.

Let us illustrate the effect of time delay on the results of Example 5, Chapter
1, where instead of (E5.7), we shall consider

$$\mu = \hat{\mu}\left(1 + \frac{K_s}{C_{HS}(t - \tau)} + \frac{C_{HS}(t - \tau)}{K_i}\right)^{-1}, \tag{2.167}$$

where C_{HS} is now evaluated at the point $t - \tau$, and τ is the time delay of the
same order of magnitude as the generation time of microorganisms. The
effect of the magnitude of the time delay on the character of the digester stationary state is shown in Fig. 2.21. Sustained oscillations appear at a certain value
of τ. Higher values of τ cause unstationary oscillations with increasing amplitudes.

2.8 Computation of Periodic Solutions

Let us consider a non-autonomous system of differential equations (2.136)
and assume that the function $\mathbf{f}$ is periodic over time t with the period T:

$$\mathbf{f}(t + T, \mathbf{x}) = \mathbf{f}(t, \mathbf{x}). \tag{2.168}$$

The theory of the existence of periodic solutions $\mathbf{x}(t)$, i.e., the solutions where

$$\mathbf{x}(t + T) = \mathbf{x}(t), \tag{2.169}$$

has been reviewed by Coddington and Levinson and by Hale [2.33], for example. A special case of a time periodic function $\mathbf{f}$ is a function which is independent of time t, i.e., an autonomous system of differential equations dependent on a parameter

$$\frac{d\mathbf{x}}{dt} = \mathbf{f}(\mathbf{x}, \alpha). \tag{2.2}$$

Using the transformation $t = T \cdot z$, we obtain a system of equations

$$\frac{d\mathbf{x}}{dz} = T\mathbf{f}(\mathbf{x}, \alpha), \tag{2.170}$$

where the solution $\mathbf{x}(z)$ has unity as its period:

$$\mathbf{x}(z + 1) = \mathbf{x}(z).$$

In a simpler form, we can write

$$\mathbf{x}(1) = \mathbf{x}(0). \tag{2.171}$$

How can the periodic solution of problem (2.2) be determined? A stable periodic trajectory can be found by dynamic simulation. The period T can be determined approximately from the simulation results. Unstable periodic solutions cannot be found in this way. In certain cases, unstable solutions can be found by integrating equations with reversed time, i.e., with a minus sign in front of the function $\mathbf{f}$ in (2.2).

In principle, two approaches can be used to solve the mixed boundary-value problem defined by Eqs. (2.170) and (2.171). The first is based on approximation of derivatives by differences on a sufficiently dense set of node points $z_0 = 0$, $z_1 = h$, $z_2 = 2h$, ..., $z_N = Nh = 1$, for example,

$$\frac{\mathbf{x}^{i+1} - \mathbf{x}^i}{h} = T\mathbf{f}(\tfrac{1}{2}(\mathbf{x}^{i+1} + \mathbf{x}^i), \alpha), \quad i = 0, 1, \ldots, N - 1. \tag{2.172}$$

Here $\mathbf{x}^i = \mathbf{x}(z_i)$. From the boundary condition (2.171), it follows that

$$\mathbf{x}^0 = \mathbf{x}^N. \tag{2.173}$$

We can fix one variable (except T) in the system of $n(N + 1)$ nonlinear algebraic equations (2.172) and (2.173) for $n(N + 1) + 1$ variables $\mathbf{x}^0, \mathbf{x}^1, \ldots, \mathbf{x}^N$, T (the fixed variable must lie on the periodic solution curve), and we can solve the resulting square system by the Newton method, for example. For systems that can be effectively integrated as initial-value problems, a more effective approach exists.

2.8.1 Transformation into an initial-value problem—the shooting method

There exists a second direct approach for determining the periodic solution. This approach is based on solving the mixed boundary-value problem for ordinary differential equations by the shooting method, e.g., [2.34].

We choose initial conditions

$$x_i(0) = \eta_i, \quad i = 1, 2, \ldots, n, \tag{2.174}$$

and a value for the period T. The system (2.170) can be integrated (for fixed α) from $z = 0$ to $z = 1$. Alternatively, integration can be performed from $z = 0$ to $z = -1$ if the solution of the system is more stable when z is negatively oriented. Only the conclusions about the stability of the periodic solution will change. We shall not discuss this alternative here.

The values of solution at $z = 1$ are obtained as a result of integration:

$$x_i(1) = \varphi_i(\eta_1, \ldots, \eta_n, T), \tag{2.175}$$

where the values are dependent on the choices of η and the period T.

Relation (2.171) must hold for any periodic solution; thus, we must satisfy n equations

$$F_i(\eta_1, \ldots, \eta_n, T) = \varphi_i(\eta_1, \ldots, \eta_n, T) - \eta_i = 0, \qquad i = 1, \ldots, n. \tag{2.176}$$

with $n + 1$ unknowns $\eta_1, \ldots, \eta_n, T$. Therefore, one variable has to be fixed. The fixed variable cannot be T, because the solution of (2.176) exists only for discrete values of T. Let us choose fixed η_k for a chosen k. Our choice will be successful if, on the trajectory of the k-th component of the required periodic solution $x_k(z)$, $z \in (0, 1)$, the chosen value actually exists, i.e., if $\eta_k = x_k(\bar{z})$ for certain $\bar{z} \in (0, 1)$. The solution of (2.176) will be obtained by applying Newton's method to the unknowns $\eta_1, \ldots, \eta_{k-1}, \eta_{k+1}, \ldots, \eta_n, T$. The Jacobian matrix contains values $\partial F_i / \partial \eta_j$ and $\partial F_i / \partial T$.

Let us differentiate (2.170) with respect to η_j and let

$$p_{ij}(z) = \frac{\partial x_i}{\partial \eta_j}. \tag{2.177}$$

We obtain variational equations ($' = d/dz$)

$$p'_{ij} = T \sum_{s=1}^{n} \frac{\partial f_i}{\partial x_s} p_{sj}, \quad i, j, = 1, 2, \ldots, n. \tag{2.178}$$

Differentiating (2.170) with respect to T, we obtain, for $q_i = \partial x_i / \partial T$,

$$q'_i = f_i + T \sum_{s=1}^{n} \frac{\partial f_i}{\partial x_s} q_s, \quad i = 1, \ldots, n. \tag{2.179}$$

Initial conditions for these equations are these (cf., the choice of (2.174))

$$p_{ij}(0) = \delta_{ij}, \quad q_i(0) = 0, \tag{2.180}$$

where δ_{ij} is the Kronecker delta. For the elements of the Jacobian matrix of the system (2.176),

$$\frac{\partial F_i}{\partial \eta_i} = p_{ij}(1) - \delta_{ij}, \quad \frac{\partial F_i}{\partial T} = q_i(1) = f_i(\mathbf{x}(1), \alpha). \tag{2.181}$$

From the latter equation, it follows that Eqs. (2.179) need not be integrated.

The column $\partial F_i / \partial \eta_k$ is redundant for the above algorithm. Since we cannot *a priori* estimate the value of η_k which will certainly lie on the desired periodic solution curve, we must use another approach. We evaluate the Newton increment by the aid of the entire Jacobian matrix ($\partial F_i / \partial \eta_j$, $j = 1, \ldots, n$, $\partial F_i / \partial T$; $i = 1, \ldots, n$), which is of dimension $n \times (n + 1)$. The Gauss–Jordan elimination method [2.5] with the pivot selection with preferences (cf., subroutine GAUSE in [2.4], see Appendix A) can be used to solve the system of n linear algebraic equations for $n + 1$ increments. Finally, just one nonzero element remains in each of n columns, and one column is filled (corresponding to the variable which is not varied in the given iteration). By means of preferences, we can assume that the fixed variable will never be the last variable, i.e., T. Moreover, we may prefer certain variables over others so that their values will remain unchanged over the course of several iteration steps. Results of this kind of calculation for Example 3, Chapter 1 are given in Table 2.14. The initial-value problems were solved by the Merson method, cf., (2.138). The values of the parameters in Table 2.14 correspond to the bifurcation diagram in Fig. 2.19. The point $Da_1 = 0.2$, $Da_2 = 0.2$ lies in the region where stationary solutions are unstable.

2.8.2 Stability of periodic solutions

We shall assume that a periodic solution has been determined, i.e., we know the values $\bar{\eta}_1, \ldots, \bar{\eta}_n, \bar{T}$. For each cycle along the periodic trajectory (limit cycle) and for fixed $T = \bar{T}$,

$$\mathbf{x}^{next}(0) = \mathbf{\eta}^{next} = \mathbf{\varphi}(\eta_1, \ldots, \eta_n, \bar{T}). \tag{2.182}$$

Evidently, (2.182) can be understood as an iteration process in variables $\eta_1, \ldots, \eta_n$; a single integration of Eqs. (2.170) from $z = 0$ to $z = 1$ corresponds to a single iteration. The point $\bar{\eta}_1, \ldots, \bar{\eta}_n$ is then a fixed point of this iteration process. If the constructed iteration process converges in the neighborhood near the point $\bar{\eta}_1, \ldots, \bar{\eta}_n$, the resulting periodic solution will be asymptotically orbitally stable. The convergence of the iteration process and, also, the stability of the periodic solution will be determined by the eigenvalues of the linearized mapping $\mathbf{\varphi}$ at the point $\bar{\eta}_1, \ldots, \bar{\eta}_n$, i.e., by the eigenvalues λ of the matrix

$$B = \left\{ \frac{\partial \varphi_i}{\partial \eta_j} \right\} = \{ p_{ij}(1) \}. \tag{2.183}$$

Table 2.14 Computation of the periodic solution in Example 3, Chapter 1. $\gamma = 1000$, $B = 12$, $\beta_1 = \beta_2 = 2$, $\theta_{c1} = \theta_{c2} = 0$, $\Lambda = 0.8$, $Da_1 = 0.2$, $Da_2 = 0.2$. Newton's method with preferences and pivoting was used for solving the system of equations (2.176).

Iteration	η_1	η_2	η_3	η_4	T	$(\sum f_i^2)^{1/2}$
0	0.50000	1.20000	0.70000	0.70000	3.70000	$1.8\,E-1$
1	0.54216	1.20000	0.76188	0.76338	3.69075	$3.1\,E-2$
2	0.57195	1.20000	0.79456	0.76102	3.73862	$2.2\,E-2$
3	0.57387	1.21676	0.79456	0.77240	3.74734	$1.1\,E-4$
4	0.57388	1.21685	0.79456	0.77276	3.74730	$3.3\,E-10$
5[a]	0.57388	1.21685	0.79456	0.77246	3.74730	$1.9\,E-18$
0	0.70000	1.50000	1.00000	1.00000	4.00000	$1.8\,E-1$
1	0.68980	1.50000	0.88347	0.97016	3.75813	$4.3\,E-3$
2	0.68824	1.50000	0.88174	0.96924	3.74737	$7.8\,E-5$
3	0.68824	1.50000	0.88174	0.96924	3.74730	$5.0\,E-9$
4	0.68824	1.50000	0.88174	0.96924	3.74730	$7.4\,E-18$
0	1.00000	2.50000	1.00000	1.00000	2.50000	$1.1\,E-0$
1	0.63123	2.50000	0.73332	1.49466	2.56045	$1.4\,E-0$
2	0.15158	2.50000	0.18964	1.61598	7.86505	$1.2\,E-0$
4	0.63306	2.50000	0.73136	1.35304	7.50726	$4.5\,E-2$
7[b]	0.63274	2.50000	0.73094	1.34623	7.49460	$4.6\,E-13$
0	1.50000	2.50000	1.50000	2.00000	3.00000	$2.0\,E-0$
1	0.57344	2.50000	0.65670	1.59769	2.79702	$1.4\,E-0$
6	0.22349	2.50000	0.26241	1.63071	11.26408	$8.3\,E-1$
11[c]	0.63274	2.50000	0.73094	1.34623	11.24190	$1.1\,E-14$

z	0.0	0.1	0.2	0.3	0.4	0.5	0.6	0.7	0.8	0.9	1.0
x_1	0.57	0.54	0.53	0.55	0.63	0.98	0.92	0.82	0.72	0.63	0.57
θ_1	1.22	1.26	1.43	1.75	2.48	5.30	3.40	2.28	1.63	1.32	1.22
x_2	0.79	0.75	0.72	0.71	0.73	0.83	0.93	0.93	0.90	0.85	0.79
θ_2	0.77	0.80	0.91	1.08	1.34	2.02	2.08	1.48	1.06	0.84	0.77

[a] Resulting periodic solution.

[b] Periodic solution with double period.

[c] Periodic solution with triple period.

Here $p_{ij}(1)$ are the values of the variational variables at the point $z = 1$ for the choice $\boldsymbol{\eta} = \overline{\boldsymbol{\eta}}$.

These eigenvalues are called multipliers, and the numbers $\mu_i = \log \lambda_i$ are called characteristic exponents [2.33]. A periodic solution is stable if $|\lambda_i| < 1$ for each λ_i with the exception of one λ,

$$\lambda_1 = 1. \tag{2.184}$$

Equality (2.184) follows from the Floquet theory [2.33]. The stability of a limit cycle changes when some λ crosses the unit circle. The eigenvalues of the matrix $\boldsymbol{B}$ can be determined, e.g., by means of the Lin–Bairstow method applied to the characteristic polynomial (see [2.5]).

2.8.3 Continuation of periodic solutions

Let us discuss an algorithm for obtaining a dependence of periodic solutions of (2.1) on a parameter α. If the value of η_k for a chosen k is fixed, the DERPAR approach may be used (cf., Sect. 2.2). Equations (2.176) take the form

$$F_i(\eta_1, \ldots, \eta_n, T, \alpha) = 0, \quad i = 1, 2, \ldots, n. \tag{2.185}$$

When we fix one η_k, we obtain n equations for $n + 1$ unknowns, $\eta_1, \ldots, \eta_{k-1}$, $\eta_{k+1}, \ldots, \eta_n$, T, α. This is the set of equations appropriate for using DERPAR (cf., Appendix A). In addition to n columns of derivatives (2.181) for $j \neq k$, we also need a column of derivatives $\partial F_i/\partial \alpha$. The column can be obtained by integrating variational equations for functions $r_i = \partial x_i/\partial \alpha$:

$$r_i' = T \sum_{s=1}^{n} \frac{\partial f_i}{\partial x_s} r_s + T \frac{\partial f_i}{\partial \alpha}, \quad i = 1, 2, \ldots, n. \tag{2.186}$$

Initial conditions evidently take the form

$$r_i(0) = 0, \quad i = 1, \ldots, n. \tag{2.187}$$

Then

$$\frac{\partial F_i}{\partial \alpha} = r_i(1), \quad i = 1, \ldots, n. \tag{2.188}$$

Thus we may proceed in the standard way, repeating the subroutine DERPAR until the fixed value of η_k "disappears" from the course of the periodic solution. To prevent the disappearance, we have to modify the algorithm so that it will continually indicate whether the chosen η_k remains on the solution curve. The main steps of the modified algorithm DERPER [2.35] can be summarized as follows

(a) k is fixed;

(b) the initial approximation of $\eta_1, \ldots, \eta_n$, T, α and necessary controlling parameters for the DERPAR subroutine are chosen;

(c) the subroutine DERPAR with fixed value η_k is called; if a periodic solution is not found, step (b) is repeated for another initial approximation of $\eta_1, \ldots, \eta_n$, T, α;

(d) when the subroutine DERPAR proceeds in a standard way in $n + 1$ variables $\eta_1, \ldots, \eta_{k-1}, \eta_{k+1}, \ldots, \eta_n$, T, α, the following inequality is tested in each step:

$$\eta_k^- < \eta_k < \eta_k^+, \tag{2.189}$$

where η_k^-, η_k^+ are defined as follows:

$$\eta_k^- = \tfrac{1}{2}[(1 + \omega_1)x_k^- + (1 - \omega_1)x_k^+],$$

$$\eta_k^+ = \tfrac{1}{2}[(1 - \omega_1)x_k^- + (1 + \omega_1)x_k^+]. \tag{2.190}$$

The values of x_k^- and x_k^+ are the minimum and maximum values of $x_k(z)$ on the interval $z \in [0, z_1] \cup [z_2, 1]$ (where $x_k(z)$ is monotonous). The value of $\omega_1 \in (0, 1)$ is usually set at 0.5.

If inequality (2.189) is satisfied, the value of η_k remains unchanged, and an advantage of the multistep (Adams) predictor formula of the subroutine DERPAR can be utilized. If inequality (2.189) is not satisfied, η_k is changed according to formula

$$\eta_k \doteq 1/2(x_k^{++} + x_k^{--}).\tag{2.191}$$

The values of x_k^{++} and x_k^{--} are equal to upper and lower values of $x_k(z)$ on the monotonous part of $x_k(z)$ with the maximum change in x_k, i.e.,

$$x_k^{++} = \max_{z \in [z_3, z_4]} x_k(z), \quad x_k^{--} = \min_{z \in [z_3, z_4]} x_k(z);\tag{2.192}$$

$x_k(z)$ is monotonous on $[z_3, z_4]$ and there is no other monotonous part of $x_k(z)$ with a greater change in x_k.

The remaining components $\eta_1, \ldots, \eta_{k-1}, \eta_{k+1}, \ldots, \eta_n$ are re-evaluated using the latest known profiles of $x_1(z), \ldots, x_{k-1}(z), x_{k+1}(z), \ldots, x_n(z)$. The multistep predictor in the subroutine DERPAR must be restarted from the first-order formula.

The above procedure of changing η_k is used if

$$|x_k^+ - x_k^-| < \omega_2 |x_k^{++} - x_k^{--}|.\tag{2.193}$$

The value of ω_2 is chosen from the interval $(0.2, 0.5)$. If inequality (2.193) is satisfied, re-evaluation of η_k and of the remaining components is made according to (2.191) and according to the last profiles of $x_i(z)$, respectively. Results of computations for Example 4, Chapter 1 are presented in Table 2.15. Passage

Table 2.15 Example 4, Chapter 1: Continuation of periodic solutions depending on the value of the parameter $D_1 = D_1^{12} = D_1^{21}$, $n = 2$, $D_2^{12} = D_2^{21} = 10\,D_1$, $A = 2$, $B = 5.9$, $x_1 = c_1^1$, $x_2 = c_2^1$, $x_3 = c_1^2$, $x_4 = c_2^2$. The value of $\eta_1 = x_1(0) = 4.06$ is fixed (it was not necessary to re-evaluate the value).

Point No.	Arc Length $\bar{z}$	η_2	η_3	η_4	T	D_1	$\lambda_i, i = 2, 3, 4$
0	0.00	1.6056	2.0042	1.8878	11.898	1.2052	$0.368; -0.0002 \pm 0.0099i$
1	0.02	1.6047	2.0137	1.8859	11.881	1.2054	$0.036; -0.017 \pm 0.030i$
2	0.04	1.6038	2.0241	1.8839	11.864	1.2056	$-0.384; \pm 0.0025i$
3	0.06	1.6029	2.0351	1.8817	11.847	1.2059	$-0.757; 0.007; -0.007$
4	0.08	1.6018	2.0465	1.8794	11.831	1.2063	$-1.116; 0.006; -0.006$
34	0.68	1.5560	2.5067	1.7734	11.474	1.2254	$-1.143; 0.004; -0.004$
35	0.70	1.5540	2.5245	1.7691	11.466	1.2257	$-0.779; 0.004; -0.004$
39	0.78	1.5461	2.5977	1.7511	11.441	1.2264	$0.697; \pm 0.0017i$
40	0.80	1.5441	2.6164	1.7465	11.436	1.2265	$1.047; 0.001; -0.001$
55	1.10	1.5152	2.8799	1.6809	11.508	1.2228	$1.033; 0.012; -0.012$
56	1.12	1.5139	2.8912	1.6780	11.524	1.2228	$0.643; 0.007; -0.007$
59	1.18	1.5110	2.9182	1.6712	11.578	1.2235	$0.615; \pm 0.008i$
60	1.20	1.5103	2.9250	1.6694	11.596	1.2239	$-1.046; \pm 0.006i$

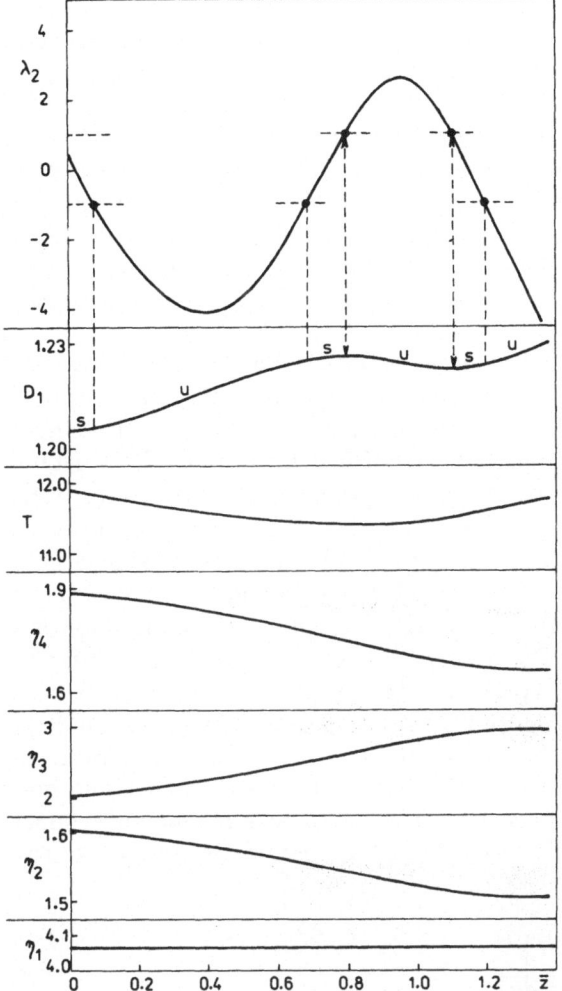

Figure 2.22 Dependence of periodic solutions on the value of the parameter D_1. For a further description, see Table 2.15. The limit points and bifurcation points are depicted: s—stable; u—unstable; $\bar{z}$—arc length of the solution locus, cf., subroutine DERPAR, Section 2.2.

through a limit point along the dependence curve of periodic solutions on the parameter is shown in Table 2.15. It is useful to evaluate multipliers at each point of the computed branches of periodic solutions. Computed values of multipliers which determine the stability of solutions are also given in Table 2.15. The dependence of periodic solutions on the parameter D_1 for Example 4, Chapter 1 are plotted in Fig. 2.22.

2.8.4 Bifurcation of periodic solutions

The stability of periodic solutions on one branch may change with the variation of α; one or two multipliers may move outside the unique circle (in the complex plane).

If one multiplier passes through the unique circle at -1 on the real axis, the originally stable periodic solution becomes unstable and a branch of periodic solutions with a two-fold period branches off at this point. Multipliers on the new branch are equal to squares of the original multipliers. With respect to the orientation of the parameter variation, the branching can be either supercritical (branching of stable periodic solutions) or subcritical (branching of unstable periodic solutions). Both supercritical and subcritical branchings are schematically shown in Fig. 2.23. The bifurcation, where a multiplier intersects the unit circle through -1, will be called the -1 bifurcation (the terms "Brunovsky bifurcation" and "double period" bifurcation have also been used in the literature). If the original branch of periodic solutions is unstable, the new branch of periodic solutions with two-fold period will also be unstable (cf., case 3 in Fig. 2.23). The -1 bifurcation often occurs repeatedly and its structure is schematically sketched as case 4 in Fig. 2.23. Localizing the point of the -1 bifurcation over the course of the continuation of periodic solutions is a relatively simple task. A trial-and-error technique, e.g., the bisection method, may be used. The direct iteration method for locating the -1 bifurcation based on decomposition is described in [2.36].

The second characteristic situation occurs when the multiplier intersects the unit circle through $+1$. This situation is not as simple as the -1 case. The typical $+1$ case corresponds to a limit point along the dependence curve of periodic solutions on a parameter (cf., case 5 in Fig. 2.23). If the continuation algorithm DERPER passes through the limit point, one multiplier intersects the unit circle at $+1$. We may encounter more complicated behavior of periodic-solution branches in the neighborhood of the point with the multiplier $+1$, e.g., when there is an inherent symmetry. The last type of bifurcation from the branch of periodic solutions occurs when two complex-conjugate

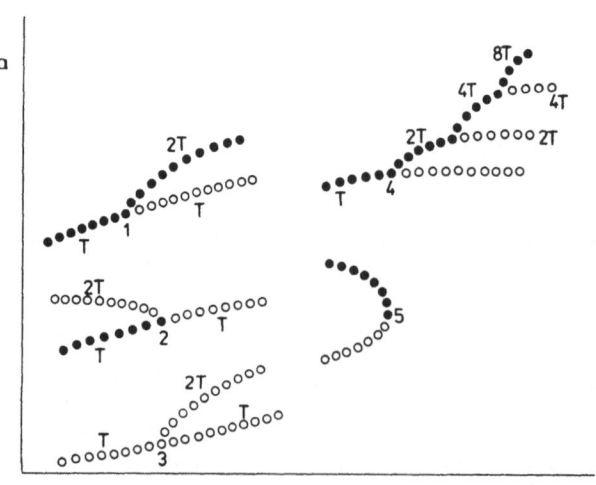

Figure 2.23 Schematic presentation of various types of bifurcations from periodic solutions; a—amplitude of periodic solution. ●●●● stable periodic solution; OOO unstable periodic solution; 1—supercritical -1 bifurcation; 2—subcritical -1 bifurcation; 3—-1 bifurcation from unstable branch; 4—pitchfork bifurcation (repeated -1 bifurcations); 5—limit point ($+1$ bifurcation).

eigenvalues intersect the unit circle. The originally stable branch of periodic solutions becomes unstable and a stable or unstable torus may appear at the bifurcation point. The continuation algorithm DERPER passes through such point when moving along the original branch of periodic solutions. Intersection of multipliers with the unit circle can be detected.

Examples of the change of stability at individual bifurcation points are shown in the solution diagram plotted in Fig. 2.22 for Example 4.

2.9 Chaotic Attractors

Lorenz has demonstrated [2.37] that very simple low-dimensional dynamical systems can display "chaotic" or "turbulent" behavior. Attractors which display such behavior have been called "strange attractors" by Ruelle and Takens [2.38].

Lorenz [2.37] numerically studied the following system which he obtained by truncating equations that describe Bénard convection:

$$\dot{x} = -10x + 10y,$$

$$\dot{y} = -xz + 28x - y,$$

$$\dot{z} = xy - \tfrac{8}{3}z. \tag{2.194}$$

His conclusions and those of later studies agree with the results obtained through a study of a number of other systems. These results are summarized as follows:

(a) The solutions of the equations asymptotically tend toward a *set with complicated structure called a "strange (chaotic) attractor."* Locally this set looks like the product of a manifold and a Cantor set.

(b) The solutions exhibit a *sensitive dependence on the initial condition.* A small error in the initial condition generally grows exponentially over time.

It is now well known that even simple systems as, for example, iterated maps of the interval [2.39, 2.40] yield chaotic attractors. A typical sequence of events leading to chaotic behavior in such systems may be described in the following way.

1. When the characteristic parameter is varied, a cascade of bifurcations doubling the period of the original periodic solution occurs.

2. The sequence of bifurcation points (-1 bifurcations) is convergent and ends in the accumulation point. The behavior of trajectories behind this point is chaotic. The asymptotic rate of convergence of the sequence of bifurcation points to the accumulation point is universal for a certain class of maps [2.41].

3. Windows of stable periodic solutions exist inside the chaotic region; these solutions arise as a result of tangent bifurcation, i.e., together with unstable periodic solutions at the limit points (saddle-node bifurcation). Stable periodic solutions in each window again undergoes a sequence of periodic doubling bifurcations ending in an accumulation point. The chaotic attractor reappears behind the accumulation point and exists until a new tangent bifurcation occurs. The order of appearance of new periodic solutions in dependence on a parameter is again universal for a wide class of maps of the interval [2.42].

4. Almost periodic oscillations interrupted by highly nonperiodic bursts (laminar and turbulent phases) are observed close to tangent bifurcations. This behavior is called intermittency [2.43].

5. When the chaotic attractor touches its own boundary of attraction, it is destroyed but its remnant captures nearby trajectories for a long time before they are ultimately repelled. This type of trajectories is called "transient" or "metastable" chaos. Many features of the above described behavior (e.g., period doubling with the universal rate of convergence and intermittency) persist in the maps of the plane [2.45, 2.46], and also in higher dimensions [2.47], where coexistence of two (chaotic) attractors is a new phenomenon [2.48]. Other routes to turbulence, namely transition from the phase locked or quasi-periodic motions to chaos, exist in the maps of the plane and even in the one-dimensional maps of the circle [2.49, 2.50]. The transition from the quasi-periodic motion to chaos may also possess universal features [2.51, 2.52]. We may expect that other types of transitions will exist in systems of higher dimensions.

The diffeomorphisms are related through Poincaré maps (cf. later) to differential equations (flows) and, therefore, similar types of behavior may often be observed. Period doubling sequences with the universal convergence rate as well as intermittency and metastable chaos have been described in the Lorenz model [2.43, 2.44, 2.53] and in many other systems [2.54–2.56]. The transition to chaos via quasi-periodic or phase-locked flow (corresponding attractor is a torus) have also been found [2.57–2.59], but this transition is much more complicated than the period doubling route, and is still not well understood. Another type of transition to chaos, via homoclinic orbits, is also known [2.60].

There have been a number of experimental observations of chaotic behavior in physical, chemical, biological, and mechanical systems. For example, in chemical systems chaotic oscillations have been observed in the catalytic reaction of the hydrogen oxidation on nickel [2.61] and in the Belousov–Zhabotinski reaction [2.62, 2.63]. A review of experimental observations of chaos in nonlinear systems is given in [2.64]. Many of the above discussed features of the transitions to chaos known in one-dimensional maps were observed in experiments.

2.9.1 Characterization of chaotic attractors

There are three commonly used types of characterization of a strange attractor or chaotic behavior with the help of a computer. The first possibility is to compute Liapunov exponents [2.65], the second possible method for characterizing the strange attractor is to compute the power spectra [2.66], the third possibility is to study the structure of the attractor in the phase space by means of Poincaré sections. The chaotic attractor is characterized by at least one positive one-dimensional Liapunov exponent (from the set of all one-dimensional Liapunov exponents λ_i), by a power spectrum containing broad band noise or a Cantor set like structure of the Poincaré map.

We shall further discuss the numerical technique for computing Liapunov exponents, proposed by Benettin *et al.* [2.67] and by Shimada and Nagashima [2.68]. The results of the computations will be illustrated using Example 4, Chapter 1, i.e., the case of two coupled reaction-diffusion cells with the Brussellator kinetic scheme. The results of power spectra and Poincaré map computation (cf., Appendix C) will be presented as well and the usefulness of the algorithm for the continuation of periodic solutions will be stressed.

2.9.2 Liapunov exponents

The Liapunov exponents characterize the asymptotic orbital instability of dynamic systems [2.67–2.68]. Further, we will discuss the definition and some properties of Liapunov exponents. Let the phase space M be formed by an n-dimensional differentiable variety with the Riemann metric. Let the variety induce a norm $\| \cdot \|$ of the vector $\mathbf{v} \in T_x M$ on every tangent space $T_x M$. Let us denote the flow on M as $\varphi_{\mathbf{x}}^t$, $t \in R^1$ (where $\mathbf{x}$ denotes the initial state). The one-dimensional Liapunov exponent $\lambda(\mathbf{x}, \mathbf{v})$ is defined as

$$\lambda(\mathbf{x}, \mathbf{v}) = \lim_{t \to \infty} \frac{1}{t} \ln \frac{\|d\varphi_{\mathbf{x}}^t(\mathbf{v})\|}{\|\mathbf{v}\|}. \tag{2.195}$$

The ratio $\|d\varphi_{\mathbf{x}}^t(\mathbf{v})\|/\|\mathbf{v}\|$ describes the growth rate of the norm of the vector $\mathbf{v}$ along the trajectory $\varphi_{\mathbf{x}}^t$, $t \in R^1$. The number $\lambda(\mathbf{x}, \mathbf{v})$ is a measure of the divergence rate of trajectories in the $\mathbf{v}$ direction. If $\lambda(\mathbf{x}, \mathbf{v}) = \lambda_1 \neq 0$, we have (for sufficiently high t)

$$\frac{1}{t} \ln \frac{\|d\varphi_{\mathbf{x}}^t(\mathbf{v})\|}{\|\mathbf{v}\|} \sim \lambda_1 \Rightarrow \|d\varphi_{\mathbf{x}}^t(\mathbf{v})\| \sim \|\mathbf{v}\| e^{\lambda_1 t}, \tag{2.196}$$

i.e., the trajectories diverge exponentially for $\lambda_1 > 0$. The Liapunov exponent $\lambda(\mathbf{x}, \mathbf{v})$ depends on $\mathbf{x} \in M$ and $\mathbf{v} \in T_x M$. For fixed $\mathbf{x}$, the exponent $\lambda(\mathbf{x}, \mathbf{v})$ can acquire only a finite number of mutually different values on $T_x M$, i.e., the values $\lambda_1, \lambda_2, \ldots, \lambda_k$, $k \leq n$. Such a basis $\mathbf{e}_1, \mathbf{e}_2, \ldots, \mathbf{e}_n$ of the space $T_x M$ may be chosen so that $\lambda(\mathbf{x}, \mathbf{e}_i) = \lambda_i$, $i = 1, 2, \ldots, n$. The Liapunov exponents characterize the trajectory behavior in the neighborhood of the given trajectory just as the matrix eigenvalues characterize the behavior in the neighborhood of a saddle point. The dependence of $\lambda(\mathbf{x}, \mathbf{v})$ on $\mathbf{x}$ is more complex. The limit (2.195) exists for nearly all $\mathbf{v} \in M$ (with respect to Lebesgue measure) [2.65].

The k-dimensional Liapunov exponent $\lambda^{(k)}(\mathbf{x}, e^{(k)})$, $1 \le k \le n$, is defined in the following way:

$$\lambda^{(k)}(\mathbf{x}, e^{(k)}) = \lim_{t \to \infty} \frac{1}{t} \ln \frac{\|d\varphi_{\mathbf{x}}^t(\mathbf{e}_1) \Lambda \ldots \Lambda d\varphi_{\mathbf{x}}^t(\mathbf{e}_k)\|}{\|\mathbf{e}_1 \Lambda \mathbf{e}_2 \Lambda \ldots \Lambda \mathbf{e}_k\|} . \tag{2.197}$$

Here $e^{(k)}$ denotes a k-dimensional subspace of the space $T_x M$ with the basis $\mathbf{e}_1, \mathbf{e}_2, \ldots, \mathbf{e}_k$. The norm of the k-vector $\mathbf{e}_1 \Lambda \ldots \Lambda \mathbf{e}_k$ corresponds to the volume of the k-dimensional parallelepiped formed by the vectors $\mathbf{e}_1, \ldots, \mathbf{e}_k$. The k-dimensional Liapunov exponent describes the changes in volume of the parallelepiped during its course along the trajectory defined by the flow φ^t.

The k-dimensional Liapunov exponent may acquire, at most, $\binom{n}{k}$ different values [2.65]. Each value $\lambda^{(k)}(\mathbf{x}, e^{(k)})$ may be obtained in the following way: From the set of one-dimensional exponents $\lambda_1, \lambda_2, \ldots, \lambda_n$, a k-tuple $\lambda_{i_1}, \lambda_{i_2}, \ldots, \lambda_{i_k}(i_1, i_2, \ldots, i_k$ mutually differ) can be chosen so that $\lambda^{(k)}(\mathbf{x}, e^{(k)}) = \lambda_{i_1} + \lambda_{i_2} + \cdots + \lambda_{i_k}$. If the basis $(\mathbf{e}_1, \ldots, \mathbf{e}_n)$ of the space $T_x M$ is chosen at random, the expression (2.197) for $k = 1, 2, \ldots, n$ converges to the maximum value of the k-dimensional Liapunov exponent with probability 1. Thus, if we obtain the Liapunov exponents $\lambda_{max}^{(1)}, \lambda_{max}^{(2)}, \ldots, \lambda_{max}^{(n)}$ as a result of numerical calculations, all one-dimensional Liapunov exponents can be obtained as $\lambda_{max}^{(1)} = \lambda_1$, $\lambda_{max}^{(2)} - \lambda_{max}^{(1)} = \lambda_2$, $\lambda_{max}^{(3)} - \lambda_{max}^{(2)} = \lambda_3$, etc. If all one-dimensional Liapunov exponents are negative, then all k-dimensional exponents ($1 \le k \le n$) are negative as well. The translation along the trajectory described by the flow causes the volume of every k-dimensional parallelepiped to converge to zero. The trajectories are then attracted into the set which is of topological dimension zero. If, for example, one Liapunov exponent is positive and the second one is zero, the volume of the parallelepiped increases in one direction only and does not vary significantly in the other direction. The topological dimension of the attractor is two and the tangent vectors determine the position of the tangent plane in relation to the attractor in the phase space.

The numerical method for the computation of the Liapunov exponents proposed by Shimada and Nagashima [2.68] may be summarized as follows:

Let us consider the system described by a set of n differential equations

$$\dot{\mathbf{x}} = \mathbf{f}(\mathbf{x}) \tag{2.198}$$

with initial conditions

$$\mathbf{x}(0) = \mathbf{x}_0. \tag{2.199}$$

The solution of Eq. (2.198) under the initial condition given by Eq. (2.199) is written as

$$\mathbf{x}(t) = \varphi^t \mathbf{x}_0, \tag{2.200}$$

where φ^t is the map which describes the evolution of all phase points over time. The time-evolution equations for the first variation of the orbit obey the set of non-autonomous linear differential equations

$$\delta \dot{\mathbf{x}} = (\delta \mathbf{x}\dot{)} = \frac{\partial \mathbf{f}(\varphi^t \mathbf{x}_0)}{\partial \mathbf{x}} \delta \mathbf{x}. \tag{2.201}$$

The solution of Eq. (2.201) can be written in the form

$$\delta \mathbf{x}(t) = U^t_{\mathbf{x}_0} \delta \mathbf{x}_0 = (d\varphi^t \mathbf{x}_0) \delta \mathbf{x}_0 \tag{2.202}$$

where $U^t_{\mathbf{x}_0}$ is the fundamental matrix [2.68] of Eq. (2.201) and $\delta \mathbf{x}_0$ is an initial deviation at $t = 0$. Initial conditions for $U^t_{\mathbf{x}_0}$ are chosen in the form of unit matrix:

$$U^{t=0}_{\mathbf{x}_0} = \begin{bmatrix} 1 & & & 0 \\ & 1 & & \\ & & 1 & \\ & & & 1 \\ 0 & & & 1 \end{bmatrix}. \tag{2.203}$$

The fundamental matrix satisfies the chain rule:

$$U^{t+s}_{\mathbf{x}_0} = U^t_{\varphi^s \mathbf{x}_0} \cdot U^s_{\mathbf{x}_0} \tag{2.204}$$

The initial conditions for $U^t_{\varphi^s \mathbf{x}_0}$ are again represented by a unit matrix:

$$U^{t=0}_{\varphi^s \mathbf{x}_0} = \begin{bmatrix} 1 & & 0 \\ & 1 & \\ & & 1 \\ 0 & & \ddots & 1 \end{bmatrix}. \tag{2.205}$$

The asymptotic behavior of a small deviation is described by the asymptotic behavior of the fundamental matrix for $t \to \infty$, which is characterized by the exponents defined by Eq. (2.197). Equation (2.197) can be rewritten in the form

$$\lambda^{(k)}(\mathbf{x}_0, e^{(k)}) = \lim_{t \to \infty} \frac{1}{t} \ln \frac{\| U^t_{\mathbf{x}_0} \mathbf{e}_1 \wedge U^t_{\mathbf{x}_0} \mathbf{e}_2 \wedge \ldots \wedge U^t_{\mathbf{x}_0} \mathbf{e}_k \|}{\| \mathbf{e}_1 \wedge \mathbf{e}_2 \wedge \ldots \wedge \mathbf{e}_k \|}. \tag{2.206}$$

When the Liapunov exponents are evaluated directly by integrating variational equations (on the basis of definition (2.206)), numerical problems arise. The variational equations have an exponentially divergent solution, and overflow problems in computer calculations occur. To remedy this problem, Shimada and Nagashima [2.68] have devised the renormalization scheme based on the transformation of the basis. The basis is renormalized using the Gramm–Schmidt orthonormalization:

$$\mathbf{e}_1^{j+1} = \frac{U_{\mathbf{x}_0}^\tau \mathbf{e}_1^j}{\| U_{\mathbf{x}_0}^\tau \mathbf{e}_1^j \|},$$

$$\mathbf{e}_2^{j+1} = \frac{U_{\mathbf{x}_0}^\tau \mathbf{e}_2^j - (\mathbf{e}_1^{j+1} \cdot U_{\mathbf{x}_0}^\tau \mathbf{e}_2^j)\mathbf{e}_1^{j+1}}{\| U_{\mathbf{x}_0}^\tau \mathbf{e}_2^j - (\mathbf{e}_1^{j+1} \cdot U_{\mathbf{x}_0}^\tau \mathbf{e}_2^j)\mathbf{e}_1^{j+1} \|},$$

$$\mathbf{e}_3^{j+1} = \frac{U_{\mathbf{x}_0}^\tau \mathbf{e}_3^j - (\mathbf{e}_1^{j+1} \cdot U_{\mathbf{x}_0}^\tau \mathbf{e}_3^j)\mathbf{e}_1^{j+1} - (\mathbf{e}_2^{j+1} \cdot U_{\mathbf{x}_0}^\tau \mathbf{e}_3^j) \cdot \mathbf{e}_2^{j+1}}{\| U_{\mathbf{x}_0}^\tau \mathbf{e}_3^j - (\mathbf{e}_1^{j+1} U_{\mathbf{x}_0}^\tau \mathbf{e}_3^j) \cdot \mathbf{e}_1^{j+1} - (\mathbf{e}_2^{j+1} \cdot U_{\mathbf{x}_0}^\tau \mathbf{e}_3^j) \cdot \mathbf{e}_2^{j+1} \|},$$

$$\vdots$$

$$\mathbf{e}_k^{j+1} = \frac{U_{\mathbf{x}_0}^\tau \mathbf{e}_k^j - (\mathbf{e}_1^{j+1} \cdot U_{\mathbf{x}_0}^\tau \mathbf{e}_k^j)\mathbf{e}_1^{j+1} \cdots - (\mathbf{e}_{k-1}^{j+1} \cdot U_{\mathbf{x}_0}^\tau \mathbf{e}_k^j) \cdot \mathbf{e}_{k-1}^{j+1}}{\| U_{\mathbf{x}_0}^\tau \mathbf{e}_k^j - (\mathbf{e}_1^{j+1} \cdot U_{\mathbf{x}_0}^\tau \mathbf{e}_k^j) \cdot \mathbf{e}_1^{j+3} \cdots - (\mathbf{e}_{k-1}^{j+1} \cdot U_{\mathbf{x}_0}^\tau \mathbf{e}_k^i)\mathbf{e}_{k-1}^{j+1} \|}. \tag{2.207}$$

Using the chain rule given by Eq. (2.204) and changing the basis $\{U_{\mathbf{x}_j}^\tau \mathbf{e}_i^j\}_i$ into $\{\mathbf{e}_i^{j+1}\}_i$ ($j = 0, 1, \ldots, n-1$ and $i = 1, 2, \ldots, k$) we obtain

$$\lim_{n \to \infty} \frac{1}{n\tau} \ln \frac{\| \Lambda_i U_{\mathbf{x}_0}^{n\tau} \mathbf{e}_i^0 \|}{\| \Lambda_i \mathbf{e}_i^0 \|} = \lim_{n \to \infty} \frac{1}{n\tau} \sum_{j=0}^{n-1} \ln \frac{\| \Lambda_i U_{\mathbf{x}_j}^\tau \mathbf{e}_i^j \|}{\| \Lambda_i \mathbf{e}_i^j \|}. \tag{2.208}$$

The procedure of changing bases is justified by the following property of the exterior product: if $\{\mathbf{e}_i\}$ and $\{\mathbf{c}_i\}$ generate the same k-dimensional subspace, the relation

$$\frac{\| \Lambda_i U \mathbf{e}_i \|}{\| \Lambda_i \mathbf{e}_i \|} = \frac{\| \Lambda_i U \mathbf{c}_i \|}{\| \Lambda_i \mathbf{c}_i \|}$$

holds.

The Liapunov exponents obtained by means of the above method are schematically shown in Fig. 2.24 in dependence on the characteristic parameter. Positive values of the exponents correspond to chaotic solutions, while a zero value corresponds to periodic solutions (stable limit cycle). Continuous evaluation of the Liapunov exponents depending on the chosen characteristic parameter can be thus utilized when describing the dependence of chaotic solutions on the parameter.

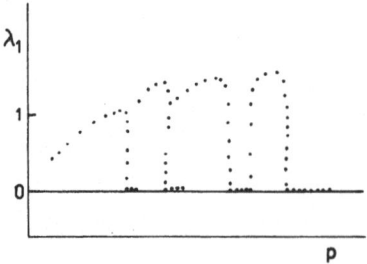

Figure 2.24 Dependence of the maximal Liapunov exponent on the characteristic parameter in a chaotic region with periodic windows.

The Liapunov exponents do not depend either on the initial choice of basis in the subspace $e^{(k)}$ or on the position of x_0 in the state space, as was illustrated for the Lorenz model in [2.68].

Now we shall present results of analysis and characterization of chaotic behavior observed in Example 4, Chapter 1 (two coupled reaction cells with mutual mass exchange).

The concentrations of reaction components x, y in two coupled cells (1, 2) with mass exchange linearly proportional to the concentration difference in the cells are described by a set of four coupled nonlinear ordinary differential equations (where the variables x_1, y_1, x_2, y_2 correspond to c_1^1, c_2^1, c_1^2, c_2^2, respectively):

$$\dot{x}_1 = A - (B + 1)x_1 + x_1^2 y_1 + D_1(x_2 - x_1),$$
$$\dot{y}_1 = Bx_1 - x_1^2 y_1 + D_2(y_2 - y_1),$$
$$\dot{x}_2 = A - (B + 1)x_2 + x_2^2 y_2 + D_1(x_1 - x_2),$$
$$\dot{y}_2 = Bx_2 - x_2^2 y_2 + D_2(y_1 - y_2). \tag{2.209}$$

Two separate subsystems for (x_1, y_1) and (x_2, y_2) when $D_1 = D_2 = 0$ can have either one stable steady state or an unstable steady state surrounded by a stable limit cycle, which bifurcates via the Hopf bifurcation when one of the parameters (A, B) is varied. However, nonhomogeneous steady-state solutions $(x_1 \neq x_2, y_1 \neq y_2)$ and nonhomogeneous limit cycles may exist for the system with coupling, $D_1, D_2 \neq 0$. The course of oscillations in the nonhomogeneous limit cycle differ in both cells. The nonhomogeneous limit cycle can be either of anti-phase type (oscillations are delayed by half a phase between oscillations in the cells) or in-phase type (nearly zero phase shift). From the limit cycle of the in-phase type may start a sequence of bifurcations doubling the period of oscillations into the chaotic region. All reported computations have used values of parameters $A = 2$, $B = 5.9$. We may follow the behavior of the system for varying D_1 for the constant ratio $D_1/D_2 = 0.1$. Hence, only the intensity of mutual interactions varies. The results of simulation are schematically shown in Fig. 2.25. The interval of the parameter D_1 is

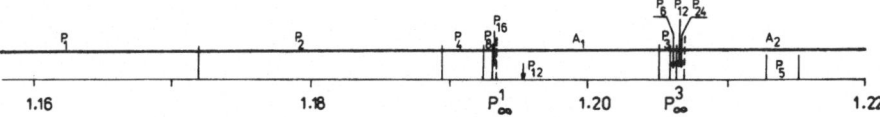

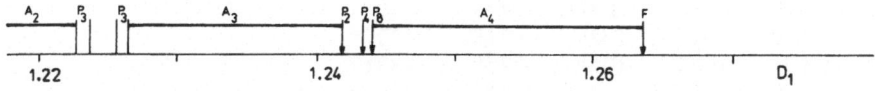

Figure 2.25 Schematic picture of chaotic region $(1.16 < D_1 < 1.27)$. P_i—region of a stable periodic solution with i fixed points on the Poincaré map; A_j—region of aperiodic behavior.

divided into subintervals exhibiting periodic (P_i) and chaotic (A_j) behavior (where i denotes the number of fixed points of the given limit cycle in a Poincaré map, cf. Appendix C, and j denotes the order of appearance of chaotic regions). With increasing D_1, a sequence of subharmonic bifurcations appears first (the sequence is followed up to the point P_{16}). This probably infinite sequence appears to reach the accumulation point P_∞^1. Then the chaotic region A_1 appears; the periodic solution P_{12}, with a period approximately 12 times longer than the period P_1, exists in this region. The region of the periodic solution P_3 appears next; this solution bifurcates into a solution with double period (P_6), and via further bifurcations, solutions with periods P_{12} and P_{24} arise. This sequence of bifurcations probably again ends at the accumulation point P_∞^3. Further on, the chaotic region A_2 appears. In A_2 the periodic solution P_5 exists. Two periodic solutions P_3 appear in the interval between A_2 and A_3. The region between these two P_3 solutions is also chaotic. Another periodic interval contains solutions P_2, P_4, and P_8. The solution is again chaotic in the interval A_4. At point F, the chaotic attractor suddenly disappears without any bifurcation sequence. The homogeneous periodic solution with the same amplitude of oscillations in both cells and with zero phase shift is then the only periodic solution available for higher values of D_1.

The system exhibits strong hysteresis. The homogeneous periodic solution is stable throughout the aperiodic region. Different initial conditions can lead to different solutions, to homogeneous periodic solution or to solutions shown in Fig. 2.25. Thus, for example, the initial condition $(x_1, y_1, x_2, y_2) = (2.1, 2.9, 0.5, 0.5)$ leads to a homogeneous periodic solution, while the initial condition $(2.1, 2.9, 1.9, 3.0)$ leads to a nonhomogeneous type of oscillations (periodic or chaotic).

The full set of one-dimensional Liapunov exponents is shown in Table 2.16 for chosen values of D_1. In all cases, the exponents are of the type $(+, 0, -, -)$, i.e., only one of the exponents is positive. The flow thus expands in one dimension. Different initial conditions used in numerical integration give approximately the same values of λ (cf., the first two rows of Table 2.16). Also the values of the exponents are independent of the choice of the basis e_i. We have chosen either the natural basis $(1, 0, 0, 0)$ $(0, 1, 0, 0)$, $(0, 0, 1, 0)$, $(0, 0, 0, 1)$, or

Table 2.16 One-dimensional Liapunov exponents.

No.	D_1	λ_1	λ_1	λ_3	λ_4
1	1.20	0.078	0.00	-2.68	-30.87
2	1.20	0.079	0.00	-2.68	-30.89
3	1.21	0.108	0.00	-2.69	-30.67
4	1.26	0.200	0.00	-2.82	-32.17

the base (1, 1, 1, 1,) (1, 1, 1, 0), (1, 1, 0, 0), (1, 0, 0, 0). Initial conditions in cases 1, 3, and 4 (Table 2.16) are (2.1, 2.9, 1.9, 3.1), and in case 2 they are (2.1, 2.9, 0.5, 6.0). The integration time is 4000 time units (cases 1 and 2) and 2000 time units (cases 3 and 4). The time step for orthogonalization is 0.5 time units on the computer ICL 4-72. Longer integration times give more accurate values of λ.

The Runge–Kutta–Merson method of integration of equations in variations is used. The required accuracy of the obtained solution is 0.0001. The computation is sensitive to the choice of the time of renormalization, τ. The parallelepiped becomes too flat and small for higher values of τ, especially the n-dimensional Liapunov exponent is subject to significant error. The error is due to the limited accuracy of numbers and the inevitable rounding errors in the computer. With double precision operations τ must be chosen so as $\tau <$ $\ln 10^{-14}/\lambda^{(n)}$, because the volume of parallelepiped cannot be determined more accurately than 10^{-14}.

2.9.3 Power spectra

Power spectra of solution trajectories can be computed by the Fourier transformation or the fast Fourier transformation method (FFT). Standard routines are available on most computers. The routine for FFT is, for example, described in detail in [2.66].

Examples of computed power spectra for chosen values of D_1 in (2.209) are shown in Figs. 2.26(a)–(g). The spectra in Figs 2.26(a and b) are taken from region A_1 (cf., Fig. 2.25), the spectrum shown in Fig. 2.26(c) is taken from region A_2; Fig. 2.26(d) shows a power spectrum corresponding to the region between two periodic ranges P_3, and Fig. 2.26(e) illustrates the spectrum from region A_3. Figs 2.26(f) and (g) show the power spectra from the region A_4; Fig. 2.26(f) illustrates the spectrum belonging to the left-hand boundary of A_4, i.e., very close to P_8, while Fig. 2.26(g) shows the power spectrum close to the point F, where the chaotic attractor disappears.

From the qualitative point of view, we may observe that the spectra can be divided into two groups. The first group contains spectra whose sharp peaks appear above the level of broad band noise (the noise level is approximately five orders of magnitude higher than in the periodic case). The spectra shown in Figs. 2.26(a), (d), (f), and, in part, in Fig. 2.26(b) belong to this group. The second group is formed by flat spectra without sharp peaks similar to broad band noise; this group contains spectra shown in Figs. 2.26(c), (e), and (g).

Similar types of spectra have been observed and discussed by Farmer *et al.*, [2.70] for the Rössler model. These authors call the first group of spectra "phase coherent" and the second type of spectra "phase incoherent". The

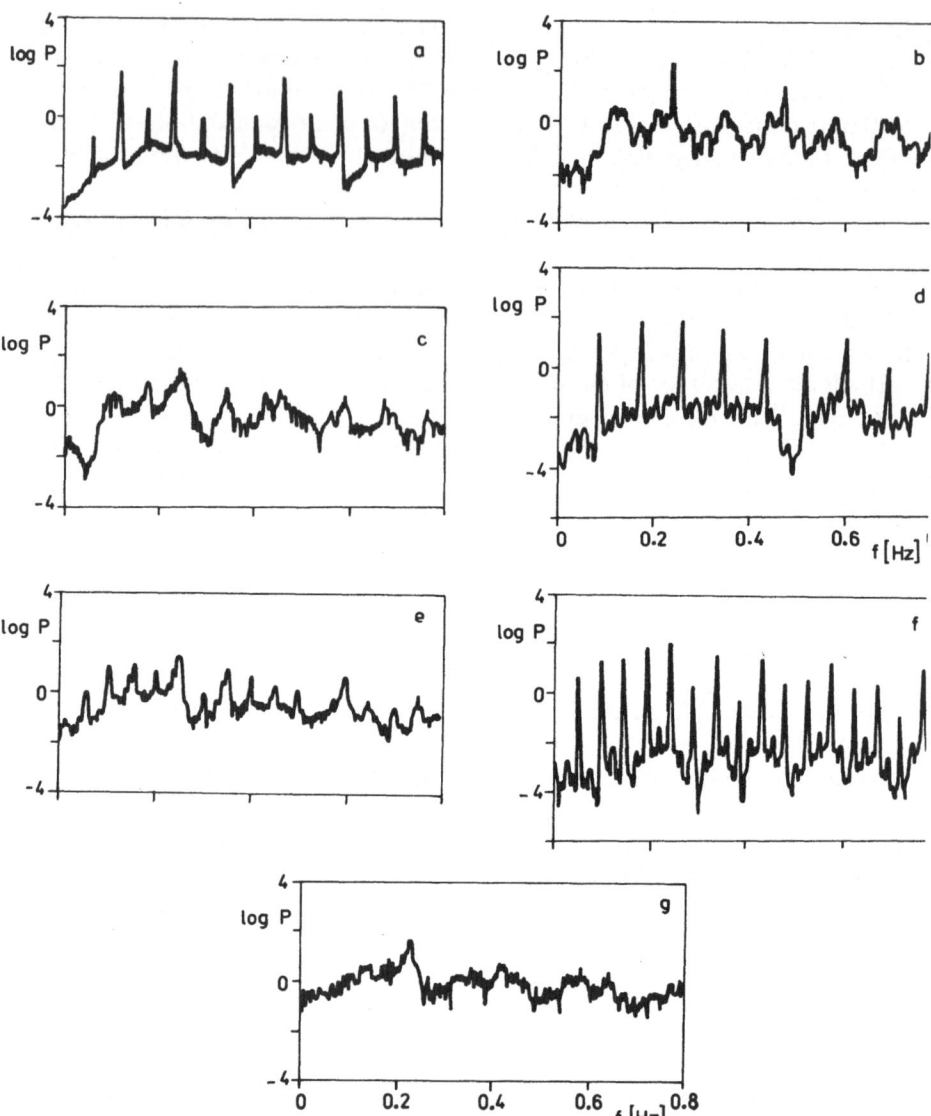

Figure 2.26 Power spectra—log P vs. f [Hz]; fast Fourier transformation from 16 384 points, sampling period $\Delta t = 0.1$. Parameter values: $A = 2$, $B = 5.9$, $D_1/D_2 = 0.1$. (a) Phase-coherent chaotic attractor with sharp peaks corresponding to an unstable limit cycle with period four; $D_1 = 1.1940$. (b) Phase-coherent chaotic attractor (lower level of phase coherence than (a)) with sharp peaks corresponding to an unstable limit cycle with period one; $D_1 = 1.20$. (c) Phase-incoherent chaotic attractor (wandering between unstable limit cycles with periods two and three); $D_1 = 1.21$. (d) Phase-coherent chaotic attractor with sharp peaks corresponding to an unstable limit cycle with period three; $D_1 = 1.225$. (e) Phase-incoherent chaotic attractor; $D_1 = 1.235$. (f) Phase-coherent chaotic attractor (connected with an unstable limit cycle with a period which is a multiple of period two); $D_1 = 1.245$. (g) Phase-incoherent chaotic attractor (located close to the point where the chaotic attractor becomes unstable); $D_1 = 1.263$.

differences between the two types of spectra are explained in [2.70] on the basis of the observation that mixing occurs in the first case mainly in the direction transversal to the flow, while in the second case, mixing in the flow direction is important.

In our case the phase-coherent attractor appears in almost the entire region A_1, and it reappears in the neighborhood of windows of periodic solutions.

The phase-incoherent attractor exists in regions A_2, A_3, and A_4 (except at the margins of regions close to periodic attractors). The phase coherence occurs close to every point of accumulation, i.e., there exists some mechanism which does not allow trajectories to leave the neighborhood of the set formed by now unstable limit cycles generated by the sequence of period doubling bifurcations behind the accumulation point. The flow on such attractor is nearly periodic which is reflected in periodic and noisy components of the power spectra. The chaotic attractor expands further from the accumulation point, and the periodic components in the spectra are lost.

2.9.4 The Poincaré map

The Poincaré map (cf. Appendix C) can help us study the structure of the chaotic attractor. In our case (a four-dimensional dynamical system), this map corresponds to the cross section of the attractor in three-dimensional subspace. In (2.209) we have set $y_2 = \text{const} = 2.5$. The cross section in the subspace $(x_1\ y_1, x_2)$ has been projected into the plane (x_1, x_2), cf. Fig. 2.27(a)–(e). The cross section in this plane looks like a continuous curve (or several continuous parts)—this shows that the attractor is two-dimensional. The cross section is analogous to the spectrum of the Liapunov exponent (cf., Table 2.16) which is of type $(+, 0, -, -)$, and it implies that the topological dimension of the attractor is two.

We can obtain many unstable periodic solutions throughout the chaotic region by applying the continuation routine DERPER, cf. Section 2.8.3. This allows us to construct a "framework" of the strange attractor. It is believed that the strange attractor contains infinitely many unstable periodic orbits. When we plot some of them in terms of the Poincaré map, we obtain the "framework." This is shown in Fig. 2.28 ($D_1 = 1.21$). This figure can be compared with the chaotic attractor shown in Fig. 2.27(c).

In certain simpler situations, Poincaré maps may be used for approximate evaluation of the Liapunov exponents [2.69]. This method has been used to confirm that experimentally observed aperiodic behavior in a chemical system is really chaotic [2.63, 2.64].

More detailed description of various bifurcation phenomena in two coupled cells (the convergence rate of the period doubling bifurcation sequences, dependences of periodic solutions on D_1, transient chaos at the point F in

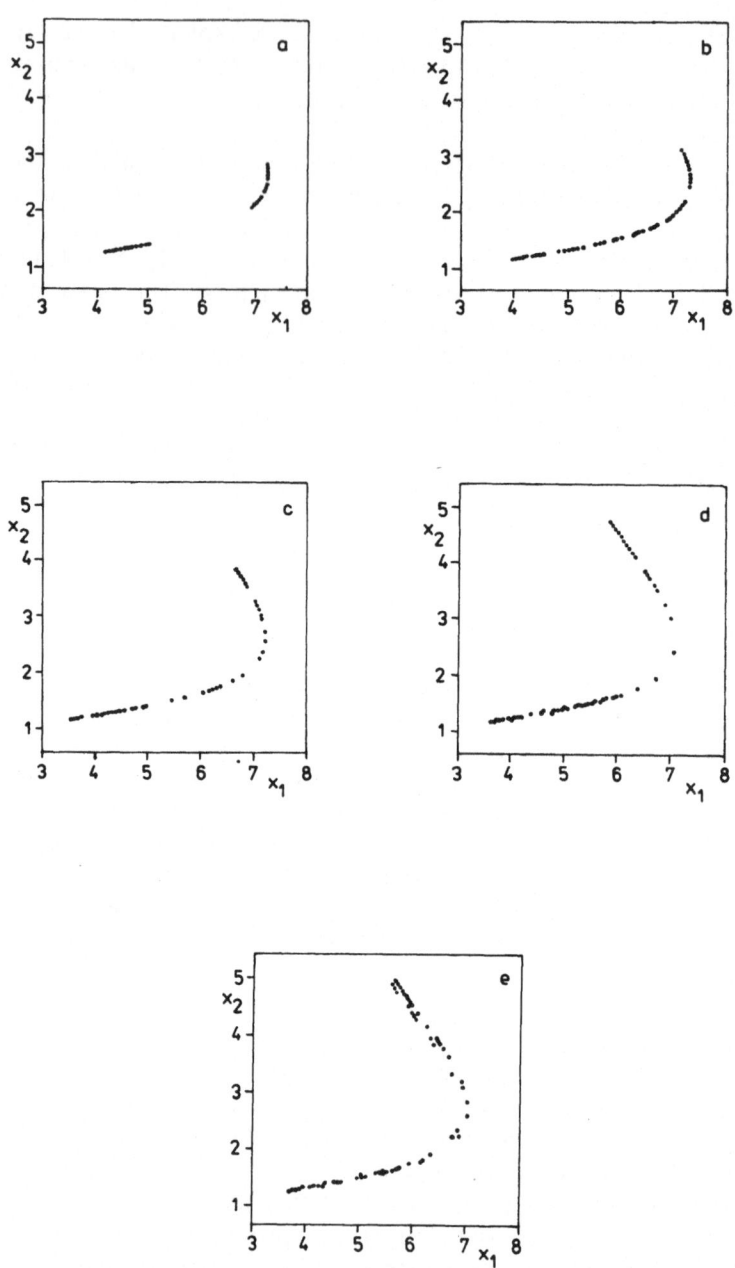

Figure 2.27 Projection of the Poincaré map into the plane (x_1, x_2). y_2 is fixed at $y_2 = 2.5$. Each figure contains approximately 100 points of intersection of the phase flow with a cross section of codimension one. Parameter values: $A = 2, B = 5.9, D_1/D_2 = 0.1, a - D_1 = 1.194, b - D_1 = 1.2, c - D_1 = 1.21, d - D_1 = 1.24, e - D_1 = 1.263.$

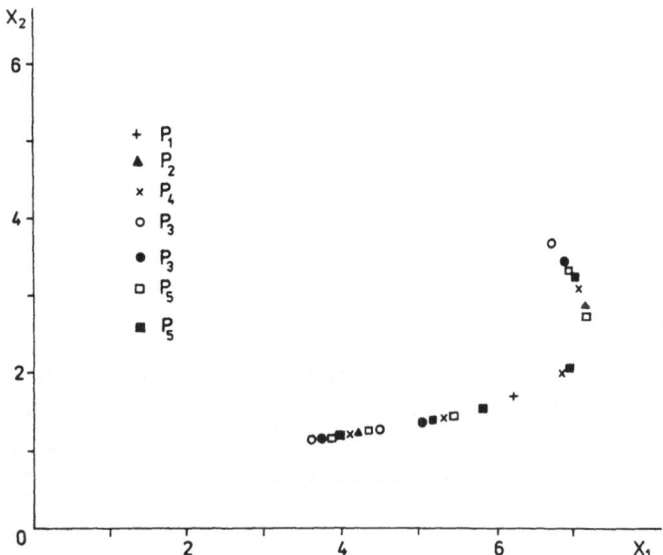

Figure 2.28 "Framework" of the chaotic attractor; Poincaré map of several unstable periodic solutions located within the chaotic attractor; Example 4, Chapter 1, $A = 2, B = 5.9, D_1/D_2 = 0.1$, $D_1 = 1.21$; P_i denotes the periodic solution with i fixed points in the Poincaré map.

Fig. 2.25, the fractal dimension of chaotic attractors, etc.) is discussed in [2.56]. The transition from the quasi-periodicity to chaos for the same problem is discussed in [2.57], and other applications of the DERPER continuation routine are described in [2.71, 2.72]. A number of review articles and books dealing with chaotic behavior appeared recently [2.73–2.76].

CHAPTER 3

Multiplicity and Stability in Distributed-Parameter Systems (DPS)

Here Eqs. (2.1) refer to an infinite-dimensional (e.g., Banach) space, contrary to the case of lumped-parameter systems where only finite-dimensional problems occur. When formulating equations for LPS, we used a general formulation of an evolution problem in a finite-dimensional (Euclidean) space. Here we shall consider a special type of distributed system, which can be written in the form

$$\frac{\partial \mathbf{X}}{\partial t} = \mathbf{F}\left(\frac{\partial^2 \mathbf{X}}{\partial z^2}, \frac{\partial \mathbf{X}}{\partial z}, \mathbf{X}, z, A\right), \tag{3.1}$$

where $\mathbf{X}(z, t)$ is a vector function of time t and of spatial coordinate z, A denotes parameters. We shall consider z to be one-dimensional, even though a number of papers dealing with the problems in higher dimensions have recently appeared in the literature.

In our further discussion, we shall concentrate mainly on the system of two parabolic partial differential equations of the form

$$\frac{\partial x}{\partial t} = \frac{D_x}{L^2} \frac{\partial^2 x}{\partial z^2} + f(x, y), \tag{3.2}$$

$$\frac{\partial y}{\partial t} = \frac{D_y}{L^2} \frac{\partial^2 y}{\partial z^2} + g(x, y). \tag{3.3}$$

Equations (3.2) and (3.3) can describe, for example, reaction diffusion systems, where one-dimensional diffusion in the direction of the coordinate z takes place and the reaction mechanism is described by functions $f(x, y)$ and $g(x, y)$. The parameter L describes the system dimension, and D_x and D_y are diffusion coefficients.

We usually search for a solution in the half-strip $t \in [0, \infty) \times z \in [0, 1]$, where boundary conditions at $z = 0$ and $z = 1$ are defined. We shall consider mainly linear boundary conditions of three types:

(a) fixed boundary conditions (Dirichlet type):

$$\text{BC 1: } x(0, t) = x(1, t) = \bar{x}, \quad y(0, t) = y(1, t) = \bar{y}, \tag{3.4}$$

(b) zero-flux boundary conditions (Neumann type):

$$\text{BC 2}: \frac{\partial x(0, t)}{\partial z} = \frac{\partial x(1, t)}{\partial z} = 0, \quad \frac{\partial y(0, t)}{\partial z} = \frac{\partial y(1, t)}{\partial z} = 0, \tag{3.5}$$

(c) boundary conditions of the third kind (of Robin type):

$$\text{BC 3}: \alpha_{x0} \frac{\partial x(0, t)}{\partial z} + \beta_{x0} x(0, t) = \gamma_{x0}, \, \alpha_{x1} \frac{\partial x(1, t)}{\partial z} + \beta_{x1} x(1, t) = \gamma_{x1},$$

$$\alpha_{y0} \frac{\partial y(0, t)}{\partial z} + \beta_{y0} y(0, t) = \gamma_{y0}, \, \alpha_{y1} \frac{\partial y(1, t)}{\partial z} + \beta_{y1} y(1, t) = \gamma_{y1}. \tag{3.6}$$

The values at the boundaries $\bar{x}$, $\bar{y}$ often describe a solution of a system without diffusion, i.e., the solution of the system

$$f(\bar{x}, \bar{y}) = 0, \quad g(\bar{x}, \bar{y}) = 0. \tag{3.7}$$

The boundary conditions BC 2 introduce an evident symmetry into the problem defined by Eqs. (3.2) and (3.3). The property "composing of solutions" follows from this symmetry. Let us have the solution $x(z, t)$ (similarly, $y(z, t)$) for a given L. We shall construct the solution $\tilde{x}(\tilde{z}, t)$ in the following way:

$$\tilde{z} \in [0, 0.5]: \tilde{x}\left(\tilde{z} = \frac{z}{2}, t\right) = x(z, t),$$

$$\tilde{z} \in [0.5, 1]: \tilde{x}\left(\tilde{z} = \frac{1}{2} + \frac{z}{2}, t\right) = x(1 - z, t). \tag{3.8}$$

We can easily verify that $\tilde{x}(\tilde{z}, t)$ is again a solution of (3.2) with BC 2 for $\tilde{L} = 2L$. Solutions for the lengths $3L$, $4L$, etc., may be found in an analogous way. The procedure is schematically shown in Fig. 3.1 for one t (the same procedure holds for stationary solutions of (3.2) and (3.3)).

Examples 6–11 in Section 1.4 will be used to illustrate the described numerical techniques. The reader may compare the formulation of the problem given above with the relations given for individual examples.

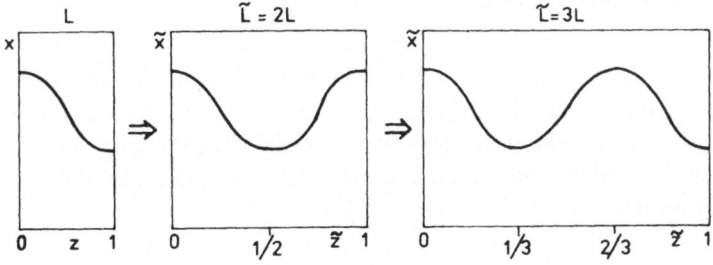

Figure 3.1 Process of composing solutions for BC 2.

In Chapter 2 we have seen that the theory, numerical methods, and terminology of bifurcation problem analysis are reasonably well established in LPS. The situation is different in distributed parameter systems. The bifurcation theory is available for linear problems or it appears in journal literature on nonlinear analysis. The work of Henry [1.84] appears to be one of the first systematic studies of the analytical approach to bifurcation problems in nonlinear DPS.

The most common approach to DPS systems consists of transforming distributed-parameter to lumped-parameter systems. The Galerkin approximation, collocation methods, and high-order difference methods [3.1, 3.2] are most often used. Spectral methods [3.3] are also used for transforming DPS into LPS. The errors of approximation usually decrease with increasing approximation fineness. Finer approximations, however, lead to LPS of higher dimensions. The difference method and the shooting method can be used when fine approximations are required. They usually give simpler and more universal algorithms of transformation of DPS into LPS.

3.1 Steady-State Solutions—Methods for Solving Nonlinear Boundary-Value Problems

Several texts describe the methods for solving nonlinear boundary-value problems in detail. Among them we can name the books by Fox [3.4], Keller [3.5], Roberts and Shipman [3.6], and Kubíček and Hlaváček [3.7]. Let us briefly discuss just two typical and, generally, the most useful types of approaches for solving nonlinear boundary-value problems—finite-difference methods and shooting methods.

The steady-state solution of (3.2) and (3.3) satisfies ($' = d/dz$)

$$x'' = -\frac{L^2}{D_x} f(x, y), \quad y'' = -\frac{L^2}{D_y} g(x, y), \tag{3.9}$$

together with the corresponding boundary conditions independent of t.

3.1.1 Finite-difference methods

The set of mesh points (usually equidistant) z_i, $z_i = ih$, $i = 0, 1, \ldots, n$ (h is the chosen step size of the set, $h = 1/n$) is defined on the interval $z \in [0, 1]$. The approximation of the solution $x_i \sim x(z_i)$, $y_i \sim y(z_i)$ is then constructed at the mesh points. The derivatives in (3.9) can be approximated by means of difference formulas, the second derivative, for example, by three-point approximation

$$y_i'' \sim y''(z_i) \sim \frac{y_{i-1} - 2y_i + y_{i+1}}{h^2}. \tag{3.10}$$

A similar approximation can be used for x. After substituting into (3.9) for $i = 1, 2, \ldots, n - 1$, we obtain the following system of nonlinear algebraic equations:

$$x_{i-1} - 2x_i + x_{i+1} + \frac{L^2 h^2}{D_x} f(x_i, y_i) = 0, \tag{3.11}$$

$$y_{i-1} - 2y_i + y_{i+1} + \frac{L^2 h^2}{D_y} g(x_i, y_i) = 0. \tag{3.12}$$

For $i = 1, 2, \ldots, n - 1$, as many as $2(n - 1)$ equations for $2(n + 1)$ unknowns $x_i, y_i, i = 0, 1, \ldots, n$ are thus obtained. The remaining four equations follow from the approximations of boundary conditions, e.g., for BC 1, we have

$$x_0 = \bar{x}, \quad x_n = \bar{x}, \quad y_0 = \bar{y}, \quad y_n = \bar{y}. \tag{3.13}$$

The situation is somewhat more complicated for BC 2 because the derivatives have to be approximated by a difference formula. If a two-point approximation of the type

$$x_0' \sim \frac{x_1 - x_0}{h} = 0$$

is used, the order of numerical approximation decreases from $O(h^2)$ in (3.11) and (3.12) to $O(h)$. Hence, we usually use the approximations which retain the order of accuracy, i.e., $O(h^2)$. We can use either asymmetric approximation, e.g.,

$$x_0' \sim \frac{-3x_0 + 4x_1 - x_2}{2h},$$

or the approximations based on virtual profiles at $i = -1$ and $n + 1$, where the validity of (3.11) and (3.12) is assumed at z_0 and z_n as well, i.e., for $i = 0$ and $i = n$. If, for $i = 0$, the approximations

$$x_0' \sim \frac{x_1 - x_{-1}}{2h} = 0, \quad y_0' \sim \frac{y_1 - y_{-1}}{2h} = 0,$$

are made, we obtain

$$2x_1 - 2x_0 + \frac{L^2 h^2}{D_x} f(x_0, y_0) = 0, \tag{3.14}$$

$$2y_1 - 2y_0 + \frac{L^2 h^2}{D_y} g(x_0, y_0) = 0. \tag{3.15}$$

In an analogous way, we have, for $i = n$,

$$2x_n - 2x_{n-1} + \frac{L^2 h^2}{D_x} f(x_n, y_n) = 0, \tag{3.16}$$

$$2y_n - 2y_{n-1} + \frac{L^2 h^2}{D_y} g(x_n, y_n) = 0. \tag{3.17}$$

Now let us arrange the unknowns into a vector $\mathbf{X} = (x_0, y_0, x_1, y_1, x_2 \cdots x_{n-1},$ $y_{n-1}, x_n, y_n)^T$, and organize the set of equations into this sequence (for BC 2):

1.   (3.14);
2.   (3.15);
3.   (3.11) for $i = 1$;
4.   (3.12) for $i = 1$;
5.   (3.11) for $i = 2$;
6.   (3.12) for $i = 2$;

$\vdots$

$2n - 1$. (3.11) for $i = n - 1$;
$2n$.   (3.12) for $i = n - 1$;
$2n + 1$. (3.16);
$2n + 2$. (3.17).

The reader can easily verify that the resulting "occurrence matrix" of this ordered system is five diagonal. If the Newton (or the Newton–Raphson) method is used for solving this system, the linear algebraic systems of equations in each iteration have a five-diagonal matrix of the system (Jacobian matrix of nonlinear system). The situation for BC 1 and BC 3 will be similar. Let us briefly illustrate the use of finite-difference methods with regard to Example 10, Chapter 1, where the application is somewhat more complicated. The difference approximations are in the form:

$$H_0 = 0,$$

$$F_0 = 0,$$

$$G_0 = 1,$$

$$\left.\begin{aligned}
\frac{H_i - H_{i-1}}{h} &= -\sqrt{\text{Re}}(F_i + F_{i-1}), \\[2mm]
\frac{F_{i-1} - 2F_i + F_{i+1}}{h^2} &= \sqrt{\text{Re}}\,H_i \frac{F_{i+1} - F_{i-1}}{2h} \\[1mm]
&\quad + \text{Re}(F_i^2 - G_i^2 + k), \\[2mm]
\frac{G_{i-1} - 2G_i + G_{i+1}}{h^2} &= 2\text{Re}\,F_i G_i \\[1mm]
&\quad + \sqrt{\text{Re}}\,\frac{G_{i+1} - G_{i-1}}{2h}\,H_i,
\end{aligned}\right\} \quad \text{for } i = 1, 2, \ldots, n-1,$$

$$(3.18)$$

$$\frac{H_n - H_{n-1}}{h} = -\sqrt{\text{Re}}(F_n + F_{n-1}),$$

$$H_n = 0,$$

$$F_n = 0,$$

$$G_n = S.$$

For given values of Re and S, and for the unknowns arranged as $X = (H_0, F_0, G_0, H_1, F_1, G_1, \ldots, H_{n-1}, F_{n-1}, G_{n-1}, H_n, F_n, G_n, k)^T$, we again obtain an almost seven-diagonal occurrence matrix. The total number of equations, equal to the number of variables, is $3(n - 1) + 7$. With respect to the fact that n may be large (for example, for Re $= 625$, it is necessary to choose $n = 100$–200) [3.8], the total number of equations is often high. A universal computer program based on finite-difference methods with an adaptive mesh size (depending on the prescribed error) has been published by Pereyra [3.9].

3.1.2 Quasi-linearization

The method of quasi-linearization is based on the construction of a sequence of approximations of a solution of nonlinear boundary-value problem (x^k, y^k):

$$x^{k+1} = x^k + \Delta x^k, \quad y^{k+1} = y^k + \Delta y^k. \tag{3.19}$$

The increments Δx^k and Δy^k (functions of the variable z) satisfy the linear boundary-value problem defined by

$$(\Delta x^k)'' + \frac{L^2}{D_x}\left[\frac{\partial f}{\partial x}\Delta x^k + \frac{\partial f}{\partial y}\Delta y^k\right] = -(x^k)'' - \frac{L^2}{D_x}f(x^k, y^k),$$

$$(\Delta y^k)'' + \frac{L^2}{D_y}\left[\frac{\partial g}{\partial x}\Delta x^k + \frac{\partial g}{\partial y}\Delta y^k\right] = -(y^k)'' - \frac{L^2}{D_y}g(x^k, y^k), \tag{3.20}$$

with the boundary conditions (for BC 2)

$$[\Delta x^k(0)]' = [\Delta x^k(1)]' = [\Delta y^k(0)]' = [\Delta y^k(1)]' = 0. \tag{3.21}$$

When the values x^k, y^k (previous approximation) are known, Eqs. (3.20) and (3.21) describe a linear boundary-value problem for Δx^k and Δy^k, and the problem can be solved, for example, by the method of superposition of solutions (see, e.g., [3.6]).

The courses of $x^k(z)$ and $y^k(z)$ again have to be retained on some sufficiently dense set of points z_i. The advantage of the initial-value technique is thus lost. Similarly, as in the case of finite-difference methods, we need an initial approximation of the solution x^0, y^0 on the above-mentioned net of points, i.e., in R^m, where m is large ($m = 2(n + 1)$). A systematic manipulation in the space of such dimensions is complicated. This fact has led to the development of techniques which decompose the problem and make use of the highly developed software for numerical integration of initial-value problems. This group of methods is called shooting methods.

3.1.3 Shooting methods

In shooting methods, the boundary-value problem is transformed into an initial-value problem. A number of efficient routines are available for integrating initial-value problems (e.g., Runge–Kutta methods, Gears package,

Episode, etc.). We shall present the simplest arrangement of a shooting method here; a more complete discussion is available in a number of textbooks, e.g., [3.6, 3.7].

Let us consider the stationary-state description (3.9) with BC 2 in the form

$$x'(0) = 0, \quad y'(0) = 0,$$ (3.22)

$$x'(1) = 0, \quad y'(1) = 0.$$ (3.23)

At a chosen point z, we need four initial conditions (the overall order of system (3.9) is four) to solve the initial-value problem. We make use of the given boundary conditions and choose two missing initial conditions at the point $z = 0$ (or at the point $z = 1$, if it is required by the stability of integration of the initial-value problem):

$$x(0) = \eta_1, \quad y(0) = \eta_2.$$ (3.24)

Together with conditions (3.22), we thus have a complete set of initial conditions. We can now integrate system (3.9) (rewritten in the form of four first-order differential equations) from $z = 0$ to $z = 1$. At $z = 1$, we obtain the values x, x', y, y' depending on the choice of $\boldsymbol{\eta} = (\eta_1, \eta_2)^T$. Hence, we obtain $x(1, \boldsymbol{\eta})$, $x'(1, \boldsymbol{\eta})$, $y(1, \boldsymbol{\eta})$, $y'(1, \boldsymbol{\eta})$. The remaining boundary conditions (3.23) will be satisfied if

$$R_1(\boldsymbol{\eta}) = x'(1, \boldsymbol{\eta}) = 0, \quad R_2(\boldsymbol{\eta}) = y'(1, \boldsymbol{\eta}) = 0.$$ (3.25)

We have obtained two nonlinear algebraic equations for two unknowns η_1 and η_2. These equations can be solved by any appropriate numerical method for solving nonlinear equations; difference formulas may be used to calculate required derivatives. One evaluation of residuals (3.25) requires one integration of the initial-value problem (3.9), (3.22), and (3.24) for chosen values of η_1, η_2. However, we may obtain partial derivatives $\partial R_i/\partial \eta_j$ analytically by forming variational differential equations for variational variables.

Let us differentiate (3.9) with respect to η_i and interexchange differentiation with respect to η_i and differentiation with respect to z. We obtain the differential equations

$$p_{xi}'' = -\frac{L^2}{D_x}\left[\frac{\partial f}{\partial x}p_{xi} + \frac{\partial f}{\partial y}p_{yi}\right],$$

$$p_{yi}'' = -\frac{L^2}{D_y}\left[\frac{\partial g}{\partial x}p_{xi} + \frac{\partial g}{\partial y}p_{yi}\right], \quad i = 1, 2.$$ (3.26)

Here we have denoted

$$p_{xi} = \frac{\partial x}{\partial \eta_i} \quad \text{and} \quad p_{yi} = \frac{\partial y}{\partial \eta_i}.$$ (3.27)

The initial conditions are

$$p_{x1}(0) = 1, \quad p_{x2}(0) = 0, \quad p_{y1}(0) = 0, \quad p_{y2}(0) = 1,$$
$$p'_{x1}(0) = 0, \quad p'_{x2}(0) = 0, \quad p'_{y1}(0) = 0, \quad p'_{y2}(0) = 0. \tag{3.28}$$

If we integrate (3.9) together with (3.26) with the initial conditions (3.22), (3.24), and (3.28), we can use the resulting values at $z = 1$ for $\partial R_i/\partial \eta_j$:

$$\frac{\partial R_1}{\partial \eta_1} = p'_{x1}(1), \quad \frac{\partial R_1}{\partial \eta_2} = p'_{x2}(1), \quad \frac{\partial R_2}{\partial \eta_1} = p'_{y1}(1), \quad \frac{\partial R_2}{\partial \eta_2} = p'_{y2}(1). \tag{3.29}$$

Application of the Newton's method is simple now (see Section 2.1). If it is inconvenient to derive variational equations (3.26), we can obtain the Jacobian matrix for Newton's method by means of differences, but we have to integrate the initial-value problem (3.9), (3.22), and (3.24) three times for choices (η_1, η_2), $(\eta_1 + \Delta, \eta_2)$, and $(\eta_1, \eta_2 + \Delta)$. Thus, we integrate the system of two second-order differential equations three times instead of performing one integration of six second-order differential equations where variational variables are used.

The methods with an automatic control of the integration step size (e.g., the Gear or Merson method) are advantageous for the integration.

For boundary conditions BC 1, i.e.,

$$x(0) = \bar{x}, \quad y(0) = \bar{y}, \tag{3.30}$$

$$x(1) = \bar{x}, \quad y(1) = \bar{y}, \tag{3.31}$$

two missing conditions

$$x'(0) = \eta_1, \quad y'(0) = \eta_2, \tag{3.32}$$

are chosen and the residuals (3.25) are now in the form

$$R_1(\boldsymbol{\eta}) = x(1) - \bar{x} = 0, \quad R_2(\boldsymbol{\eta}) = y(1) - \bar{y} = 0. \tag{3.33}$$

The variational equations (3.26) are the same, except that the initial conditions are changed as follows:

$$p_{x1}(0) = 0, \quad p_{x2}(0) = 0, \quad p_{y1}(0) = 0, \quad p_{y2}(0) = 0;$$
$$p'_{x1}(0) = 1, \quad p'_{x2}(0) = 0, \quad p'_{y1}(0) = 0, \quad p'_{y2}(0) = 1. \tag{3.34}$$

Subroutine SHOOT and its description for a simple version of the above shooting method are given in Appendix B. This version may be used when the integration of the corresponding initial-value problems using the shooting techniques (e.g., SHOOT) is successful. This integration is sometimes impossible (or very difficult) because of inherent instability of the initial-value problems with respect to the chosen initial conditions. In this case we must use either difference methods or multiple shooting methods [3.10, 3.11] where the original interval $z \in [0, 1]$ is divided into subintervals and initial-value problems are integrated on these subintervals.

Table 3.1 Results for Example 9, Chapter 1, shooting method. $Da = 0.12$, $B = 12$, $Pe_y = Pe_\theta = 2$, $\beta = 2$, $\theta_c = 0$, $\gamma = \infty$.

Iteration	η_1	η_2
0	0.000	0.000
1	0.157	0.739
2	0.221	1.030
3	0.234	1.093
4	0.235	1.096
5	0.235	1.096
0	0.900	4.000
1	0.896	4.087
2	0.897	4.078
3	0.897	4.077
0	0.750	4.000
3	0.619	3.145
0	0.980	2.900
4	0.985	3.213
0	0.950	3.600
3	0.937	3.693

Sometimes the initial-value problems are unstable in one direction only. This occurs in Example 9, Chapter 1, where integration from $z = 0$ to $z = 1$ cannot be performed for higher values of parameters Pe_y and Pe_θ. On the contrary, integration from $z = 1$ to $z = 0$ is successful. The missing initial (end) conditions are then chosen at $z = 1$:

$$y(1) = \eta_1, \quad \theta(1) = \eta_2.$$

The results obtained by means of the program SHOOT (see Appendix B) are given in Table 3.1. The problem has five different solutions for the chosen parameter values.

3.2 Dependence of Steady-State Solutions on a Parameter

Continuation of solutions of the operator Equation (2.2) (where $A = \alpha \in R^1$) will again be accomplished along individual branches of solutions (the solution branches have been defined in Section 1.3). The point $\mathbf{X}_0(\alpha_0)$, $\mathbf{F}(\mathbf{X}_0, \alpha_0) = \mathbf{0}$ is a regular point of the branch of solutions. The Fréchet derivative $\mathbf{F}'_x(\mathbf{X}_0, \alpha_0)$ has an inverse operator at the point $\mathbf{X}_0(\alpha_0)$, and according to the generalized implicit function theorem (e.g., [1.84]), just one branch of solutions passes through this point. The definition of limit and bifurcation points remains the same as in Section 1.3. Decker and Keller [3.12] present a more precise definition of these points based on knowledge of derivatives $\mathbf{F}'_x$ and $\mathbf{F}'_{xx}$. Moreover, they propose a continuation algorithm based on the Euler–Newton predictor-corrector combination. The algorithm proceeds along the arc–length of the solution locus when it passes through the bifurcation point as well. It seems that numerical realization of the proposed method, analogous to the procedure given in Section 2.2 for LPS, can be a difficult task in practice.

In the following, we shall deal with the continuation procedures which will not "take notice" of the bifurcation point. When passing through the bifurcation point, the procedures will probably continue along the branch which corresponds to the smooth continuation of the original branch. Numerical realization of these continuation algorithms is simple. We shall discuss the sequential use of a chosen iteration method, the use of the method of differentiation with respect to a parameter or to a boundary condition, the GPM method based on the shooting method, and the method following from the specific form of the problem under investigation.

The sequential use of some of the methods discussed in Section 3.1 is the simplest method for obtaining a parameter dependence of the solution of a nonlinear boundary-value problem. The application of the sequence method is the same as the one described for LPS, cf., Section 2.2. During the application of the method, the difficulties which arise in the neighborhood of limit and bifurcation points are similar to those in algebraic systems.

The second possible approach is the method of differentiation with respect to a parameter. Let us differentiate (3.9) formally with respect to L, for example. When certain assumptions about the continuity of solutions are satisfied, we obtain two partial differential equations

$$\frac{\partial^3 x}{\partial z^2 \, \partial L} + \frac{L^2}{D_x} \left[\frac{\partial f}{\partial x} \frac{\partial x}{\partial L} + \frac{\partial f}{\partial y} \frac{\partial y}{\partial L} \right] + \frac{2L}{D_x} f(x, y) = 0,$$

$$\frac{\partial^3 y}{\partial z^2 \, \partial L} + \frac{L^2}{D_y} \left[\frac{\partial g}{\partial x} \frac{\partial x}{\partial L} + \frac{\partial g}{\partial y} \frac{\partial y}{\partial L} \right] + \frac{2L}{D_y} g(x, y) = 0, \tag{3.35}$$

with boundary conditions, e.g., for BC 2,

$$\frac{\partial x(0, L)}{\partial z} = \frac{\partial x(1, L)}{\partial z} = 0, \quad \frac{\partial y(0, L)}{\partial z} = \frac{\partial y(1, L)}{\partial z} = 0. \tag{3.36}$$

An initial condition for some $L = L_0$ has to be chosen so that it satisfies the original boundary-value problem (3.9); hence,

$$L = L_0: x(z, L_0) = x_0(z), \ y(z, L_0) = y_0(z), \tag{3.37}$$

where $x_0(z)$ and $y_0(z)$ satisfy equations (3.9) for $L = L_0$. This solution is either known *a priori* for some limiting value of the parameter or it can be obtained by some of the methods described in Section 3.1. For integrating the partial differential equation (3.35), methods similar to the finite-difference methods for parabolic equations (e.g., [3.13, 3.14], cf., Section 3.5) may be used. The solution of (3.35) will differ from the solution of equations (3.39) due to errors of approximation of the applied finite-difference method. Therefore, it is necessary to correct the solution after a certain number of integration steps by the use of some of the methods described previously (from the point of view of compatibility, the finite-difference method will be the most convenient one). The above continuation method is successful on one smooth branch of the solution; transition through limit points is not possible. In the neighborhood of the limit points, the method of differentiation with respect to boundary conditions may be used [3.15]. Let us present only a schematic derivation of the method. First, we shall differentiate (3.9) with respect to the parameter ξ which will represent the value of $x(0)$ for BC 2 (it would correspond to $x'(0)$ for BC 1):

$$\frac{\partial^3 x}{\partial z^2 \, \partial \xi} + \frac{L^2}{D_x}\left[\frac{\partial f}{\partial x}\frac{\partial x}{\partial \xi} + \frac{\partial f}{\partial y}\frac{\partial y}{\partial \xi}\right] + \frac{2L}{D_x}f(x, y)\frac{dL}{d\xi} = 0,$$

$$\frac{\partial^3 y}{\partial z^2 \, \partial \xi} + \frac{L^2}{D_y}\left[\frac{\partial g}{\partial x}\frac{\partial x}{\partial \xi} + \frac{\partial g}{\partial y}\frac{\partial y}{\partial \xi}\right] + \frac{2L}{D_y}g(x, y)\frac{dL}{d\xi} = 0. \tag{3.38}$$

The boundary conditions are evidently in the form

$$\frac{\partial x(0, \xi)}{\partial z} = \frac{\partial x(1, \xi)}{\partial z} = \frac{\partial y(0, \xi)}{\partial z} = \frac{\partial y(1, \xi)}{\partial z} = 0, \quad x(0, \xi) = \xi, \tag{3.39}$$

and the initial conditions are analogous to (3.37):

$$\xi = \xi_0: x(z, \xi_0) = x_0(z), \quad y(z, \xi_0) = y_0(z), \quad L(\xi_0) = L_0, \tag{3.40}$$

where $x_0(z)$, $y_0(z)$ satisfy (3.9) for $L = L_0$, and, moreover, $x_0(0) = \xi_0$.

In all we have five boundary conditions (3.39), because in (3.38) we have added another unknown, $dL/d\xi$. The problem is thus correctly defined, and we may switch from the integration of (3.35) to the integration of (3.38) before the limit point on the L-dependence curve is reached. After passing the limit point, we can switch back to the integration of (3.35). The continuation with regard to (3.38) will pass through the limit point without difficulties. How do we determine when we are near a limit point? Either we can follow the value of the derivative $dL/d\xi$, or we can use some trial-and-error technique.

The above methods are of general utility. However, algorithmization is relatively complicated and computer memory requirements are high. For problems where the shooting method can be used, it is advantageous to use "decomposition" or transformation of the boundary-value problem into an initial-value problem. Then the general procedure of the type DERPAR presented in Section 2.2 may be used to solve the obtained algebraic problem. Let us describe the procedure for BC 2. Again we seek the dependence of the solution on the parameter L. By choice of initial conditions (3.24), we have a complete set of initial conditions for integrating the initial-value problem for a given L. If we consider the solution of the initial-value problem not only as a function of chosen initial conditions η_1 and η_2, but also as a function of the parameter L, we obtain:

$$R_1(\eta_1, \eta_2, L) = x'(1, \eta_1, \eta_2, L) = 0,$$

$$R_2(\eta_1, \eta_2, L) = y'(1, \eta_1, \eta_2, L) = 0. \tag{3.41}$$

To determine the dependence of a solution of the boundary-value problem (3.9), (3.22), and (3.23) on the parameter L, it suffices to obtain dependences $\eta_1(L)$, $\eta_2(L)$ which satisfy (3.41). We may use the approach described in Section 2.2 based on the DERPAR routine to solve the system (3.41). However, for its application, we also need (in addition to evaluating R_1 and R_2) the Jacobian matrix

$$\begin{pmatrix} \dfrac{\partial R_1}{\partial \eta_1}, & \dfrac{\partial R_1}{\partial \eta_2}, & \dfrac{\partial R_1}{\partial L} \\[2mm] \dfrac{\partial R_2}{\partial \eta_1}, & \dfrac{\partial R_2}{\partial \eta_2}, & \dfrac{\partial R_2}{\partial L} \end{pmatrix}. \tag{3.42}$$

Again the values of variational variables at $z = 1$ (see (3.29)) can be used to calculate the first two columns of the Jacobian matrix (3.42). Other variational variables $p_{xL} = \partial x/\partial L$ and $p_{yL} = \partial y/\partial L$ can be introduced for evaluating of the last column. Upon differentiating (3.9) with respect to L, we obtain

$$p''_{xL} = -\frac{L^2}{D_x}\left[\frac{\partial f}{\partial x} p_{xL} + \frac{\partial f}{\partial y} p_{yL}\right] - \frac{2L}{D_x} f(x, y),$$

$$p''_{yL} = -\frac{L^2}{D_y}\left[\frac{\partial g}{\partial x} p_{xL} + \frac{\partial g}{\partial y} p_{yL}\right] - \frac{2L}{D_y} g(x, y), \tag{3.43}$$

with initial conditions

$$p_{xL}(0) = p'_{xL}(0) = p_{yL}(0) = p'_{yL}(0) = 0. \tag{3.44}$$

The values of the derivatives are then

$$\frac{\partial R_1}{\partial L} = p'_{xL}(1), \quad \frac{\partial R_2}{\partial L} = p'_{yL}(1). \tag{3.45}$$

The approach based on the shooting method and variational variables is called GPM—the general parameter mapping technique [3.16, 3.17]. For every call of the subrouting FCTN in the subroutine DERPAR, we have to integrate the initial-value problem for eight second-order differential equations (3.9), (3.26), and (3.43). An efficient numerical technique for integrating initial-value problems has to be used because the integration is often repeated during the computation. The Runge–Kutta method with an automatic step-size control (e.g., the Merson modification) seems to be appropriate; the use of the Gear package or EPISODE codes may speed up the computations in some cases.

Now let us present several examples of the dependences of the solution on a parameter obtained by means of the above methods. The first example will be Example 6, Chapter 1. The dependence of the stationary solution (of the values of η_1 and η_2) on the value of the parameter L is shown in Fig. 3.2 for BC 2 and in Fig. 3.3 for BC 1 [3.18]. Figure 3.3 need not be complete; several

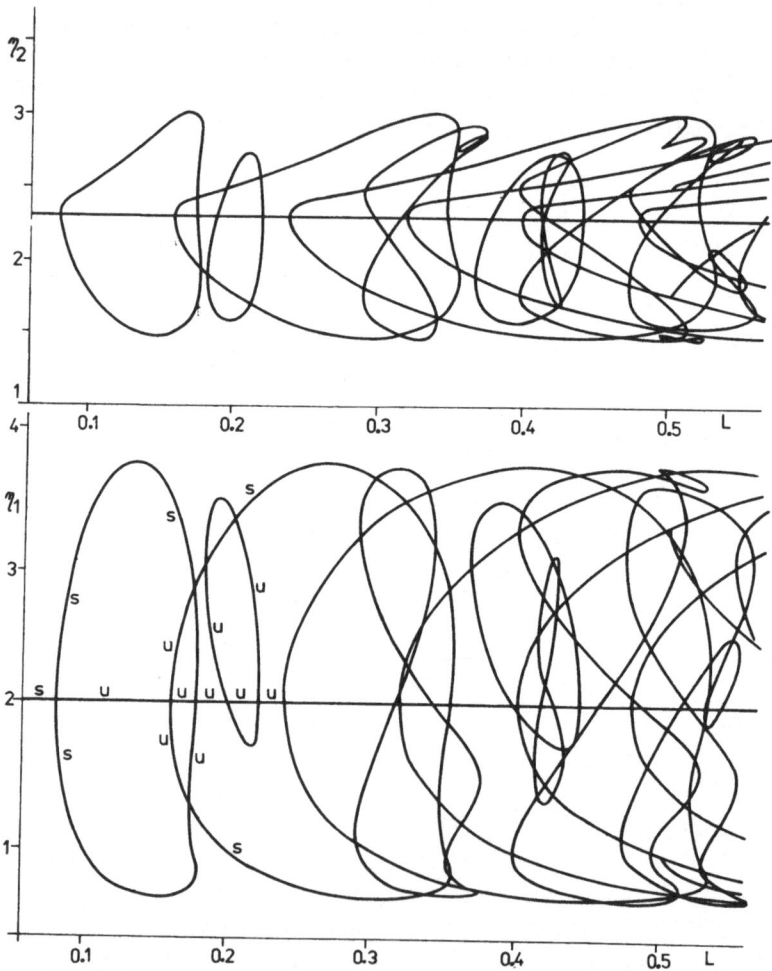

Figure 3.2 Dependence of the steady-state solutions on the parameter L; Brusselator model, Example 6, Chapter 1, $A = 2$, $B = 4.6$, BC 2, $D_x = 0.0016$, $D_y = 0.008$, $\eta_1 = x(0)$, $\eta_2 = y(0)$. s—stable, u—unstable.

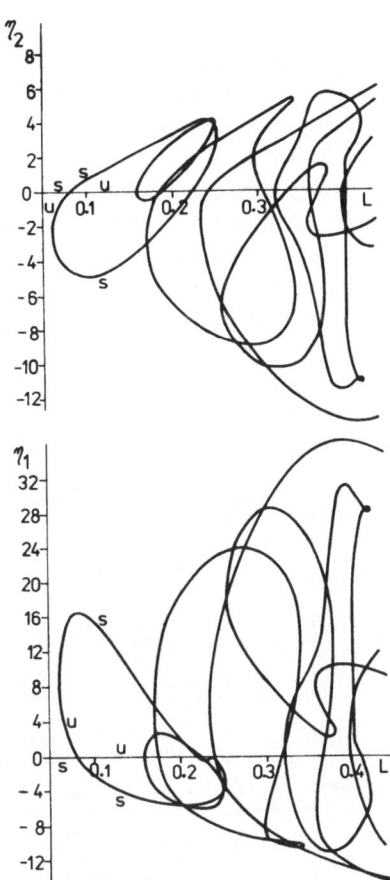

Figure 3.3 Dependence of the steady-state solutions on the parameter L; Brusselator model, Example 6, Chapter 1, $A = 2$, $B = 4.6$, BC 1, $D_x = 0.0016$, $D_y = 0.008$, $\eta_1 = x'(0)$, $\eta_2 = y'(0)$. s—stable, u—unstable.

branches of solution may be missing. However, we believe that all significant branches are presented in Fig. 3.2.

A knowledge of the trivial solution and of the value of parameter L at the point of primary real bifurcation (see Section 3.3) has been utilized in the computations. The initial points for continuation by the GPM-DERPAR technique have been chosen in the neighborhood of known points of the solution. Most solution branches have been computed in this way. The procedure of "composing solutions" (following from the symmetry of the problem) has been used for BC 2. In addition to the above-mentioned branches of solutions, branches arising through secondary bifurcation (i.e., a bifurcation from a non-trivial solution) also exist. Random generation of initial estimates has been used to determine the starting point for continuation in this case.

Dependences of steady-state solutions on a parameter for Example 7, Chapter 1 will be presented in Chapter 4.

Let us consider Example 8, Chapter 1. Three trivial solutions are possible for the chosen parameter values. They are presented in the legend to Fig. 3.4, where the dependence of the solution on the length L for BC 2 is shown. The possibility of composing solutions has again been used for the construction of Fig. 3.4.

The dependences of the solution on parameter Da for Example 9, Chapter 1, again obtained by the GPM-DERPAR technique, are shown in Figs. 3.5 and 3.6 [1.77]. From these figures, it can be seen that within a certain range of parameter values, five solutions of the given boundary-value problem exist. The dependence of the temperature and conversion profiles on parameter Da

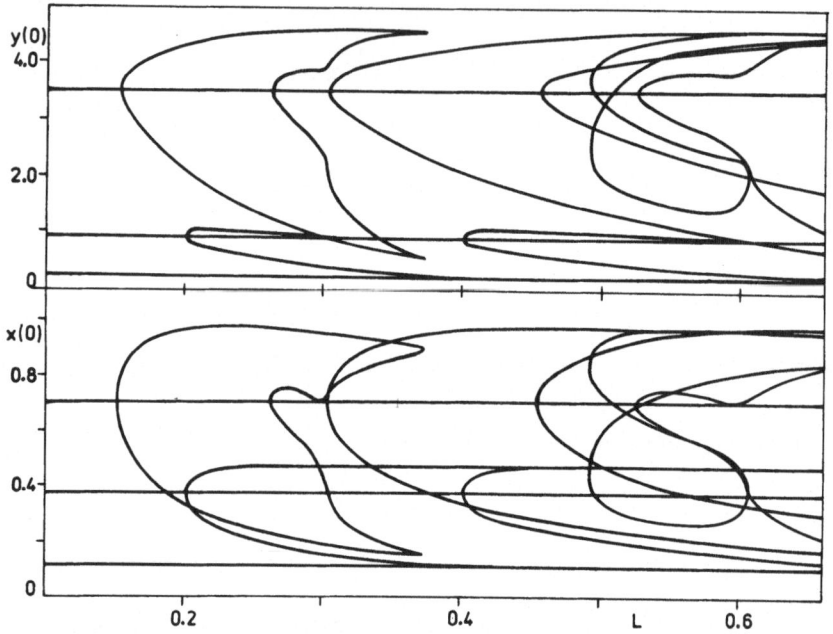

Figure 3.4 Dependence of $x(0)$ and $y(0)$ on L, Example 8, Chapter 1, BC 2, $\alpha = 12, \beta = 1.5, \gamma = 3, \delta = 1, v_0 = 0.01, D_x = 0.008, D_y = 0.004.$

Trivial steady-state solutions:

i	$\bar{x}_i$	$\bar{y}_i$	in CSTR[b]	L (bifurcation length)[a]	
1	0.1157	0.1962	stable focus	–	
2	0.3728	0.8915	saddle	0.1965	
3	0.7006	3.5106	unstable focus	0.1516	0.2630

[a] See Section 3.3.

[b] $D_x = D_y = 0.$

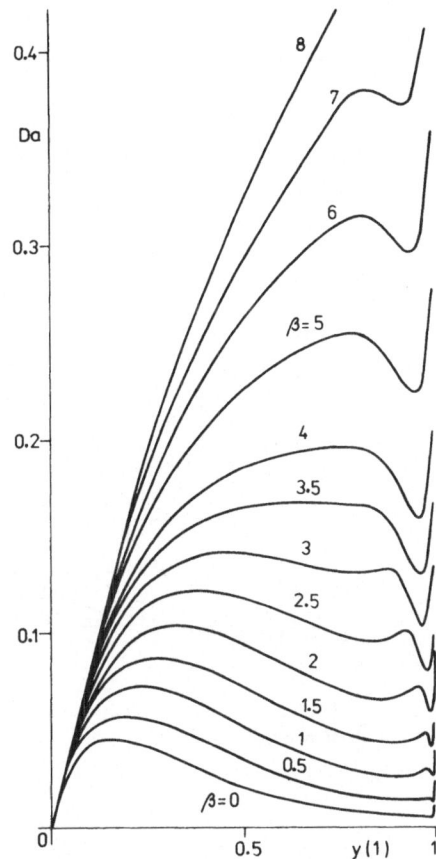

Figure 3.5 Example 9, Chapter 1, dependence of exit conversion $y(1)$ on the Damköhler number Da for various values of β; $\gamma = 20$, $B = 15$, $\text{Pe}_y = 10$, $\text{Pe}_\theta = 5$, $\theta_c = 0$.

is shown in Fig. 3.7. Five solutions of the boundary-value problem exist for Da = 0.07. Let us mention that no trivial solution exists for Example 9, Chapter 1, contrary to Examples 6, 7, and 8 of Chapter 1.

The entire line in Figs. 3.5 and 3.6 (for a fixed β) has been computed by means of the GPM-DERPAR algorithm; the starting point was $\eta_1 = y(1) = 0$, $\eta_2 = \theta(1) = 0$ for Da = 0.

The problem discussed in Example 10, Chapter 1 is more difficult because the shooting method is unstable for higher values of the Reynolds number Re. Therefore, the difference approximation (3.18) has been used sequentially for the sequence of values of Re and for $S = 0.8$ [3.8]. The dependence of the k-value on S for a fixed Re (obtained by sequential use of the difference approximation (3.18) for the sequence of S-values) is shown in Fig. 3.8. The solution had to be obtained for every branch individually, and problems arose at the limit points. A large number of man-machine interactions were required for the construction of Fig. 3.8. The results are described in more detail in [3.19], where, for the values of $S = -1, 0, 1$, all profiles $H(z)$, $F(z)$, $G(z)$ are

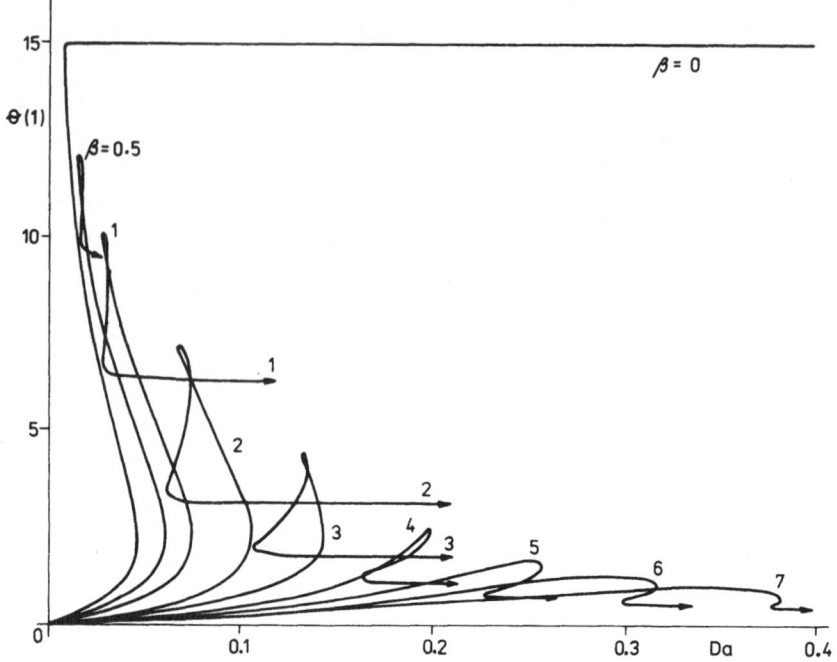

Figure 3.6 Example 9, Chapter 1, dependence of exit temperature $\theta(1)$ on Da for various values of β; $\gamma = 20$, $B = 15$, $Pe_y = 10$, $Pe_\theta = 5$, $\theta_c = 0$.

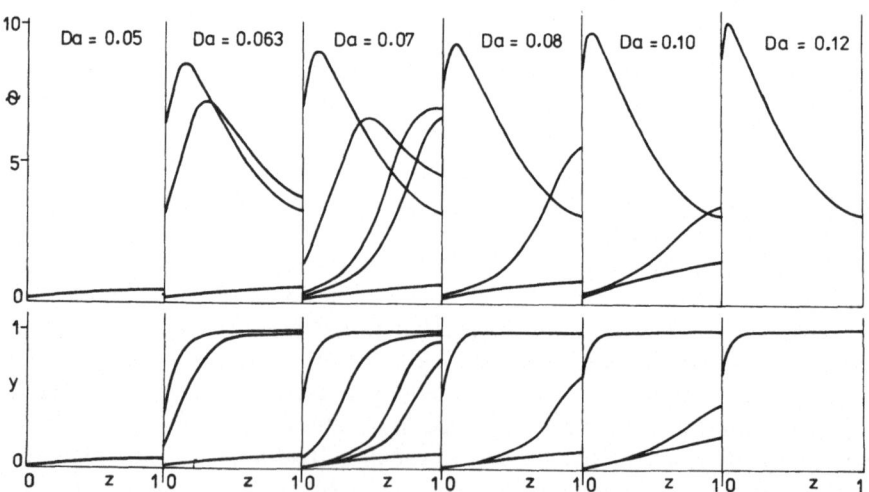

Figure 3.7 Example 9, Chapter 1, dependence of temperature and conversion profiles on Da; $\gamma = 20$, $B = 15$, $Pe_y = 10$, $Pe_\theta = 5$, $\beta = 2$, $\theta_c = 0$.

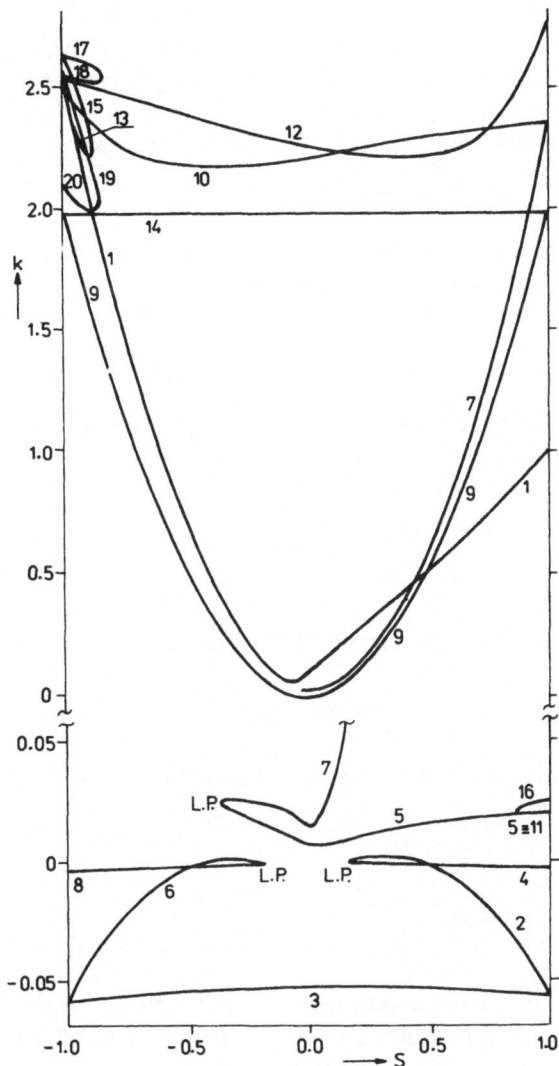

Figure 3.8 Example 10, Chapter 1, dependence of the solution on S, Re = 625.

given. An attempt to use the multiple shooting method in combination with the GPM-DERPAR algorithm for the construction of the above-mentioned dependences has shown [3.20] that the method is successful for far higher values of the Reynolds number Re than the simple shooting method.

For certain problems, we can find such transformation of variables and parameters (depending on the nature of the problem) that in a single integration of such a transformed initial-value problem (hence, for one choice of initial conditions), we can obtain one solution of the boundary-value problem for the parameter value determined at the end of the integration. Thus, we obtain one point for the picture of the dependence on this parameter. The procedure is discussed in detail in the papers of Klamkin [3.21, 3.22] and in [3.7]. Let us illustrate this procedure (sometimes called parameter mapping) on the simplified problem (3.9) where we assume $D_y \to 0$. In this case, the second of equations (3.9) can be understood as an algebraic equation

$$g(x, y) = 0 \tag{3.46}$$

in which $y(x)$ can be expressed and then introduced into the first of equations (3.9) with the following result ($' = d/dz$):

$$x'' = -\frac{L^2}{D_x} f(x, y(x)) = -\frac{L^2}{D_x} F(x). \tag{3.47}$$

Let us consider, for example, the boundary conditions BC 2, i.e.,

$$x'(0) = 0, \quad x'(1) = 0. \tag{3.48}$$

Let us introduce the substitution

$$\xi = Lz; \tag{3.49}$$

(3.47) is then expressed in the form

$$\frac{d^2x}{d\xi^2} = -\frac{1}{D_x} F(x) \tag{3.50}$$

with unchanged boundary conditions (3.48),

$$\frac{dx(0)}{d\xi} = 0, \tag{3.51}$$

$$\frac{dx(L)}{d\xi} = 0. \tag{3.52}$$

Note that the parameter L now appears in the boundary conditions. Let us now choose the value of x_0; upon using (3.51), we shall obtain a complete set of initial conditions at $\xi = 0$:

$$\xi = 0: x(0) = x_0, \quad \frac{dx(0)}{d\xi} = 0. \tag{3.53}$$

Now we integrate the initial-value problem (3.50) and (3.53) rewritten in the form of two first-order differential equations:

$$\frac{dx}{d\xi} = u,$$

$$\frac{du}{d\xi} = -\frac{1}{D_x} F(x). \tag{3.54}$$

We proceed with the integration until, at a certain point $\bar{\xi}$,

$$u(\bar{\xi}) = 0. \tag{3.55}$$

Then $L = \bar{\xi}$ and the point (x_0, L) is the point used for the dependence of the solution x on parameter L. To locate the point $\bar{\xi}$ during integration, we may use either a trial-and-error method or the following procedure: we integrate (3.54) with initial conditions (3.53) up to a point $\tilde{\xi} > 0$, where

$$\xi = \tilde{\xi}: x = \tilde{x}, \quad u = \tilde{u}. \tag{3.56}$$

At this point, we switch from the integration of (3.54) to the integration of the system

$$\frac{d\xi}{du} = -\frac{D_x}{F(x)}, \quad \frac{dx}{du} = -\frac{uD_x}{F(x)} \tag{3.57}$$

with the initial condition

$$u = \tilde{u}: \xi = \tilde{\xi}, \quad x = \tilde{x}. \tag{3.58}$$

We integrate until the value $u = 0$ is reached (hence, we integrate on the interval $[\tilde{u}, 0]$). We have exchanged the independent variable ξ for u. This can be done only in cases where the dependence $u(\xi)$ is monotonous on the interval $[\tilde{\xi}, \bar{\xi}]$, i.e., there is no point where $du/d\xi = 0$. Therefore, we choose the point $\tilde{\xi}$ only after the extremum point $du/d\xi = 0$ is crossed (the dependence $x(\xi)$ has an inflection point there).

Over the course of integration of (3.57), we have to follow $F(x)$ and determine whether or not it approaches zero. After integration on the interval $[\tilde{u}, 0]$, the system (3.57) with the initial conditions (3.58) yields the values

$$u = 0: \xi = \bar{\xi} = L, \quad x = x(\bar{\xi}) = x(z = 1). \tag{3.59}$$

3.3 Branch Points—Methods for Evaluating Real and Complex Bifurcation Points

The theoretical procedures discussed in Appendix C also hold for equations in the Banach space, i.e., for distributed-parameter problems which satisfy certain requirements [1.84]. However, the development of numerical procedures for determinating the bifurcation points is in its initial stage. We shall discuss the techniques used on example of reaction-diffusion systems.

As we mentioned at the beginning of Chapter 3, the trivial solution

$$x(z) \equiv \bar{x}, \quad y(z) \equiv \bar{y}, \tag{3.60}$$

often exists for reaction-diffusion systems. The trivial solution may be independent of values of some parameters, e.g., L, D_x, D_y. The study of the behavior

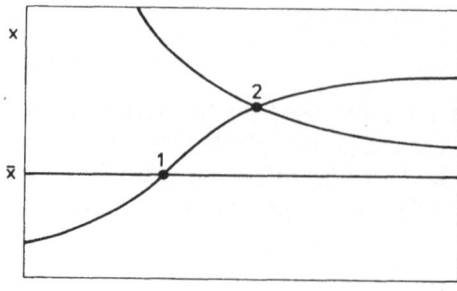

Figure 3.9 Primary (1) and secondary (2) bifurcation.

of stationary solutions in the neighborhood of this trivial solution and the questions of its stability are simpler. This study has its own terminology which has developed historically. Bifurcation of solution branches from the trivial branch is called "primary bifurcation," and bifurcation from a non-trivial branch is then a secondary bifurcation (see Fig. 3.9).

3.3.1 Primary bifurcation

The stability of the trivial solution (3.60) is determined by eigenvalues of linearized operator [1.84]

$$\mathscr{L} = \begin{pmatrix} \dfrac{D_x}{L^2}\dfrac{d^2}{dz^2} + \dfrac{\partial f(\bar{x}, \bar{y})}{\partial x}, & \dfrac{\partial f(\bar{x}, \bar{y})}{\partial y} \\ \dfrac{\partial g(\bar{x}, \bar{y})}{\partial x}, & \dfrac{D_y}{L^2}\dfrac{d^2}{dz^2} + \dfrac{\partial g(\bar{x}, \bar{y})}{\partial y} \end{pmatrix}, \tag{3.61}$$

when corresponding boundary conditions are considered.

The trivial solution is stable if all eigenvalues of the operator $\mathscr{L}$ have negative real parts, and it is unstable if at least one eigenvalue has a positive real part.

In the case of the trivial solution, the operator $\mathscr{L}$ has constant coefficients and the eigenfunctions are in the well-known form:

BC 1: $(c_n \sin n\pi z, d_n \sin n\pi z)^T$, $n = 1, 2, \ldots$, $\tag{3.62}$

BC 2: $(c_n \cos n\pi z, d_n \cos n\pi z)^T$, $n = 0, 1, 2, \ldots$, $\tag{3.63}$

where the coefficients c_n and d_n can be determined. If we introduce the eigenfunctions into characteristic equation of $\mathscr{L}$, $\mathscr{L}u = \lambda u$, we can see that the eigenvalue λ is simultaneously an eigenvalue of the matrix (for BC 1 and BC 2)

$$A_n = \begin{pmatrix} -D_x E_{(n)} + a_{11} & a_{12} \\ a_{21} & -D_y E_{(n)} + a_{22} \end{pmatrix}. \tag{3.64}$$

Here we have denoted $E_{(n)} = (n\pi/L)^2$, and the matrix A is then equal to

$$A = \begin{pmatrix} a_{11} & a_{12} \\ a_{21} & a_{22} \end{pmatrix} = \begin{pmatrix} \dfrac{\partial f(\bar{x}, \bar{y})}{\partial x} & \dfrac{\partial f(\bar{x}, \bar{y})}{\partial y} \\ \dfrac{\partial g(\bar{x}, \bar{y})}{\partial x} & \dfrac{\partial g(\bar{x}, \bar{y})}{\partial y} \end{pmatrix}. \tag{3.65}$$

The eigenvalues λ_n of the matrices A_n are the roots of characteristic equations

$$\lambda_n^2 - P(n)\lambda_n + Q(n) = 0, \tag{3.66}$$

where, evidently, the following relations hold:

$$P(n) = a_{11} + a_{22} - (D_x + D_y)E_{(n)}, \tag{3.67}$$

$$Q(n) = D_x D_y E_{(n)}^2 - (D_x a_{22} + D_y a_{11})E_{(n)} + a_{11}a_{22} - a_{12}a_{21}. \tag{3.68}$$

The necessary condition for a real bifurcation (cf., Appendix C and Section 1.3), the existence of zero eigenvalue, follows from properties of the quadratic equation (3.66):

$$Q(n) = 0. \tag{3.69}$$

For a complex bifurcation, we have

$$P(n) = 0, \quad Q(n) > 0. \tag{3.70}$$

We can take L, D_x, or D_y as bifurcation parameters. Let us take L, which is contained solely in $E_{(n)}$. The values of $E_{(n)}$ satisfying the condition of a real bifurcation (3.69), $E_{(n)}^*$, are given by the relation (in case the real solution of (3.69) exists):

$$E_{(n)1,2}^* =$$
$$\frac{D_x a_{22} + D_y a_{11} \pm [(D_x a_{22} + D_y a_{11})^2 - 4D_x D_y(a_{11}a_{22} - a_{12}a_{21})]^{1/2}}{2D_x D_y}.$$
$$\tag{3.71}$$

The right-hand side of (3.71) does not depend on n; hence, we have 0, 1, or 2 values of E^* for which the real bifurcation occurs. From $E_{(n)} = (n\pi/L)^2$ it follows that the bifurcation length is determined by the relation

$$L_{(n)1,2}^* = \frac{n\pi}{\sqrt{E_{(n)1,2}^*}} \tag{3.72}$$

for $E^* > 0$. It is evident that

$$L_{(n)}^* = nL_{(1)}^* \tag{3.73}$$

for each of the two possible groups of bifurcation lengths. In an analogous way, from (3.70), we obtain

$$L_{(1)}^+ = \pi\sqrt{\frac{D_x + D_y}{a_{11} + a_{22}}} \tag{3.74}$$

Table 3.2 Conditions for the existence of
real (L^*) and complex (L^+) bifurcation
lengths for $n = 1$.

	Exists
$ab < -1$	one L^*
$a + \rho b > 0$ $\|a - \rho b\| > 2\sqrt{\rho}$	two L^*
$a + b > 0$ $\|a - \rho b\| < 1 + \rho$	L^+
others	no bifurcation length

for a complex bifurcation length L^+ if the square root is defined and if the
condition $Q(1) > 0$ holds. Also,

$$L_{(n)}^+ = nL_{(1)}^+. \tag{3.75}$$

Let us denote

$$a = \frac{a_{11}}{\sqrt{-a_{12}a_{21}}}, \quad b = \frac{a_{22}}{\sqrt{-a_{12}a_{21}}}, \quad \rho = \frac{D_x}{D_y}, \tag{3.76}$$

and consider the case $a_{12}a_{21} < 0$. The criteria of existence of bifurcation
lengths are summarized in Table 3.2.

Let us consider the linearized operator $\mathscr{L}$ as a function of the characteristic
length L. The operator has the following properties:

(i) For every positive integer multiple of a complex bifurcation length L^+,
a pair of complex-conjugate eigenvalues of $\mathscr{L}$ crosses the imaginary
axis from the left complex half-plane to the right one.

(ii) If there exists exactly one real bifurcation length L^*, for each of its
multiples, one real eigenvalue crosses the imaginary axis as follows:

(a) from left to right for $P < 0$, or
(b) from right to left for $P > 0$.

(iii) If there exists $L_1^* < L_2^*$ (for $n = 1$), for every multiple of $L_1^*(L_2^*)$, one
real eigenvalue crosses the imaginary axis

(a) from left to right for $P(L_1^*) < 0 \ (P(L_2^*) > 0)$, or
(b) from right to left for $P(L_1^*) > 0 \ (P(L_2^*) < 0)$.

The pictures in a "parametric plane a-b" showing the existence of
individual bifurcation lengths are presented in [3.23], together with
possible ways in which the eigenvalues cross the imaginary axis.

As in Chapter 2, we may construct a bifurcation diagram of the dependence
of bifurcation points on another parameter v of the problem. A typical bi-

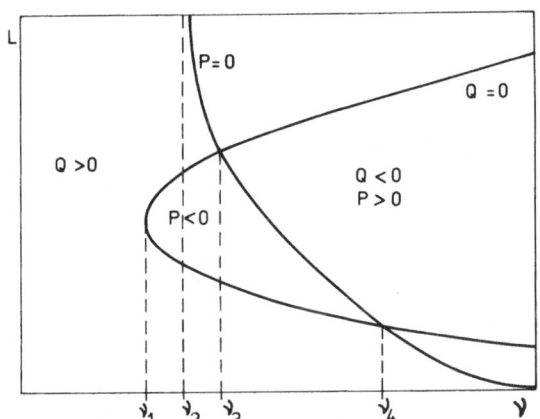

Figure 3.10 A typical bifurcation diagram.

furcation diagram (for $n = 1$) is shown in Fig. 3.10. For $v < v_1$, no bifurcation lengths exist, for $v \in (v_1, v_2) \cup (v_3, v_4)$, two real bifurcation lengths exist, and for $v \in (v_2, v_3) \cup (v_4, \cdot)$, two real and one complex bifurcation lengths exist. The bifurcation diagram for Example 6, Chapter 1, including the multiples of bifurcation lengths, is shown in Fig. 3.11.

Now let us deal with stability of a trivial solution for small values of L. For $L \to 0^+$ is the trivial solution of problem (3.2), (3.3), and (3.4), i.e., the solution for BC 1, stable. The stability of the solution for BC 2 depends on the stability of the solution for $D_x = D_y = 0$ in (3.2) and (3.3) ("well-mixed system") [1.64].

The stability is determined by eigenvalues of the matrix A in (3.65). If the solution in the well-mixed system is stable, the conclusions for BC 2 are the same as those for BC 1 for $L \to 0^+$. If the solution for the well-mixed system is unstable, the trivial solution for BC 2 is unstable for all $L > 0$. We can expect that a nontrivial branch of solutions on the dependence curve of solutions on L branches off at the first bifurcation length when the trivial solution for $L \to 0^+$ is stable and when bifurcation lengths exist. The solutions

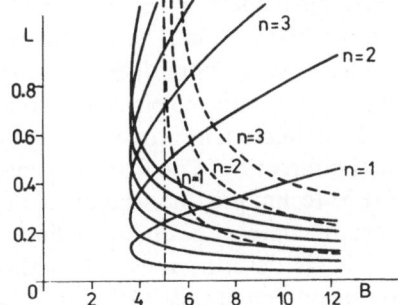

Figure 3.11 Bifurcation diagram for the model "Brusselator," Example 6, Chapter 1, $A = 2$, $D_x = 0.0016$, $D_y = 0.008$. —— $Q = 0$; ---- $P = 0$.

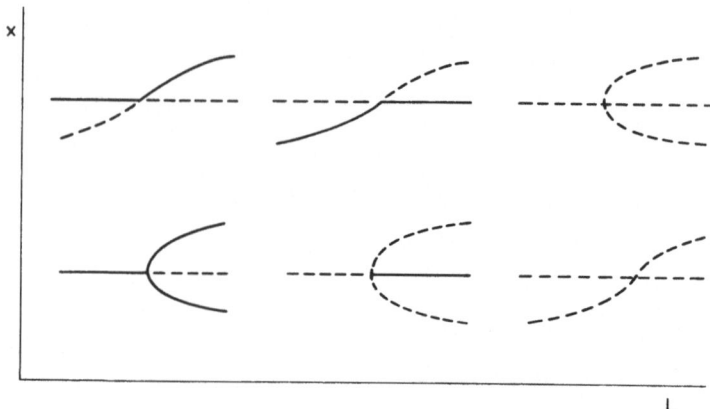

Figure 3.12 Possible cases of real bifurcation from the trivial solution. —— stable; - - - - unstable solutions.

on the nontrivial branch will be stable if the bifurcation is supercritical (cf., Section 1.3), i.e., the bifurcation occurs for $L > L^*$. Various possibilities of real branching of the nontrivial branch at the bifurcation point (primary) are sketched in Fig. 3.12. These are the only possibilities that can occur. Here we have not considered the "perpendicular bifurcation" which usually occurs in linear systems. Complex bifurcations are not considered either. Similar situations occur in complex bifurcation. The cases where a real eigenvalue of odd multiplicity or a single pair of complex eigenvalues cross the imaginary axis are the only cases which will be examined (cf., Appendix C for a discussion of other cases).

3.3.2 Secondary real bifurcation

There is no principal difference between primary and secondary bifurcation, and all conclusions regarding the stability of bifurcated branches and the possibilities of bifurcation described above are valid. Determination of the points of primary real and complex bifurcation is a simple process because the linearized operator $\mathscr{L}$ has constant coefficients and the eigenfunctions are known beforehand. The case of nontrivial solutions (i.e., the case of a secondary bifurcation) is more difficult since the coefficients of $\mathscr{L}$ depend on z and it is not easy to find the eigenfunctions.

Two different approaches may be used for computating a point of secondary real bifurcation. The first approach is based on the shooting method and the GPM technique; the second is based on finite-difference approximations for a certain complex boundary-value problem.

Let us describe the first method applied to the system of two reaction-diffusion equations (3.9) with boundary conditions BC 2. The reader is encouraged to read Section 2.3 before continuing further. After choosing initial conditions (3.24) and integrating (3.9) from $z = 0$ to $z = 1$, we obtain two residuals (3.25) dependent on the choice of L:

$$R_1(\eta_1, \eta_2, L) = x'(1, \boldsymbol{\eta}, L) = 0,$$
$$R_2(\eta_1, \eta_2, L) = y'(1, \boldsymbol{\eta}, L) = 0. \tag{3.77}$$

If we integrate the variational system (3.26), also, with initial conditions (3.28) together with (3.9), we obtain a Jacobian matrix at $z = 1$:

$$G(\eta_1, \eta_2, L) = \begin{pmatrix} \dfrac{\partial R_1}{\partial \eta_1} & \dfrac{\partial R_1}{\partial \eta_2} \\ \dfrac{\partial R_2}{\partial \eta_1} & \dfrac{\partial R_2}{\partial \eta_2} \end{pmatrix}. \tag{3.78}$$

Relation (3.29) holds for its individual elements.

Then the necessary condition for the branching is

$$R_3(\eta_1, \eta_2, L) = \det G(\eta_1, \eta_2, L) = p'_{x1}(1) \cdot p'_{y2}(1) - p'_{x2}(1) \cdot p'_{y1}(1) = 0. \tag{3.79}$$

As in Section 2.3, we may consider Eqs. (3.77) and (3.79) to be a system of three equations for determining the unknowns η_1, η_2, and L at a limit point (eventually, also, at a bifurcation point). It is necessary to determine the Jacobian matrix of the system when we apply Newton's method to the above set of three equations. We need not calculate the elements in the first two rows (for $\partial R_i / \partial L$, $i = 1$, 2, we use the values of variational variables $p'_{xL}(1)$ and $p'_{yL}(1)$ obtained through solving (3.43)). The derivatives in the third row can be either obtained from their finite-difference approximations or the following variational variables have to be introduced:

$$p_{x11} = \frac{\partial^2 x}{\partial \eta_1^2}, \quad p_{x12} = \frac{\partial^2 x}{\partial \eta_1\, \partial \eta_2}, \quad p_{x22} = \frac{\partial^2 x}{\partial \eta_2^2}, \quad p_{x1L} = \frac{\partial^2 x}{\partial \eta_1\, \partial L},$$

$$p_{x2L} = \frac{\partial^2 x}{\partial \eta_2\, \partial L}\,;$$

$$p_{y11} = \frac{\partial^2 y}{\partial \eta_1^2}, \quad p_{y12} = \frac{\partial^2 y}{\partial \eta_1\, \partial \eta_2}, \quad p_{y22} = \frac{\partial^2 y}{\partial \eta_2^2}, \quad p_{y1L} = \frac{\partial^2 y}{\partial \eta_1\, \partial L},$$

$$p_{y2L} = \frac{\partial^2 y}{\partial \eta_2\, \partial L}, \tag{3.80}$$

and we derive differential equations for these variables. Thus, for example, the differential equation for p_{x12} can be obtained upon differentiating (3.26) for p_{x1} with respect to η_2:

$$p''_{x12} = -\frac{L^2}{D_x} \left[\left(\frac{\partial^2 f}{\partial x^2} p_{x2} + \frac{\partial^2 f}{\partial x\, \partial y} p_{y2} \right) p_{x1} + \frac{\partial f}{\partial x} p_{x12} \right.$$
$$\left. + \left(\frac{\partial^2 f}{\partial y\, \partial x} p_{x2} + \frac{\partial^2 f}{\partial y^2} p_{y2} \right) p_{y1} + \frac{\partial f}{\partial y} p_{y12} \right]. \tag{3.81}$$

The remaining nine differential equations can be obtained in a similar way. The initial conditions for all variational variables (3.80) at $z = 0$ are equal to zero. A more detailed derivation is given in [3.24]. We must integrate the initial-value problem for 16 second-order differential equations to obtain one Newton iteration of the system of three nonlinear equations (3.77), (3.79). The third row of the Jacobian matrix is then evidently given by differentiating (3.79) with respect to η_1, η_2, and L:

$$\frac{\partial R_3}{\partial \eta_1} = p'_{x11}(1)p'_{y2}(1) + p'_{x1}(1) \cdot p'_{y12}(1) - p'_{x12}(1)p'_{y1}(1) - p'_{x2}(1)p'_{y11}(1),$$

$$\frac{\partial R_3}{\partial \eta_2} = p'_{x12}(1)p'_{y2}(1) + p'_{x1}(1) \cdot p'_{y22}(1) - p'_{x22}(1)p'_{y1}(1) - p'_{x2}(1)p'_{y12}(1),$$

$$\frac{\partial R_3}{\partial L} = p'_{x1L}(1)p'_{y2}(1) + p'_{x1}(1) \cdot p'_{y2L}(1) - p'_{x2L}(1)p'_{y1}(1) - p'_{x2}(1)p'_{y1L}(1).$$

$$(3.82)$$

The above relations are sufficient for determining limit points (cf., Section 2.3). To compute bifurcation points, we add another equation (see Section 2.3):

$$R_4(\eta_1, \eta_2, L) = \det \bar{G}(\eta_1, \eta_2, L) = p'_{x1}(1) \cdot p'_{yL}(1)$$

$$- p'_{xL}(1)p'_{y1}(1) = 0. \qquad (3.83)$$

Partial derivatives of R_4 with respect to η_1, η_2, and L can be obtained by using the variational variables (3.80) which are known at $z = 1$ after integrating the initial-value problem. Table 3.3 illustrates convergence of Newton's method for the limit point in Fig. 3.3, which corresponds to Example 6, Chapter 1, with BC 1 for several initial estimates. An example of the convergence rate of the Gauss–Newton method for a bifurcation point in Fig. 3.3 is shown in Table 3.4, together with the results of the method for limit points, where the convergence is slower. The bifurcation point in question is simultaneously a limit point, which explains the sensitivity of the iteration process to rounding errors and approximation errors.

An example of convergence of Newton's method for determining limit points in the case of Example 9, Chapter 1, is given in [3.24]. If we continue the limit point $(\eta_1 = y(1), \eta_2 = \theta(1), Da)^{*T}$ depending on another parameter of the problem, for example β, we shall obtain a bifurcation diagram similar to that for LPS. An example of such a bifurcation diagram is shown in Fig. 3.13. The point $\beta = 2$, $Da = 0.07$ lies in the region where five solutions of the problem exist (compare with Figs. 3.5, 3.6, and 3.7).

If the second derivatives of functions R_1, R_2 are obtained (for example, by means of variational variables (3.80) or by means of difference approximations), the point of isola formation (isola center) can be located according

Table 3.3 Example 6, Chapter 1. BC 1, $A = 2$, $B = 4.6$, $D_x = 0.0016$, $D_y = 0.008$. Newton's method for three equations analogous to (3.77), (3.79). $\eta_1 = x'(0)$, $\eta_2 = y'(0)$.

Iteration	η_1	η_2	L	R_1	R_2	R_3
0	8.000	−2.000	0.0600	$1.8\,E-1$	$-5.7\,E-2$	$-1.8\,E-2$
1	9.581	−2.367	0.0606	$3.1\,E-3$	$4.0\,E-4$	$1.0\,E-2$
2	9.315	−2.303	0.0606	$1.3\,E-3$	$-2.4\,E-4$	$4.4\,E-4$
3	9.306	−2.301	0.0606	$3.3\,E-6$	$-7.1\,E-7$	$6.5\,E-7$
4	9.306	−2.301	0.0606			
0	10.000	−2.000	0.0700	$-2.8\,E0$	$1.0\,E0$	$-2.6\,E-1$
1	8.733	−2.202	0.0616	$2.5\,E-2$	$-3.5\,E-2$	$-1.4\,E-2$
2	9.325	−2.306	0.0606	$9.8\,E-3$	$-2.4\,E-3$	$1.7\,E-3$
3	9.306	−2.301	0.0606	$4.4\,E-5$	$-9.8\,E-6$	$7.2\,E-6$
4	9.306	−2.301	0.0606	$8.3\,E-10$	$-1.9\,E-10$	$1.3\,E-10$
0	12.000	−3.000	0.0800	$-2.7\,E0$	$6.4\,E-1$	$-1.4\,E-1$
1	9.888	−2.521	0.0636	$-1.3\,E-1$	$-1.6\,E-2$	$1.4\,E-2$
2	9.387	−2.317	0.0604	$1.4\,E-2$	$-4.7\,E-4$	$4.1\,E-3$
3	9.307	−2.301	0.0606	$3.8\,E-4$	$-8.5\,E-5$	$5.8\,E-5$
4	9.306	−2.301	0.0606	$5.3\,E-8$	$-1.2\,E-8$	$8.0\,E-9$
0	6.000	−2.000	0.0500	$3.1\,E0$	$-1.2\,E0$	$4.3\,E-1$
1	−0.004	0.265	0.0206	$-4.8\,E-2$	$2.7\,E-1$	$8.6\,E-1$
2	0.322	−0.065	0.0883	$-7.4\,E-2$	$3.9\,E-2$	$-1.0\,E-1$
4	0.392	−0.109	0.0750	$1.4\,E-2$	$-4.0\,E-3$	$5.7\,E-3$
7	0.055	−0.015	0.0792	$2.7\,E-4$	$-1.0\,E-4$	$7.3\,E-5$
11[a]	0.003	−0.001	0.0798	$1.1\,E-6$	$-4.1\,E-7$	$3.0\,E-7$
0	6.000	−1.000	0.1000	$-3.4\,E0$	$1.1\,E0$	$-6.5\,E-1$
11[a]	0.001	−0.001	0.2211	$9.6\,E-9$	$-2.1\,E-8$	$-1.2\,E-9$
0	5.000	−1.000	0.0400	$1.7\,E0$	$-2.1\,E-1$	$2.5\,E-1$
11[a]	0.004	−0.001	0.0798	$1.4\,E-6$	$-5.2\,E-7$	$3.8\,E-7$

[a] Converged to a point of primary bifurcation.

to the procedure described in Section 2.4.3. A procedure similar to that described in Section 2.4.2 can also be used for determining the directions of branches of solutions intersecting at the bifurcation point.

A more detailed discussion and more extensive results of Example 9, Chapter 1, are given in [1.77].

As has already been discussed, the computational techniques based on the shooting methods are very effective when they can be used at all, i.e., if the corresponding initial-value problems are stable and, hence, integrable. Then the advantage arising from decreasing the dimensionality of the problem is great. If the shooting method cannot be used, we have to use, for example, some finite-difference method. The study of branch points for a system of difference equations would be, however, incomprehensive because the

Table 3.4 Course of iterations for Example 6, Chapter 1. BC 1, $A = 2$, $B = 4.6$, $D_x = 0.0016$, $D_y = 0.008$. The Gauss–Newton method for four equations (3.77), (3.79), (3.83).

Iteration	$\eta_1 = x'(0)$	$\eta_2 = y'(0)$	L	R_1	R_2	R_3	R_4
0	5.000	-2.500	0.160	0.5 E0	-0.2 E0	0.2 E - 1	-0.2 E - 1
1	5.829	-2.856	0.173	0.6 E - 1	0.3 E - 1	-0.8 E - 2	0.2 E0
2	6.260	-2.982	0.170	-0.6 E - 2	0.3 E - 2	-0.1 E - 3	-0.2 E - 1
3	6.274	-2.984	0.170	-0.4 E - 4	0.2 E - 4	-0.3 E - 5	-0.6 E - 4
4	6.274	-2.984	0.170	-0.2 E - 4	-0.1 E - 4	0.2 E - 5	0.1 E - 4

Newton's method for three equations (3.77), (3.79)

Iteration	η_1	η_2	L
0	5.000	-2.500	0.160
1	6.377	-3.005	0.172
2	6.641	-3.090	0.170
4	6.545	-3.060	0.170
6	6.505	-3.049	0.170
8	6.666	-3.094	0.170
10^a	6.501	-3.048	0.170

[a] $\mathbf{R} = (0.1\,E - 5,\ 0,\ 0.4\,E - 6,\ 0.2\,E0\,(1))^T$.

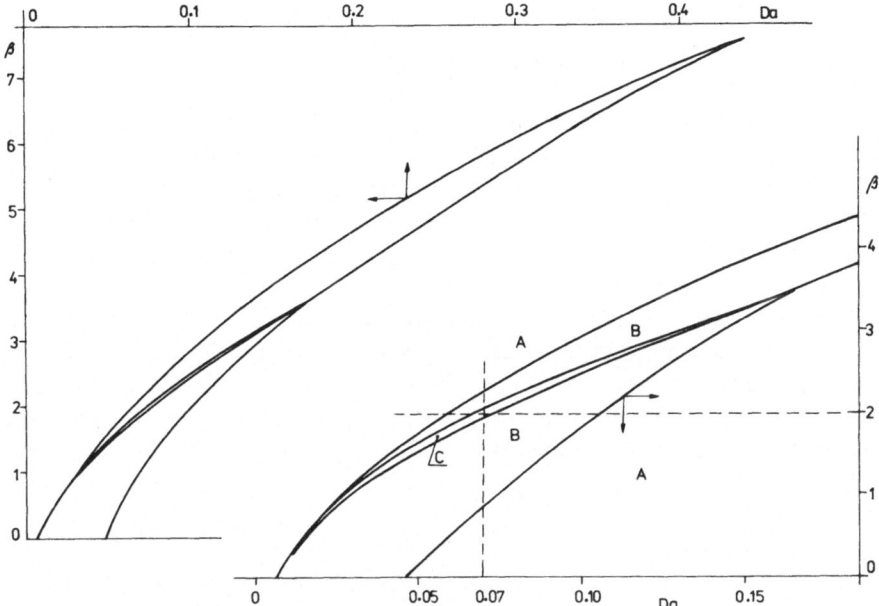

Figure 3.13 Bifurcation diagram for Example 9, Chapter 1, $B = 15$, $Pe_y = 10$, $Pe_\theta = 5$, $\theta_c = 0$, $\gamma = 20$. A—1 solution; B—3 solutions; C—5 solutions.

problem dimension is usually high. Therefore, we present here a somewhat different approach and we demonstrate its derivation using Example 10, Chapter 1 (where the shooting method fails for higher values of Re). An analogous procedure was used in [3.25] for the case of one second-order differential equation.

Let us consider the boundary-value problem given by Eqs. (E10.1)–(E10.3) with boundary conditions (E10.4). Let us denote the variables

$$f = \frac{\partial F}{\partial k}, \quad g = \frac{\partial G}{\partial k}, \quad h = \frac{\partial H}{\partial k}, \quad s = \frac{dS}{dk}. \tag{3.84}$$

Upon differentiating the system (E10.1)–(E10.3) with respect to k, we obtain the following system of differential equations:

$$f'' = \sqrt{\mathrm{Re}}[hF' + Hf'] + \mathrm{Re}(2Ff - 2Gg + 1), \tag{3.85}$$

$$g'' = 2\mathrm{Re}[Fg + fG] + \sqrt{\mathrm{Re}}[g'H + G'h], \tag{3.86}$$

$$h' = -2\sqrt{\mathrm{Re}}f. \tag{3.87}$$

At the limit point (on the dependence curve of the solution on S), the following condition must be satisfied:

$$s = \frac{dS}{dk} = 0. \tag{3.88}$$

Also, the following boundary conditions must be satisfied:

$$h(0) = f(0) = h(1) = f(1) = g(0) = g(1) = 0. \tag{3.89}$$

We have thus obtained a system of six differential equations (E10.1)–(E10.3), (3.85)–(3.87), of total order equal to 10, as well as 12 boundary conditions (E10.4) and (3.89). It seems that the problem is overdetermined. However, we have two additional undetermined parameters k and S, and we must determine their values at the limit point. Thus the problem is correctly defined.

We have used difference approximations to approximate the system of differential equations. Let us denote $\Delta = 1/n$, $z_i = i\Delta$, $i = 0, 1, \ldots, n$. Then we can approximate the derivatives F'', F', G'', G', f'', f', g'', and g' by a central three-point approximation at the points z_i, $i = 1, 2, \ldots, n - 1$. Thus, for example, (3.85) is approximated by

$$\frac{f_{i-1} - 2f_i + f_{i+1}}{\Delta^2} = \sqrt{\mathrm{Re}}\left[h_i \frac{F_{i+1} - F_{i-1}}{2\Delta} + H_i \frac{f_{i+1} - f_{i-1}}{2\Delta} \right]$$
$$+ \mathrm{Re}[2F_i f_i - 2G_i g_i + 1], \quad i = 1, \ldots, n - 1. \tag{3.90}$$

The derivatives H' and h' are approximated by a two-point difference formula centered between two neighboring node points. Approximations of (E10.3) and (3.87) thus take the form

$$\frac{H_i - H_{i-1}}{\Delta} + \sqrt{\mathrm{Re}}(F_i + F_{i-1}) = 0, \quad i = 1, 2, \ldots, n,$$

$$\frac{h_i - h_{i-1}}{\Delta} + \sqrt{\mathrm{Re}}(f_i + f_{i-1}) = 0. \tag{3.91}$$

The boundary conditions (E10.4) and (3.89) provide 12 equations

$$H_0 = F_0 = H_n = F_n = h_0 = f_0 = h_n = f_n = g_0 = g_n = 0,$$
$$G_0 = 1, \quad G_n = S. \tag{3.92}$$

We arrange the unknowns into a vector $\mathbf{X}$,

$$\mathbf{X} = (H_0, h_0, F_0, f_0, G_0, g_0, H_1, h_1, F_1, \ldots,$$
$$H_n, h_n, F_n, f_n, G_n, g_n, k, S)^T. \tag{3.93}$$

After an appropriate arrangement of the difference equations, we obtain a system of $6(n - 1) + 2$ equations with an almost 15-diagonal structural occurrence matrix (nonzero elements outside these 15 diagonals appear only in the last two columns). Newton's method was chosen for solving this system of nonlinear equations. The system of linear algebraic equations in each iteration was solved by the method of modified matrices [3.26, 3.27], which was derived for almost band matrices.

The number of iterations in Newton's method depends on the accuracy of the initial approximation of the limit point. We have no *a priori* information about the values of variables h, f, and g. In this case, we have made use of the already known dependence of the solution on the parameter S (see Fig. 3.8). Initial estimates can be computed from two neighboring branches in the neighborhood of the limit point in question. Newton's method then converged to the limit point within several (5–8) iterations. When poorer initial approximations were used, the iteration process usually diverged due to the sensitivity of solutions, because in the neighborhood of the limit point, a large number of other solutions exist (cf., Fig. 3.8).

After evaluating the limit point for a particular value of the parameter Re, we can use the results as an initial estimate for iteration for the values Re + Δ Re or Re − Δ Re. Continuing in this manner, we obtain the dependence of the limit points on the parameter Re—the bifurcation diagram. The results of continuing three limit points from Fig. 3.8 are presented in Table 3.5, resulting curves in the bifurcation diagram in the parametric plane "Re-S" are depicted in Fig. 3.14 [3.29]. The grid spacing $n = 100$ has been used to compute the results presented. For $n = 200$, the results only slightly deviate from the curves presented in Fig. 3.14. A computer with 14 decimal digits of accuracy has been used for computations. Note that the bifurcation diagram in Fig. 3.14 is not complete; there exist additional curves in this figure [3.28, 3.29].

Table 3.5 Resulting limit points for a sequence of values of Re, $n = 100$. The value of ΔRe varies from 1 to 100.

Re	S	k	Re	S	k	Re	S	k
121	−0.9980	0.01395						
124	−0.9856	0.01342						
125	−0.9807	0.01324	175	1.0945	0.01293			
200	−0.6256	0.00426	200	0.8433	0.00971	348	−0.3903	0.01991
300	−0.3983	0.00103	300	0.4408	0.00276	350	−0.3865	0.02070
400	−0.2865	0.00016	400	0.2927	0.00093	400	−0.3468	0.02366
500	−0.2215	−0.00013	500	0.2163	0.00031	500	−0.3419	0.02416
600	−0.1795	−0.00023	600	0.1704	0.00005	600	−0.3588	0.02393
625	−0.1713	−0.00024	625	0.1617	0.00002	625	−0.3636	0.02383
800	−0.1291	−0.00026	800	0.1186	−0.00011	800	−0.3872	0.02313
1000	−0.1002	−0.00023	1000	0.0908	−0.00015	1000	−0.3880	0.02278
2000	−0.0471	−0.00012	2000	0.0431	−0.00011	2000	−0.3773	0.02230
5000	−0.0198	−0.00004	5000	0.0194	−0.00004	5000	−0.3693	0.02044
10000	−0.0113	−0.00001	10000	0.0112	−0.00001	10000	−0.3583	0.01798
10000[a]	−0.0116	−0.00002	10000[a]	0.0115	−0.00001	10000[a]	−0.3759	0.02196

[a] $n = 200$.

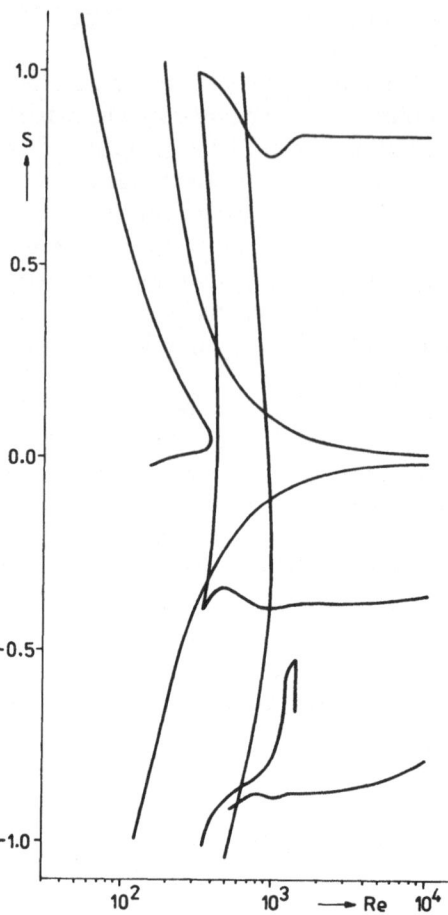

Figure 3.14 Bifurcation diagram for Example 10, Chapter 1; only some branches of the dependence of limit points on Re are presented.

3.3.3 Secondary complex bifurcation

Complex bifurcation from a trivial solution ("primary complex bifurcation") was discussed in the previous section. We shall now discuss the problem of determining the point of secondary complex bifurcation in distributed systems. We shall discuss three possible approaches. The first approach is a trial-and-error technique which is closely connected with the procedure discussed in Sections 3.4 and 4.2, i.e., dynamic simulation of (3.2) and (3.3) for a sequence of values of the considered parameter or for the value of the parameter slowly varying over time, (cf., Section 4.2). The point of complex bifurcation (from a non-trivial stationary solution) is recognized as a point where a periodic solution arises. Stable oscillatory solutions (arising, e.g., by way of supercritical bifurcation from a stable stationary solution) are obtained by dynamic simulation, unstable oscillatory solutions (arising, e.g., by way of subcritical bifurcation from a stable stationary solution) cannot practically be determined by this procedure.

The second approach is based on the approximation of DPS by means of LPS using appropriate semi-discretization (called the "method of lines"). Let us choose a set of points (for example, an equidistant set) $z_0 = 0, z_1, \ldots,$ $z_n = 1$ on the interval $z \in [0, 1]$ and denote the solution of (3.2), (3.3) at these points in the following way:

$$x_i(t) \sim x(z_i, t), \quad y_i(t) \sim y(z_i, t). \tag{3.94}$$

Upon approximating the derivatives in the z-direction by means of a central three-point difference formula, we obtain a system of ordinary differential equations ($h = 1/n$) from (3.2), (3.3):

$$\frac{dx_i}{dt} = \frac{D_x}{L^2 h^2} [x_{i-1} - 2x_i + x_{i+1}] + f(x_i, y_i),$$

$$i = 1, 2, \ldots, n-1,$$

$$\frac{dy_i}{dt} = \frac{D_y}{L^2 h^2} [y_{i-1} - 2y_i + y_{i+1}] + g(x_i, y_i). \tag{3.95}$$

The remaining equations are obtained from the boundary conditions; thus, for example, for BC 1 (3.4), we have

$$x_0 = \bar{x}, \quad x_n = \bar{x}, \quad y_0 = \bar{y}, \quad y_n = \bar{y}. \tag{3.96}$$

For BC 2, we also use approximations (3.95) for $i = 0$ and $i = n$:

$$x_{-1} = x_1, \quad x_{n+1} = x_{n-1}, \quad y_{-1} = y_1, \quad y_{n+1} = y_{n-1}. \tag{3.97}$$

Thus, we obtain $2(n-1)$ differential equations for BC 1 and $2(n+1)$ ordinary differential equations for BC 2. Now we can use the methods of analysis of complex (and also real) bifurcations for the system of ordinary differential equations discussed in Section 2.5. The question of the accuracy of approximating of DPS by LPS is very important. This accuracy evidently depends on the discretization step size h. With a decrease in step size (hence, improved accuracy of approximation), the dimension of LPS increases. Therefore, we usually have to be satisfied with approximate results obtained for small n (large h). However, for certain types of equations, we can obtain surprisingly good, qualitatively correct results even for very sparse discretization [3.30]. Let us discuss the use of the method of lines for Example 11, Chapter 1, and consider a characteristic parameter Lw (a stationary solution does not depend on this parameter). The computation thus proceeds as follows: for given values of the parameters, a stationary solution is obtained, e.g., by a shooting method or by solving a system of nonlinear algebraic equations corresponding to a stationary solution of LPS (the system arises from the application of the method of lines). For lower values of parameters $\gamma\beta$, and ϕ, these solutions differ only slightly. In this case, the system (3.95) can be schematically expressed in the form $(a = 0)$: $y_{-1} = y_1$, $\theta_{-1} = \theta_1$, $y_n = 1$, $\theta_n = 0$:

$$\mathrm{Lw}\frac{dy_i}{dt} = F(y_{i-1}, y_i, y_{i+1}, \theta_i, \phi), \quad i = 0, 1, \ldots, n - 1, \tag{3.98}$$

$$\frac{d\theta_i}{dt} = G(\theta_{i-1}, \theta_i, \theta_{i+1}, y_i, \phi), \quad i = 0, 1, \ldots, n - 1. \tag{3.99}$$

Let us arrange the variables into a vector $\mathbf{X} = (y_0, y_1, \ldots, y_{n-1}, \theta_0, \theta_1, \ldots, \theta_{n-1})^T$, and equations (3.98), (3.99) consecutively. The occurrence matrix of the Jacobian matrix $\mathbf{J}$ of the right-hand sides then takes the form (e.g., for $n = 5$):

$$\begin{bmatrix}
y_0 & y_1 & y_2 & y_3 & y_4 & \theta_0 & \theta_1 & \theta_2 & \theta_3 & \theta_4 \\
x & x & & & & & x & & & \\
x & x & x & & & & & x & & \\
& x & x & x & & & & & x & \\
& & x & x & x & & & & & x \\
& & & x & x & & & & & x \\
x & & & & & x & x & & & \\
& x & & & & x & x & x & & \\
& & x & & & & x & x & x & \\
& & & x & & & & x & x & x \\
& & & & x & & & & x & x
\end{bmatrix}
\begin{array}{l}
\\
i = 0 \quad (3.98) \\
i = 1 \\
i = 2 \\
i = 3 \\
i = 4 \\
i = 0 \quad (3.99) \\
i = 1 \\
i = 2 \\
i = 3 \\
i = 4 \quad (3.100)
\end{array}$$

After the stationary solution is obtained, we can compute the values of elements of the Jacobian matrix (3.100) (they do not depend on Lw). For individual values of Lw, we then obtain the Jacobian matrix when dividing the first n rows of the matrix calculated beforehand by the value of Lw. The eigenvalues of this matrix indicate the stability of the stationary solution in question. If a trial-and-error technique is used (for a fixed ϕ) to obtain the values of Lw^+ for which a pair of complex-conjugate eigenvalues pass from the left to the right complex half-plane, we have one point for the bifurcation diagram "ϕ-Lw." Such a bifurcation diagram is shown for $a = 0$ in Fig. 3.15 (no real bifurcation occurs for chosen parameter values). Computations reported below have determined the accuracy of Lw obtained for the choice $n = 5$, cf. Fig. 3.15 [1.102]. Then, in the upper part of the figure, periodic solutions exist. The periodic solutions, which are stable, can be found to be stable oscillations obtained by numerical simulation of the transient equations. Stable and unstable limit cycles can be found using procedures analogous to the procedures for LPS, hence, by means of a transformation into a boundary-value problem. Jensen and Ray [3.31] and Heinemann and Poore [3.32, 3.33] have studied in detail the application of the method of lines to the

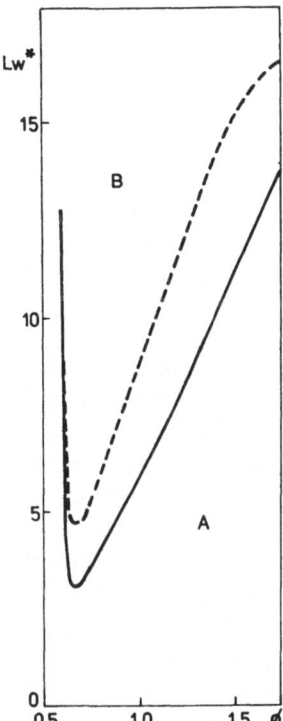

Figure 3.15 Bifurcation diagram for Example 11, Chapter 1, $a = 0, \gamma = 20, \beta = 0.2,$ ----- $n = 5$, method of lines. —— shooting method, cf. Eqs. (3.103). A—stable steady-state solutions; B—unstable steady-state solutions and stable periodic solutions.

problem of determination of the Hopf bifurcation points in Example 9, Chapter 1.

Recently an iterative method has been developed to determine the points of complex bifurcation in DPS. The method is based on principles similar to those for LPS described in Section 2.5.2 [3.34]. We shall briefly summarize the principle ideas of the method.

Let us consider the problem (3.2), (3.3) with boundary conditions BC 2 given by Eq. (3.5). The linearized operator of the right-hand sides of Eqs. (3.2) and (3.3) evaluated for the steady-state solution in question has a purely imaginary eigenvalue $\lambda = 0 + i\omega$ at the point of complex bifurcation. Hence, the system of equations (3.101) holds:

$$p_R'' + \frac{L^2}{D_x}\left[\frac{\partial f}{\partial x}p_R + \frac{\partial f}{\partial y}q_R\right] = -\omega p_I,$$

$$p_I'' + \frac{L^2}{D_x}\left[\frac{\partial f}{\partial x}p_I + \frac{\partial f}{\partial y}q_I\right] = \omega p_R, \qquad (3.101)$$

$$q_R'' + \frac{L^2}{D_y}\left[\frac{\partial g}{\partial x}p_R + \frac{\partial g}{\partial y}q_R\right] = -\omega q_I,$$

$$q_I'' + \frac{L^2}{D_y}\left[\frac{\partial g}{\partial x}p_I + \frac{\partial g}{\partial y}q_I\right] = \omega q_R.$$

Here we have denoted the eigenfunction of the linearized operator corresponding to the eigenvalue λ as (p, q). Real and imaginary components $p = p_R + ip_I, q = q_R + iq_I$ are expressed separately. The boundary conditions evidently take the form

$$p_R'(0) = p_I'(0) = q_R'(0) = q_I'(0) = 0,$$
$$p_R'(1) = p_I'(1) = q_R'(1) = q_I'(1) = 0. \tag{3.102a}$$

The eigenfunctions (p, q) are determined except for two real constants. This means that we can choose two fixed values on these functions, for example:

$$p_R(0) = 1, \quad q_R(0) = 1. \tag{3.102b}$$

Thus we have obtained a nonlinear boundary-value problem (3.9), (3.101) of overall order equal to 12 with as many as 14 boundary conditions (3.5), (3.102a), and (3.102b) and with two parameters (ω and L), which must be determined at the point of complex bifurcation. The problem is well determined and can be solved, for example, by the shooting method or by difference methods.

We shall illustrate the above algorithm for the location of the Hopf bifurcation point in Example 11, Chapter 1 [3.34]. The shooting method is used in this case. We denote $E = \exp(\theta/(1 + \theta/\gamma)$, $E_1 = E/(1 + \theta/\gamma)^2$ and consider the case where $a = 0$. Then the set of six second-order differential equations to be solved is in the following form:

$$y'' - \phi^2 yE = 0 \tag{3.103a}$$

$$\theta'' + \phi^2\gamma\beta yE = 0 \tag{3.103b}$$

$$\frac{1}{Lw}[p_R'' - \phi^2 Ep_R - \phi^2 yE_1 q_R] = -\omega p_I \tag{3.103c}$$

$$\frac{1}{Lw}[p_I'' - \phi^2 Ep_I - \phi^2 yE_1 q_I] = \omega p_R \tag{3.103d}$$

$$q_R'' + \phi^2\gamma\beta Ep_R + \phi^2\gamma\beta yE_1 q_R = -\omega q_I \tag{3.103e}$$

$$q_I'' + \phi^2\gamma\beta Ep_I + \phi^2\gamma\beta yE_1 q_I = \omega q_R. \tag{3.103f}$$

Equations (3.103a) and (3.103b) describe steady states of Eqs. (E11.1) and (E11.2), Eqs. (3.103c–f) correspond to Eqs. (3.101). Boundary conditions corresponding to (E11.3) and (E11.4) are

$$y'(0) = \theta'(0) = p_R'(0) = p_I'(0) = q_R'(0) = q_I'(0) = 0 \tag{3.103g}$$

$$y(1) = 1, \theta(1) = 0, p_R(1)p_I(1) = q_R(1) = q_I(1) = 0. \tag{3.103h}$$

Let us transform the boundary-value problem into the initial-value problem at $z = 0$ by choosing six missing initial conditions

$$y(0) = \eta_1, \quad \theta(0) = \eta_2, \quad p_R(0) = \eta_3, \quad q_R(0) = \eta_4, \quad p_I(0) = \eta_5,$$
$$q_I(0) = \eta_6. \tag{3.103i}$$

Table 3.6 Course of Newton's iteration for determination of a Hopf bifurcation point in the problem of heat and mass transfer in a porous catalyst particle, cf. Eqs. (3.103). ($a = 0, \gamma = 20, \beta = 0.2, \phi = 1$)

Iteration	η_1	η_2	η_5	η_6	Lw	ω	$\sum\limits_{i=1}^{6} F_i^2$
0	0.0200	4.0000	16.0000	5.0000	5.0000	5.0000	4.2 E5
1	0.0304	3.8783	8.6553	−164.84	4.3185	5.0243	3.0 E3
2	0.0334	3.8663	1.3145	−41.630	4.4471	4.8374	6.3 E2
3	0.0336	3.8658	1.0359	−49.417	6.5241	2.0723	2.6 E0
4	0.0336	3.8658	1.2361	−46.191	6.2455	2.3378	2.1 E − 2
5	0.0336	3.8658	1.1950	−45.001	5.9707	2.4464	3.6 E − 8
6	0.0336	3.8658	1.1952	−45.006	5.9759	2.4490	1.9 E − 16

After integrating Eqs. (3.103a–f) with initial conditions (3.103g, i) from $z = 0$ to $z = 1$ we obtain six residuals

$$F_1 = y(1) - 1, \quad F_2 = \theta(1), \quad F_3 = p_R(1), \quad F_4 = q_R(1), \quad F_5 = p_I(1),$$

$$F_6 = q_I(1)$$

and the following set of six nonlinear equations is to be solved

$$F_i(\eta_1, \ldots, \eta_6, \text{Lw}, \omega) = 0, \quad i = 1, 2, \ldots, 6. \tag{3.103j}$$

According to (3.102b) we fix $\eta_3 = \eta_4 = 1$ and solve the system (3.103j) for six unknowns $\eta_1, \eta_2, \eta_5, \eta_6$, Lw, ω by Newton's method (cf. Section 3.1.3). The course of iterations is presented in Table 3.6. If we compute the values of (critical) Lewis number Lw at the Hopf bifurcation points in dependence on another parameter of the problem, e.g. the parameter ϕ, we obtain the bifurcation diagram, cf. Fig. 3.15. These results may be compared with the results obtained by the method of lines, cf. Fig. 3.15.

More detailed results together with the results for Example 9, Chapter 1, are presented in [3.34].

3.4 Methods for Transient Simulation of Parabolic Equations—Finite-Difference Methods

A number of methods exist for numerical solution of (3.2) and (3.3). These include the method of lines, described in the previous section, the method of perpendicular lines, the method of spline functions, and the method of finite elements. However, difference methods are most often used. There exist a large number of monographs describing difference methods in detail [3.35, 3.36, 3.37, 3.38]. Here we shall discuss only the questions of effectiveness and algorithmization of difference methods.

In the half-band $z \in [0, 1] \times t \in [0, \infty)$, we shall choose an equidistant set of node points

$$(z_i, t_j), \quad z_i = ih, \quad i = 0, 1, \ldots, n, \quad h = 1/n;$$
$$t_j = jk, \quad j = 0, 1, \ldots \quad k > 0. \tag{3.104}$$

The value of a solution x, y at the point (z_i, t_j) will be denoted

$$x_i^j \sim x(z_i, t_j), \quad y_i^j \sim y(z_i, t_j). \tag{3.105}$$

The values x_i^j and y_i^j approximate the solution at the point (z_i, t_j).

The simplest and most often used difference approximation of (3.2), (3.3) is the six-point Crank–Nicolson formula ($w = \frac{1}{2}$):

$$\frac{x_i^{j+1} - x_i^j}{k} = D_x w \frac{x_{i-1}^{j+1} - 2x_i^{j+1} + x_{i+1}^{j+1}}{L^2 h^2}$$

$$+ D_x(1 - w) \frac{x_{i-1}^j - 2x_i^j + x_{i+1}^j}{L^2 h^2} + f_i^{j+w}, \tag{3.106}$$

$$\frac{y_i^{j+1} - y_i^j}{k} = D_y w \frac{y_{i-1}^{j+1} - 2y_i^{j+1} + y_{i+1}^{j+1}}{L^2 h^2}$$

$$+ D_y(1 - w) \frac{y_{i-1}^j - 2y_i^j + y_{i+1}^j}{L^2 h^2} + g_i^{j+w}, \tag{3.107}$$

which is a central three-point formula in the direction z and a two-point formula in the direction t. The Crank–Nicolson formula in the form (3.106), (3.107) includes a simple explicit method ($w = 0$) and a purely implicit approximation ($w = 1$). Schematic illustrations of the node points used in individual cases are shown in Fig. 3.16. In addition to (3.106) and (3.107), we also have approximations of boundary conditions, e.g., for BC 1, we have

$$x_0^{j+1} = \bar{x}, \quad x_n^{j+1} = \bar{x}, \quad y_0^{j+1} = \bar{y}, \quad y_n^{j+1} = \bar{y}. \tag{3.108}$$

The explicit method ($w = 0$) must satisfy the condition of numerical stability which, for BC 1, takes the form

$$\frac{D_x k}{L^2 h^2} \leq \frac{1}{2}; \quad \frac{D_y k}{L^2 h^2} \leq \frac{1}{2}. \tag{3.109}$$

If conditions (3.109) are not satisfied, the method cannot be used because, after several steps in the t direction, rounding errors will invalidate the results.

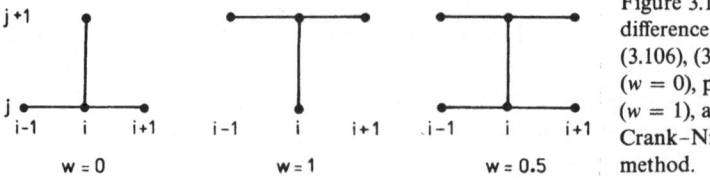

Figure 3.16 Finite-difference approximations (3.106), (3.107); explicit ($w = 0$), purely implicit ($w = 1$), and Crank–Nicolson ($w = \frac{1}{2}$) method.

An advantage of the explicit method follows from the fact that a new profile $j + 1$ can be calculated recurrently, point after point, from the old profile (the term "profile" will be used for a vector of all variables with the same index j; the values of the solution for one profile lie in a rectangular set on a horizontal line). The nonlinear term can be easily evaluated, for example, as $f_i^j = f$ (x_i^j, y_i^j), and the values x_i^j, y_i^j are already known. The method described by (3.106), (3.107) is a one-step method in the t direction. This means that the "old" profile j can be forgotten as soon as the "new" profile $j + 1$ is computed; the profile $j + 1$ will then play the role of an "old" profile for the new profile $j + 2$, and so on. This property allows the possibility of a simple variation of the step size k during integration, depending on the maximal prescribed error. A new profile for $w \neq 0$ satisfies the system of equations which are generally nonlinear in the case where the nonlinear term is approximated by

$$f_i^{j+w} \sim f(wx_i^{j+1} + (1 - w)x_i^j, \quad wy_i^{j+1} + (1 - w)y_i^j) \tag{3.110}$$

(in an analogous way for g^{j+w}). This system of nonlinear algebraic equations has a band occurrence matrix (for an appropriate arrangement of equations and variables) and can be simply solved by Newton's method. Usually one or two iterations are sufficient, since the profile j serves as a very good initial approximation for the iteration. Even though Newton's method usually converges very quickly, a number of methods which do not require iteration have been developed. Let us summarize some of them.

3.4.1 Nonlinearity approximation

The method of simple linearization
The nonlinear terms are approximated by

$$f_i^{j+w} \sim a^j + b^j x_i^{j+1} + c^j y_i^{j+1}, \tag{3.111}$$

where a^j, b^j, c^j are coefficients which depend on the values of solution on the jth ("old") profile. Thus, for example, for $w = 1$, the term $(x_i^{j+1})^2$ can be rewritten in the form $x_i^{j+1} \cdot x_i^j$. Some further possibilities are

$$(x_i^{j+1})^3 \sim (x_i^j)^2 \cdot x_i^{j+1},$$
$$(x_i^{j+1/2})^2 \sim \tfrac{1}{4}(x_i^j + x_i^{j+1})^2 \sim \tfrac{1}{4}[(x_i^j)^2 + 2x_i^j x_i^{j+1} + x_i^j x_i^{j+1}], \text{ etc.}$$

The simplest frequently used method of linearization is

$$f_i^{j+w} \sim f_i^j. \tag{3.112}$$

Taylor expansion
The Taylor expansion up to first-order terms is used for approximation of (3.111).

$$f_i^{j+w} \sim f_i^j + \frac{\partial f}{\partial x}\bigg|_i^j (x_i^{j+w} - x_i^j) + \frac{\partial f}{\partial y}\bigg|_i^j (y_i^{j+w} - y_i^j). \tag{3.113}$$

Here we already have to evaluate derivatives of nonlinear terms, and x_i^{j+w} is approximated by $w \cdot x_i^{j+1} + (1 - w)x_i^j$ (similarly for y_i^{j+w}).

In both cases, from the method of simple linearization as well as from Taylor expansion, we obtain a system of linear algebraic equations of a band structure for a new profile. A similar result is obtained when the extrapolation technique is used.

Extrapolation technique

The value of f_i^{j+w} is obtained by extrapolation from several old profiles (retained for this purpose). Thus, for $w = 1$, we have

$$f_i^{j+1} \sim 2f_i^j - f_i^{j-1}, \tag{3.114}$$

$$f_i^{j+1} \sim 3f_i^j - 3f_i^{j-1} + f_i^{j-2}, \tag{3.115}$$

when two or three old profiles are used. At the point $t = 0$, where only one old profile is available (an initial condition), we have to use, for example, the simple linearization technique.

Predictor-corrector technique

In this case, explicit and implicit methods are combined. The profile $j + w$ is computed by means of an explicit method, the values obtained are introduced into nonlinear functions f and g, and a system of linear algebraic equations is obtained for a new profile $j + 1$. The profile $j + w$, computed by an explicit method, was used for evaluation of nonlinearities only. After being introduced into f and g, it can either be discarded or used for an automatic control of time step k.

3.4.2 Automatic control of time step k

Since the evaluation of a single profile can often be relatively time consuming (for large n), it is appropriate to control the step size k in agreement with the required accuracy. The required step size may greatly vary over the course of integration, and thus an automatic control of the time step size is useful. We can use several methods.

1. We may compare the values of a solution obtained with a step size k and with two steps of the length $k/2$. The error estimate may then be used in the form

$$E = \|\mathbf{x}^{j+1(k)} - \mathbf{x}^{j+1(k/2)}\|. \tag{3.116}$$

We can also use the Richardson extrapolation. Let us note that the error estimate is obtained after several solutions of linear algebraic systems. This technique is, however, only rarely used; if at all, it is used only after several time steps.

2. We may use the difference between the extrapolated values $\hat{\mathbf{f}}^{j+w}$ and the computed values $\mathbf{f}^{j+w}$. The error estimate (here u denotes $u = (x, y)$) then

takes the form:

$$E' = \|\mathbf{f}(w\mathbf{u}^{j+1} + (1 - w)\mathbf{u}^j) - \hat{\mathbf{f}}^{j+w}\|. \tag{3.117}$$

A similar estimate can be used for the predictor-corrector technique, where $\hat{\mathbf{x}}^{j+w}$ denotes a predicted value and $E = \|w\mathbf{x}^{j+1} + (1 - w)\mathbf{x}^j - \hat{\mathbf{x}}^{j+w}\|$.

3. For the methods of simple linearization and of the Taylor expansion, the expressions for the error estimate can be used:

$$E' = \|\mathbf{f}(w\mathbf{u}^{j+1} + (1 - w)\mathbf{u}^j) - a^j - b^j\mathbf{x}^{j+1} - c^j\mathbf{y}^{j+1}\|, \tag{3.118a}$$

$$E' = \left\| \mathbf{f}(w\mathbf{u}^{j+1} + (1 - w)\mathbf{u}^j) - \mathbf{f}(\mathbf{x}^j) - w\left[\frac{\partial \mathbf{f}}{\partial \mathbf{x}}\Big|^j (\mathbf{x}^{j+1} - \mathbf{x}^j) \right.\right.$$
$$\left.\left. + \frac{\partial \mathbf{f}}{\partial \mathbf{y}}\Big|^j (\mathbf{y}^{j+1} - \mathbf{y}^j)\right]\right\|. \tag{3.118b}$$

If ε is a prescribed error, a profile $j + 1$ is accepted for $E < \varepsilon$; depending on the ratio E/ε, the step size k may be increased for the next profile. For $E > \varepsilon$, the profile $j + 1$ is not accepted and integration of a new profile is repeated for a shorter step size k.

3.4.3 Automatic control of spatial step size h, equidistant net

In principle, there are two different approaches to automatic control of the net in the z-direction. One consists of the adaptive variation of n while keeping a net equidistant, and the second includes the formation of an adaptive non-equidistant net. Let us deal with the former approach first.

To determine whether a given n (or h) is sufficiently large (or small) with respect to the prescribed accuracy ε, we must estimate an approximation error for the z direction.

An evaluation of a new profile with the step sizes h and $h/2$ and a comparison of results could be used as an error estimate, but we would have to interpolate the old profile. The latter and the less demanding method, is to use the remaining terms of the difference formulas used for the approximation. In the case of (3.106), (3.107), we have used the following approximation of the second derivative:

$$\frac{\partial^2 x(z_i, t_j)}{\partial z^2} = \frac{x_{i-1}^j - 2x_i^j + x_{i+1}^j}{h^2} - \frac{1}{12} h^2 \frac{\partial^4 x(\xi, t_j)}{\partial z^4} \tag{3.119}$$

for $\xi \in (z_{i-1}, z_{i+1})$. If we succeed in approximating the fourth derivative of the solution based on our knowledge of the jth profile, we can use the a priori estimate of the approximation error given by (3.116). We can make use of a difference formula

$$\frac{\partial^4 x(\xi, t_j)}{\partial z^4} \sim \frac{1}{h^4} [x_{i-2}^j - 4x_{i-1}^j + 6x_i^j - 4x_{i+1}^j + x_{i+2}^j] \tag{3.120}$$

for $\xi \in (z_{i-1}, z_{i+1})$ and $i = 2, 3, \ldots, n - 2$. The error estimate of the approximation in the z-direction is then given by

$$E_x = \max_{i = 2, 3, \ldots, n-2} \frac{1}{12h^2} |x^j_{i-2} - 4x^j_{i-1} + 6x^j_i - 4x^j_{i+1} + x^j_{i+2}|. \quad (3.121)$$

The result is analogous for E_y; for $i = 1$ and $i = n - 1$, we obtain identical results when using an asymmetric formula for approximating the fourth derivative. The estimate (3.121) thus includes the boundaries of the interval. If we denote $E = \max(E_x, E_y)$, we can use the following strategy:

1. $E > \varepsilon$ the step size h is too large and is halved. The values of the solution of the net points between the original net points are obtained by interpolation, e.g.,

$$x^j_{i+1/2} = \tfrac{1}{16}[-x^j_{i-1} + 9x^j_i + 9x^j_{i+1} - x^j_{i+2}], \quad i = 1, 2, \ldots, n - 2,$$
$$x^j_{1/2} = \tfrac{1}{8}[3x^j_0 + 6x^j_1 - x^j_2]. \quad (3.122)$$

The interpolation for $x^j_{n-1/2}$ is organized in a similar way. After renumbering the net points, $2n \to n, h/2 \to h$, we can proceed with the computations for this new value of h.

2. $E < \varepsilon/\omega$, ω is usually chosen between 4 and 8: the step size h is too small and can be doubled. The points with indeces 1, 3, 5, $\ldots$, $n - 1$ are deleted, renumbering of the net points is accomplished, $2h \to h, n/2 \to n$, and the computation continues with this new value of h.

3. In all other cases, the step size h remains unchanged and the computation continues as it would without step-size control.

It is useful to prescribe a minimal value of n (maximal h) and maximal n (minimal h) for the computation.

3.4.4 Adaptive nonequidistant net

Let us consider a nonequidistant net of node points $z_0 = 0, z_1, \ldots, z_n = 1$ and the values of solution at these points $x^j_i, y^j_i, i = 0, 1, \ldots, n$. Instead of the formula $(x^j_i)'' \sim (x^j_{i-1} - 2x^j_i + x^j_{i+1})/h^2$ used for a constant step size h, we use a three-point formula to approximate the second derivative (similarly for y):

$$\frac{\partial^2 x(z_i, t_j)}{\partial z^2} \sim \frac{2}{(z_i - z_{i-1})(z_{i+1} - z_{i-1})} x^j_{i-1} - \frac{2}{(z_i - z_{i-1})(z_{i+1} - z_i)} x^j_i$$
$$+ \frac{2}{(z_{i+1} - z_{i-1})(z_{i+1} - z_i)} x^j_{i+1}. \quad (3.123)$$

The difference approximations on the right-hand sides of (3.106) and (3.107) are changed according to (3.123), the computation remains otherwise unchanged. The estimate of the approximation error in the z-direction is now performed on each subinterval (z_{i-1}, z_i) separately. Eigenberger [3.39] has proposed the following procedure for computing the estimate. The value of the solution at the point $(z_{i-1} + z_i)/2$ is obtained by interpolation from the values at the points z_{i-2}, z_{i-1}, z_i:

$$x_{i-1/2}^{j(1)} = \frac{-(z_i - z_{i-1})^2}{4(z_i - z_{i-2})(z_{i-1} - z_{i-2})} \cdot x_{i-2}^j + \frac{z_i + z_{i-1} - 2z_{i-2}}{4(z_{i-1} - z_{i-2})} x_{i-1}^j$$

$$+ \frac{z_i + z_{i-1} - 2z_{i-2}}{4(z_i - z_{i-2})} x_i^j. \tag{3.124}$$

The same value is obtained by interpolation from the points z_{i-1}, z_i, z_{i+1}; hence,

$$x_{i-1/2}^{j(2)} = \frac{2z_{i+1} - z_i - z_{i-1}}{4(z_{i+1} - z_{i-1})} x_{i-1}^j + \frac{2z_{i+1} - z_i - z_{i-1}}{4(z_{i+1} - z_i)} x_i^j$$

$$- \frac{(z_i - z_{i-1})^2}{4(z_{i+1} - z_{i-1})(z_{i+1} - z_i)} x_{i+1}^j. \tag{3.125}$$

The value of an error in the interval $z \in (z_{i-1}, z_i)$ is then defined as

$$E_x^i = |x_{i-1/2}^{j(1)} - x_{i-1/2}^{j(2)}|. \tag{3.126}$$

The same procedure is followed for the variable y and the error is defined as $E^i = \max(E_x^i, E_y^i)$. Now the following strategy is used.

1. $E^i > \varepsilon$. Another node point $(z_{i-1} + z_i)/2$ is added between the points z_{i-1} and z_i. The values at this point are obtained by interpolation from the neighboring four node points $z_{i-2}, z_{i-1}, z_i,$ and z_{i+1}. This is done for $i = 1, 2, \ldots, n$. Then renumbering of the node points is accomplished and the computation proceeds.

2. $E^i < \varepsilon/\omega$ for $i = k$ and $i = k + 1$. The point z_k is deleted and a recalculation of E^i for neighboring indices i is performed. The node points are renumbered and the computation proceeds. The value of ω is usually chosen between 10 and 20.

3. In all other cases, there are no changes in the set of node points.

As in the previous control strategy, it is useful to prescribe maximal and minimal values for the difference $z_i - z_{i-1}$. However, we can also use some other procedure to estimate the approximation error between two node points, e.g., a procedure similar to that described for an equidistant set.

Remarks
A three-point approximation was used above for approximating in the z-direction. If we want to obtain higher accuracy for a lower number of node points, we can use an approximation with a greater number of node points.

Thus, for example, a five-point approximation of the second derivative takes the form (an equidistant net)

$$\frac{\partial^2 x(z_i, t_j)}{\partial z^2} \sim \frac{1}{24h^2}[-2x_{i-2}^j + 32x_{i-1}^j - 60x_i^j + 32x_{i+1}^j - 2x_{i+2}^j],$$

$$i = 2, 3, \ldots, n - 2, \tag{3.127}$$

$$\frac{\partial^2 x(z_1, t_j)}{\partial z^2} \sim \frac{1}{24h^2}[22x_0^j - 40x_1^j + 12x_2^j + 8x_3^j - 2x_4^j]. \tag{3.128}$$

Similarly, we can integrate with higher accuracy in the t-direction, e.g., using the Runge–Kutta method on the basis of the method of lines. The problems with numerical stability may be encountered, particularly in so-called stiff parabolic systems. For such problems, we usually must use a purely implicit method or a method based on the A-stability concept for ordinary differential equations (cf., Section 2.7).

Here we have not dealt with the solution of equations in higher-dimensional space. The problems of high dimensionality of the system are becoming important. Most often, the so-called ADI–alternating-direction implicit methods are used for computing dynamic systems, cf., Mitchell [3.36]. A method of over-relaxation type is used for obtaining the stationary solution. An attempt was made [3.40] to determine an optimal value of the relaxation factor for a nonlinear elliptic problem.

CHAPTER 4

Development of Quasi-stationary Patterns with Changing Parameter

Let us consider a dynamical system which depends on a parameter A:

$$\frac{d\mathbf{X}}{dt} = \mathbf{F}(\mathbf{X}, A) \tag{4.1}$$

with specified initial conditions. Up until now, we have considered the parameter A to be fixed and independent of time t. Let us now consider that the parameter A varies very slowly (on the time scale of evolution of dissipative structures—patterns— described by Eq. (4.1)) over time, according to either a given functional dependence

$$A = \varphi(t) \tag{4.2}$$

or a prescribed differential equation

$$\frac{dA}{dt} = \psi(t, A). \tag{4.3}$$

In both cases, we assume that the changes of parameter A over time are very slow in comparison to the changes of state variables $\mathbf{X}$. Usually we assume that the right-hand side of (4.3) is independent of $\mathbf{X}$; hence, in both cases, we are dealing with a gradual control of system (4.1) by means of the parameter A. Thus, we study special cases of the general evolutionary problem well known in physical and biological systems. The results will be represented in evolution diagrams.

If (4.3) can be pre-solved, the form of (4.2) is obtained again. Two of the most common forms of variation of a parameter over time are "linear growth,"

$$A = A_0 + tA_1, \tag{4.4}$$

and "exponential growth,"

$$\frac{dA}{dt} = cA, \tag{4.5}$$

i.e.,

$$A = A_0 \exp(ct), \tag{4.6}$$

where c is a scalar variable. Exponential growth describes the situation where the parameter value keeps increasing by a certain percentage in a given time interval. The rate of change of the parameter is constant for linear growth.

The main behavior under quasi-stationary conditions can be described as movement along a chosen stable branch of solutions (steady state, periodic, chaotic). The behavior of a system is regular in these situations, i.e., we do not observe "jumps" in behavior. The jumps may occur at bifurcation and limit points only. Examples of possible types of movement across bifurcation and limit points are given in Fig. 4.1. More complex branching from periodic solutions (including, for example, the appearence of chaotic solutions) are not included.

Case (a) (Fig. 4.1) corresponds to a simple bifurcation point. A solution continues after the bifurcation point along a stable branch. In case (b), where the bifurcation point is also a limit point, the branch on which the solution continues is chosen at random (from two possibilities). In real physical or biological situations, the problem has to be formulated stochastically; the character of distribution of fluctuations of state variables will determine the probability of the choice of individual branches of solutions, cf., [1.95, 4.1].

Case (c) is very interesting because after crossing the limit point the system will evolve into the closest stable state or, in other words, into a state in whose domain of attraction the limit point belongs. (This state can be the stable steady-state, periodic, or chaotic solution.) Stochastic formulation corresponds to the solution of "the time of the first passage problem," cf., [4.1].

Case (d) corresponds to the Hopf bifurcation of a stable periodic solution from a stable stationary solution.

The behavior in case (e) is analogous to that in the case of the limit point (c). In case (f), a limit point lies on the dependence curve of periodic solutions on a parameter; the behavior of the system after crossing this point is analogous to the behavior in case (c).

The total number of possible phase portraits observed during evolution in a given range of parameter values (if the number is countable) can often be determined on the basis of the theory of catastrophes [1.1] or on the basis of the more general singularity theory [1.30]. In problems of higher codimension, we can follow situations where a higher number of parameters vary simultaneously, and we can observe more complex and far richer types of behavior, cf., [1.3].

The numerical realization of quasi-stationary behavior is not complicated. It arises through the use of dynamic simulation methods (i.e., methods described in Section 2.7 for LPS and in Section 3.4 for DPS) with a slow variation of the chosen parameter during computation.

There is one problem with numerical calculations. If the rate of change of the considered parameter is too high, both phenomena, i.e., transient behavior and the variation of the parameter, may occur along similar time scales, and we do not observe the quasi-stationary behavior, but one of the large number

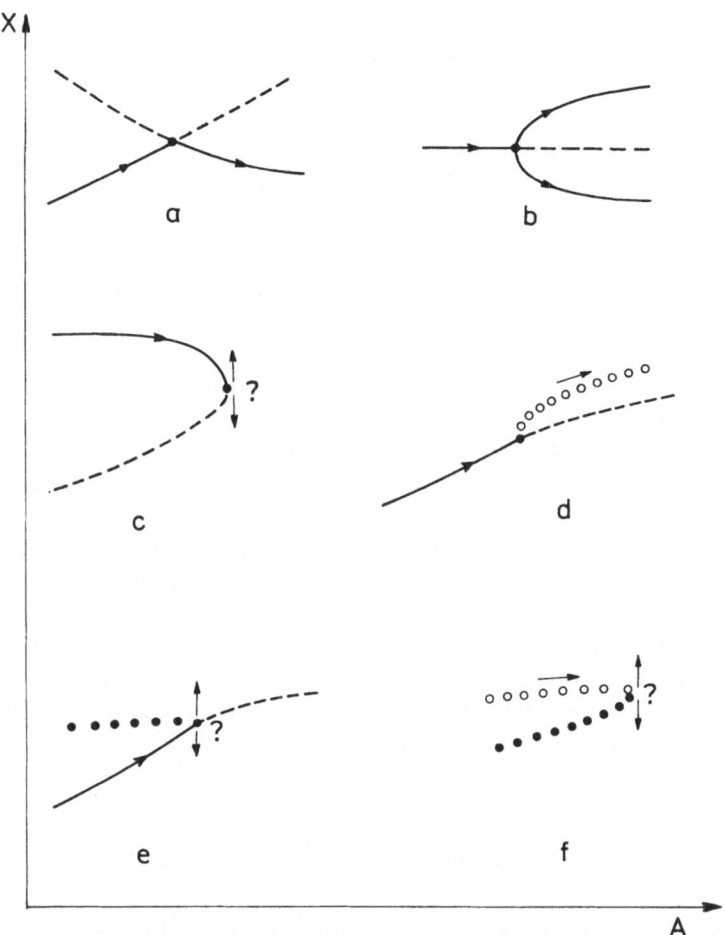

Figure 4.1 Examples of possible crossing of bifurcation and limit points during quasi-stationary behavior. —— stable; – – – – unstable; ○○○ periodic stable; ●●● periodic unstable.

of possibilities of transient behavior. If the rate of change of the parameter is chosen too low, the required computation time increases with the increasing time interval necessary for one passage of the chosen parameter interval. We have to reach a compromise based on a choice of a rate of change of the parameter slow enough so that it will not affect stable stationary or periodic solutions; during transitions at the limit points, we may, however, observe jumps to different branches of solutions for various rates of change of the parameter. In such situations, we can use a trial-and-error technique based on our knowledge of the stationary bifurcation diagram.

Among transitions from one branch of solutions to another one, we may observe transitions which are always realized in the same way. We call them deterministic (cf cases (a) and (d) in Fig. 4.1). Transitions where the following branch of solutions is chosen more or less at random are called stochastic (cf., cases (b), (c), (e), and (f) in Fig. 4.1). In the latter case, a stochastic formulation, together with a chosen probability distribution of fluctuations of state variables, will determine the probabilities of individual transitions.

The behavior is, however, more or less deterministic if we know the overall picture of solutions obtained by continuation. For example, the behavior in a "stochastic case" is also deterministic if we know that after crossing the limit point in case (c), Fig. 4.1, only one stable solution exists for higher values of the parameter. At the bifurcation points, during transitions to another branch which is not connected to the considered bifurcation point, we may sometimes observe a transient phenomenon, where the transient phase of transition to the following branch is relatively long. The transition is not always monotonous; it can also be oscillatory (focal or spiral type).

In engineering practice, we may often meet the problem of whether we can reach the chosen stable solution by realizable variations of chosen parameters in the case where several stable solutions of the problem exist. Hence, we can introduce a notion of attainability of a stable state by means of slow variations of a parameter. Sometimes it is very difficult to control initial conditions for state variables externally and the required stable state cannot be obtained directly. However, it can be realized when we follow the evolution of a system corresponding to the variation of another parameter. Let us present several examples of quasi-stationary behavior.

4.1 Quasi-stationary Behavior in LPS—Examples

Let us follow quasi-stationary behavior of a solution of the problem given in Example 3, Chapter 1. First we shall choose a problem where only a unique steady-state solution, together with complex bifurcations, exists. This is the case described in Table 2.9 with the bifurcation diagram in the plane "Da_1-Da_2" shown in Fig. 2.18. A stationary solution was chosen as the initial condition for the value of the parameter $Da_{10} = 0$ (Da_1 is the considered evolutionary parameter).

The dependence of the amplitude of a periodic solution on the parameter for the linear growth-rate relation (4.4),

$$Da_1 = Da_{10} + at \tag{4.7}$$

is presented in Fig. 4.2. The cases for each of two values of a are given in the figure. For the higher value of a, we can observe that the initial parts of the dependences of amplitudes on the parameter Da_1 are distorted at the point of complex bifurcation, i.e., the system begins to oscillate a little later rather than immediately after the point of complex bifurcation has been passed. The

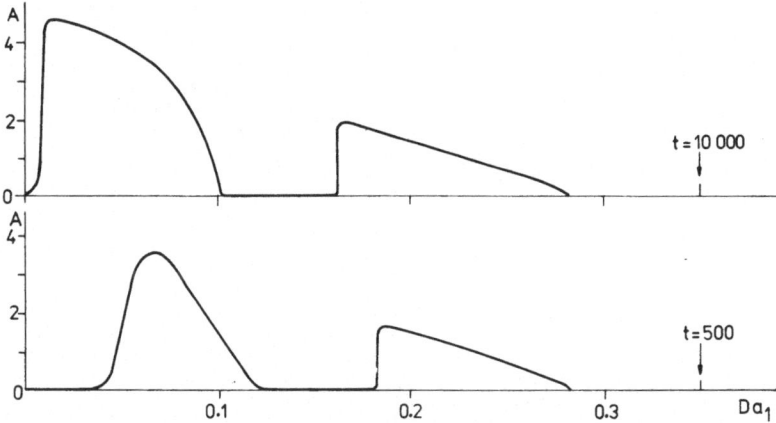

Figure 4.2 Quasi-stationary behavior in Example 3, Chapter 1, $\gamma = 1000$, $B = 12$, $\beta_1 = \beta_2 = 2$, $\theta_{c1} = \theta_{c2} = 0$, $\Lambda = 0.8$, $Da_2 = 0.2$. Dependence of the amplitude A of periodic solutions on Da_1 over the course of transient simulation; Da_1 is varied according to (4.7), $Da_{10} = 0$.

situation is better for the slower rate of change of the parameter value: the points of complex bifurcation ($Da_1^+ = 0.0956, 0.1574, 0.2730$, cf., Table 2.9) are already well approximated. However, direct evaluation of the points of complex bifurcation, as was described in Section 2.5, is a more practical technique, particularly with respect to the fact that the dynamic simulations for low values of a are time consuming.

The case in which multiple steady-state solutions coexist is sketched in Fig. 4.3. Reparametrization of the problem

$$Da_1 = Da_2 = k_0 \vartheta, \beta_1 = \alpha_1 \vartheta, \quad \beta_2 = \alpha_2 \vartheta \tag{4.8}$$

is performed (ϑ denotes residence time). The time change of ϑ is described by relation (4.6) in the form

$$\vartheta = \vartheta_0 \exp(at). \tag{4.9}$$

The rate of change is linear in Fig. 4.3 because the scale is logarithmic.

The value of $a = \pm 0.01$ has been found to be an acceptable rate of change of ϑ, which has still enabled us to obtain a true quasi-stationary picture of subsequent evolution of solutions. In the Fig. 4.3a, the upper parts (branches) of dependences of steady-state solutions are inaccessible when only the residence time varies. On the contrary, in the Fig. 4.3.b, any stable branch of solutions, with the exception of a small stable region on the "bow," can be reached when we vary ϑ forward and backward. Transitions between branches at the limit points are schematically shown in the figure.

More interesting behavior can be found in the system of two coupled reactors described in Example 4, Chapter 1. Let us take the values of parameters of

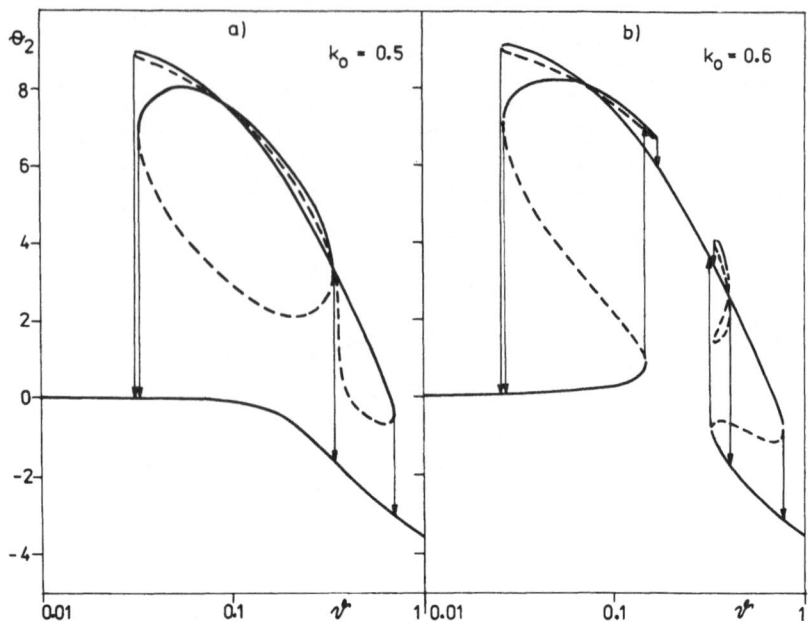

Figure 4.3 Quasi-stationary behavior in Example 3, Chapter 1; the case with multiple steady-state solutions; $\gamma = 20$, $B = 10$, $\beta_1 = \beta_2 = 1$, $\theta_{c1} = \theta_{c2} = -5$, $\Lambda = 1$; ϑ is varied according to (4.9). —— stable; ---- unstable steady-state solution.

the bifurcation diagram shown in Fig. 2.20: $A = 2$, $D_i = D_i^{12} = D_i^{21}$, $i = 1, 2$; $D_1/D_2 = 0.1$. Let us follow the dynamic behavior of solutions, which depend on the parameter D_1, by numerical simulation. Initial conditions

$$t = 0: c_1^1 = 2.1, \quad c_2^1 = 2.9, \quad c_1^2 = 1.9, \quad c_2^2 = 3.0 \tag{4.10}$$

are used in all simulations. The results of simulations for increasing and descreasing D_1,

$$D_1 = D_{10} 2^{\pm at}, \tag{4.11}$$

are shown in the evolution diagram in Fig. 4.4 [1.92, 1.93]. Simulation originally begins (at $t = 0$) at a homogeneous ($c_i^1 = c_i^2$, $i = 1, 2$) periodic solution. When D_1 increases, the homogeneous periodic solution changes into a nonhomogeneous stationary solution; then a nonhomogeneous periodic solution appears, and it bifurcates further into an aperiodic (chaotic) regime (here we have passed through a small region of subharmonic bifurcations in the parameter space), and finally, the homogeneous periodic solution arises again. When D_1 decreases, the homogeneous periodic solution does not evolve into an aperiodic solution, but changes directly into a nonhomogeneous stationary solution and then into a nonhomogeneous periodic solution, which for sufficiently small D_1 evolves again into a homogeneous periodic

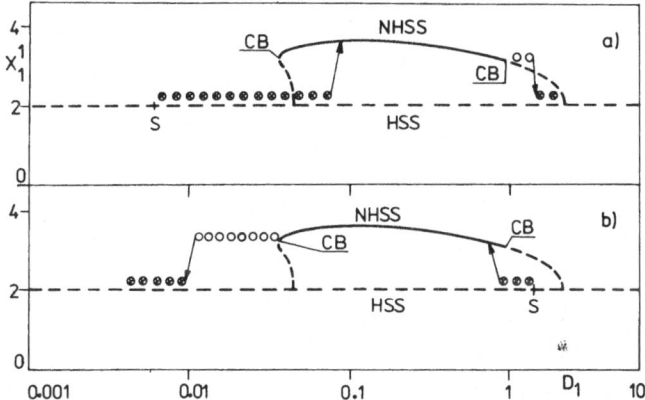

Figure 4.4 Evolution diagrams for the evolution of stationary and periodic solutions with slow changes of D_1, Example 4, Chapter 1, $A = 2$, $B = 6.0$, $D_1/D_2 = 0.1$, two coupled cells, $a = 0.02$ in (4.11); (a) increase of D_1, (b) decrease of D_1; S—initial state at $t = 0$; $\bigcirc$—nonhomogeneous periodic solution; $\otimes$—homogeneous periodic solution; NHSS—nonhomogeneous stationary state; HSS—homogeneous stationary state; CB—complex bifurcation point.

solution. Hence, there exist two interesting regions of the values of the parameter D_1: the first one, where homogeneous and nonhomogeneous periodic solutions coexist, and the other, where a homogeneous periodic solution and an aperiodic (chaotic) solution coexist. The dimensions of the regions of coexistence of particular solutions depend on the value of the parameter B, e.g., for $B = 5.5$, the region is smaller than for $B = 5.9$. The domain of attraction of the homogeneous periodic solution appears to be small, which is why larger perturbations will cause transitions to other solutions, i.e., either to a nonhomogeneous periodic solution or to an aperiodic (chaotic) solution.

Let us follow the dynamic behavior of solutions for $B = 5.9$ in the vicinity of the complex bifurcation point for $D_1^+ \doteq 0.9$. For $D_1 > D_1^+$, a stable nonhomogeneous periodic solution coexists with a stable homogeneous periodic solution. A further increase of D_1 causes an increase in the amplitude of the nonhomogeneous solution first and the loss of its stability afterwards. Stable solutions with more complex characteristics appear. In Fig. 4.5, the D_1-dependence of the values of the variable c_1^2 at subsequent local maxima is shown. The behavior is similar to that known from studies of difference equations ("pitchfork bifurcations"). The time variables for individual regimes are shown in Figures 4.6(a)–(f), where the unbroken line corresponds to c_1^2 and dashed line to c_1^1. The course of periodic concentration oscillations in both cells is synchronized. For $D_1 = 1.16$, we can observe one-point cycle oscillations (Fig. 4.6(a)), for $D_1 = 1.18$, two-point cycle oscillations (Fig. 4.6(b)), for $D_1 = 1.19$, four-point cycle oscillations (Fig. 4.6(c)), and for $D_1 = 1.1925$, eight-point cycle oscillations (Fig. 4.6(d)). The periods of consecutively existing solutions are $4, 8.5, 19$, and 33.8, i.e., approximately in the ratio $1:2:4:8$ (or $2^n, n = 0, 1, 2, 3$).

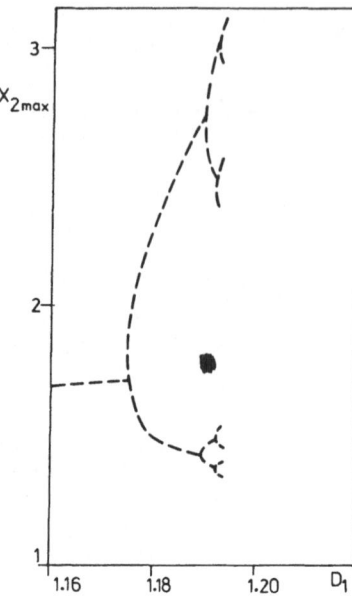

Figure 4.5 Dependence of $c_{1 max}^2 = x_{2 max}$ on $D_1 \cdot A = 2$, $B = 5.9$, $D_1/D_2 = 0.1$.

We have not been able to locate any solutions with higher periods, since, for $D_1 = 1.1935$, an aperiodic solution already arises (see Fig. 4.6(e) for $D_1 = 1.22$). In the aperiodic region, we can observe repetition of the profiles after an odd number of oscillations. The regions with the "period three" are shown in Fig. 4.6(e). Here the aperiodic regime corresponds, in principle, to wandering of the concentration trajectory between unstable periodic trajectories. The periodicities of these trajectories can determine the instantaneous spectral characteristics of the chaotic solutions. For a sufficiently high value of D_1, the aperiodic solution becomes unstable and a homogeneous periodic solution arises (cf., Fig. 4.6(f)).

4.2 Quasi-stationary Behavior in DPS—Examples

Let us consider Example 7, Chapter 1 [4.2]. The dependence of representative values of steady-state solutions for BC 2 on the length L for one chosen set of parameters is given in Fig. 4.7. The method of "composing solutions" has been used; Fig. 4.7 is not necessarily complete. The stability of solutions was tested by using dynamic simulation.

The model used in the example has been formulated to describe the process of differentiation in Hydra [1.97]. The process of differentiation does not occur under steady-state conditions, but in the transient situation of growth, where

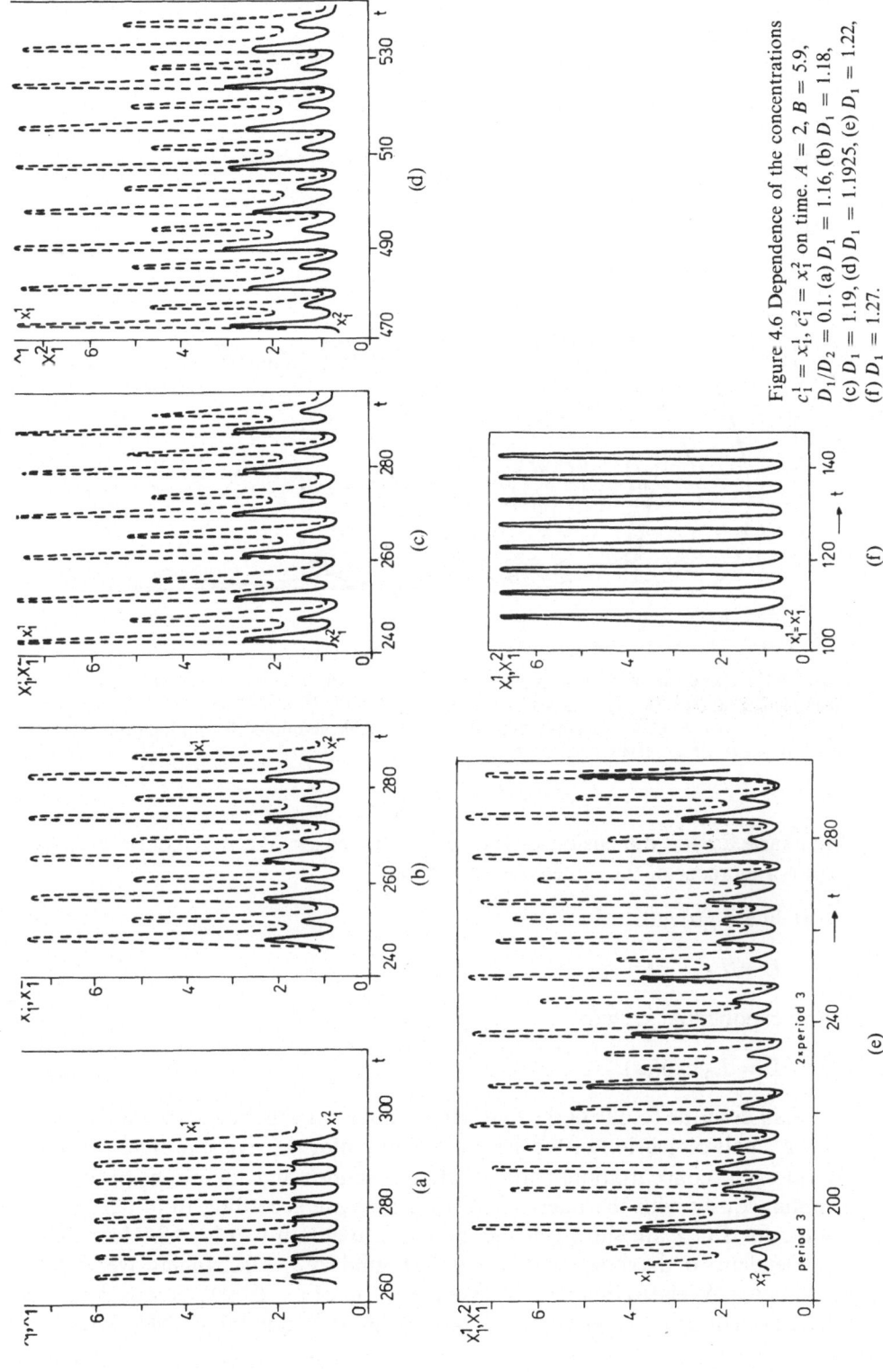

Figure 4.6 Dependence of the concentrations $c_i^1 = x_1^1$, $c_i^2 = x_1^2$ on time. $A = 2$, $B = 5.9$, $D_1/D_2 = 0.1$. (a) $D_1 = 1.16$, (b) $D_1 = 1.18$, (c) $D_1 = 1.19$, (d) $D_1 = 1.22$, (e) $D_1 = 1.1925$, (f) $D_1 = 1.27$.

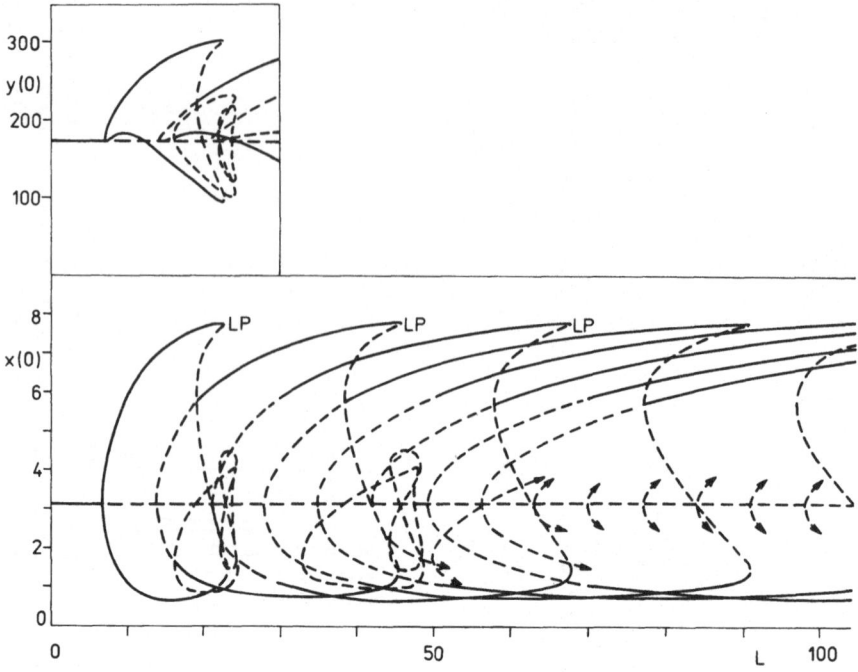

Figure 4.7 Dependence of the steady-state concentrations $x(0)$, $y(0)$ on the characteristic length; Example 7, Chapter 1, BC 2, $\mu = 0.0035$, $v = 0.0045$, $\rho_0 = 6.10^{-4}$, $c = 0.05$, $c' = 0.025$, $D_x = 0.01$, $D_y = 0.45$, $\rho = \rho' = 3.2$. ⎯⎯ stable solutions; ⎯ ⎯ ⎯ unstable solutions. Several limit points are denoted as LP, $L_1^* = 7.0349$, $L_2^* = 23.7548$.

the characteristic dimension of the system increases. We may consider two types of increase of the dimension over time:

(a) linear growth:

$$L = L_0 + at;$$ (4.12)

(b) exponential growth

$$L = L_0 \exp(at).$$ (4.13)

We shall describe results of the following simulation experiment. Let us begin with a small dimension of the system where uniform (trivial) profiles of the morphogen concentrations are stable. Reaction and diffusion proceed simultaneously with the linear dimension increase according to (4.12). The results for BC 2 are shown in the evolution diagram in Fig. 4.8. Here the L-dependence of the concentration x at the boundary $z = 0$ changing over time is presented. A relatively slow change of L over time ($a = 0.0002$) is considered. Characteristic spatial concentration profiles are shown in the Fig. 4.8 as well.

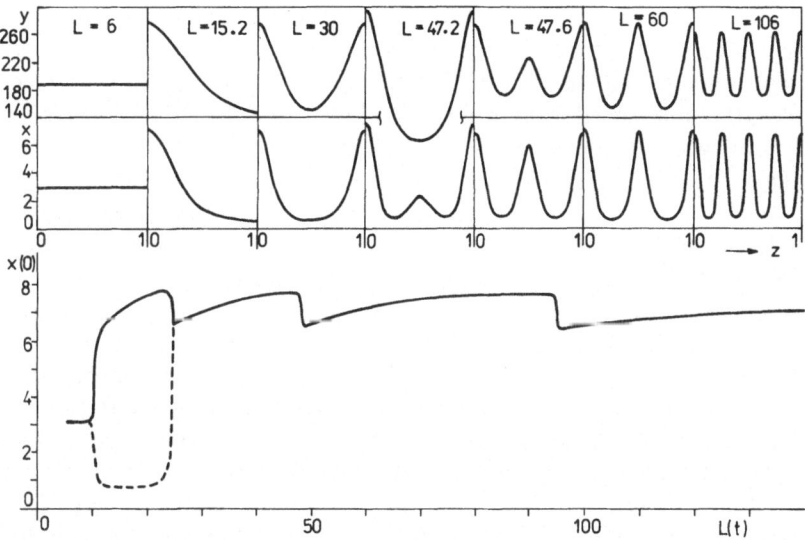

Figure 4.8 Evolution diagram—concentration of the morphogen x at the boundary during linear growth (4.12), Example 7, Chapter 1, BC 2, $a = 0.0002$, $L_0 = 6$. 48 space increments have been used for numerical approximation. Time step $\Delta t = 125$. For other parameters see legend to Fig. 4.7. Profiles of x and y for several values of L are also presented.

From Fig. 4.8, we can see that after the loss of stability, the uniform concentration profile changes abruptly into the non-symmetric spatial concentration profile which belongs to the first closed curve in Fig. 4.7. The form of the concentration profile remains essentially the same over its stability region. There are two possibilities for the course of profiles $x_{1,2}(z)$ over time: the cases $x_1(0) > x_1(1)$ and $x_2(1) > x_2(0)$, where the profiles $x_{1,2}(z)$ are mutual mirror images. The choice of one of these profiles from the two possibilities is random, and in simulation the choice depends on the chosen numerical constants and round-off errors. For example, the upper trajectory was calculated for $\Delta t = 125$, while the lower one resulted from numerical calculation with $\Delta t = 250$, cf., Fig. 4.8. In an actual physico-chemical or biological system, the distribution of probabilities of concentration fluctuations, magnified in the neighborhood of L_1^*, will determine the probability of the appearance of $x_1(z)$ or $x_2(z)$. The profile with one maximum changes into the profile with two maxima at the value $L \doteq 25$, and the profile with three maxima appears at $L \doteq 50$. At $L \doteq 95$, profiles with five maxima occur. When comparing Fig. 4.8 to Fig. 4.7, we can see that the manner in which the profiles change with the changing L (which depends on t) consists of qualitatively different parts. One type of parts can be characterized as "traveling" along the stable branch of steady-state solutions. Between such parts, we may observe the transient transition from one branch to the following one. Such a transition is enforced by the disappearance of the former branch. Therefore, it can exist at limit points only. The changes of profiles during such a transition process are also shown in Fig. 4.8. From the Fig. 4.8, we may conclude that the pattern of concentration profiles is derived for multiples of L from stable elementary profiles (the first closed loop in Fig. 4.7). This is true for BC 2 and for a relatively slow growth rate. The case $a = 0.0002$ used here can be considered to be quasistationary. Very similar results have been obtained for exponential growth (4.13) with $L_0 = 6$ and $a = 0.00002$. For $a = 0.002$ and the linear growth rate of (4.12), both processes (transient behavior and the change in L over time) occur on a similar time scale. Therefore, there is no correspondence between transient results and steady-state results (Fig. 4.7) for low L. This correspondence appears for $L = 60$ and for higher values of L.

The situation for BC 1 is not as simple as for BC 2 because composing solutions from elementary patterns is not possible. Two cases of evolution diagrams with linear growth rate are presented in Figs. 4.9 and 4.10. The cases differ in the space increment Δz used in the difference approximation of the partial differential equations. The choice of the first nontrivial branch after $L > L_1^*$ again occurs at random. The profiles at various characteristic lengths are drawn in the Figs. 4.9 and 4.10. The corresponding steady-state diagram has not been calculated. We can see that the dependences in Figs. 4.9 and 4.10 coincide for $L > 80$. As for BC 2, we can observe that the number of maxima increases when a transition from one branch to another occurs. In the linear growth-rate relation (4.12), the time interval of the existence of a particular

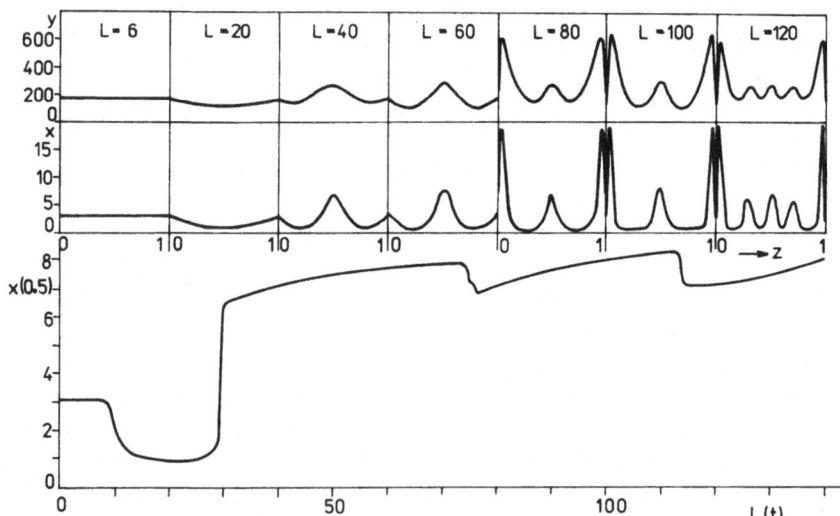

Figure 4.9 Evolution diagram—concentration of the morphogen x in the center of the system, i.e., $x(0.5)$ during linear growth (4.12), Example 7, Chapter 1, BC 1, $L_0 = 6$, $a = 0.0002$. 48 space increments have been used for numerical approximation. Time step length $\Delta t = 125$. For other parameters see legend to Fig. 4.7. Profiles x and y for several values of L are also presented (all profiles are symmetric).

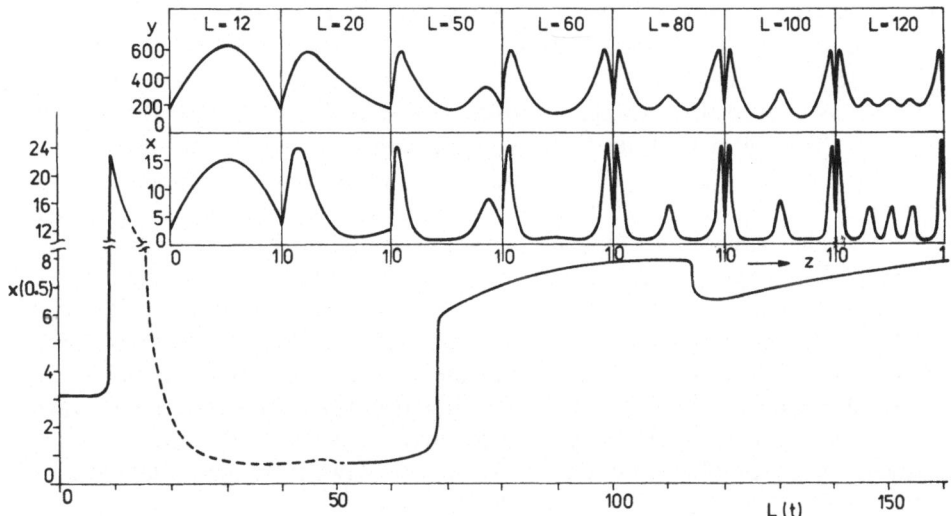

Figure 4.10 Evolution diagram—the same situation as in Fig. 4.9; the number of space increments is 80, the development in growth is changed. ——symmetric profiles; ----asymmetric profiles.

spatial profile of the morphogen doubles with every increase of the maxima number on the profile. For exponential growth rate (4.13), the time periods do not depend on the number of maxima.

In Fig. 4.11, the characteristic length dependence of the steady-state concentrations of the components x, y at the boundary $(x(0), y(0))$ is shown for BC 2 [1.64]. In this case, two bifurcation lengths, $L_1^* = 0.1115$ and $L_2^* = 0.1583$, can be observed, and here $2L_1^* > L_2^*$. The trivial (homogeneous) solution is stable for the interval of lengths $L \in (0, L_1^*)$, unstable for $L \in (L_1^*, L_2^*)$, stable again for $L \in (L_2^*, 2L_1^*)$, unstable for $L \in (2L_1^*, 2L_2^*)$, stable for $L \in (2L_2^*, 3L_1^*)$, and finally unstable for $L \in (3L_1^*, \infty)$. When we simulate behavior of spatial profiles under conditions of the slow growth rate, the homogeneous (trivial) profile becomes unstable at $L = L_1^*$; it changes into a spatial structure with the wave number 1 (with one maximum) which will be stable between L_1^* and L_2^* and slowly change its gradient. At $L = L_2^*$, the spatial structure will change smoothly into the homogeneous solution which will

Figure 4.11 The dependence of the steady-state solution on length, Example 6, Chapter 1, $D_x = 0.0016$, $D_y = 0.008$, $A = 2$, $B = 3.7$, BC 2; s—stable; u—unstable.

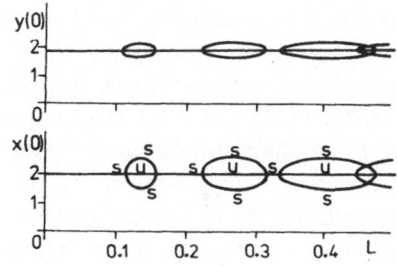

become stable and be established in the system. At $L = 2L_1^*$, the stable spatial structure with wave number 2 arises and will be retained by the system until $L = 2L_2^*$ is reached, etc.

Let us choose the parameter values $A = 2$, $B = 5.45$ for the Brusselator model (Example 6, Chapter 1). In the system without concentration gradients (LPS), i.e., for $D_x = D_y = 0$, a limit-cycle solution surrounding a unique unstable steady state will arise, with the period of oscillations $T_p = 3.84$. Further, let us choose diffusion coefficients in the distributed system, $D_x = 0.008$, $D_y = 0.004$. Then we can calculate the complex bifurcation length $L^+ = 0.51302$, cf., Section 3.3.1. Let us first consider the case of Dirichlet boundary conditions (BC 1). For small values of L, the trivial solution is stable. When the characteristic dimension of the system is increased beyond L^+, a periodic solution with a spatial profile similar to the eigenfunction with wave number 1 appears; a further increase in length causes the subsequent appearance of more complicated solutions. The course of two stable periodic solutions for $L = 2.25$ and $L = 4.5$ is presented in [4.3, 4.4].

When we consider the Neumann boundary conditions (BC 2), we take into account that the trivial solution in the distributed system is unstable, and thus the solutions bifurcating from this solution will also be unstable if no change of stability (e.g., secondary bifurcation) occurs. Apparently, this is what happens in our case.

A homogeneous (spatially uniform) periodic solution appears to be stable for $L \in (0, 2)$; close to $L = L^+$, a stable nonhomogeneous solution appears in the form similar to the corresponding eigenfunction, $x(z) \sim A_x \cos n\pi z$, $n = 1$. With a further increase in length, solutions with higher wave numbers appear at $2L^+$, $3L^+$. In Figs. 4.12(a), (b), (c), and (d) such solutions are illustrated. With respect to the symmetry properties of the original partial differential equations and BC 2, the solutions for $n = 2, 3, 4$ can be composed from the solution for $n = 1$ because the corresponding lengths are multiples of $L = 1.125$ (similar to the composition principle for steady-state solutions).

If the characteristic length L is increased, a large number of coexisting periodic solutions with different characteristics of spatial profiles (e.g., symmetric, asymmetric, composable from elementary solutions, mirror images along a central axis, profiles with large concentration gradients, etc.) can be calculated. For example, for $L = 4.5$, 15 stable periodic solutions (all with period $T_p = 3.64$, but with varying regions of attraction with respect to the initial conditions) may be observed. Convergence to the periodic solutions is rather slow, depending on the approximation and rounding errors. More detailed double precision computations have shown that true values of T_p approach the value for the lumped parameter system, $T_p = 3.84$.

From the point of view of comparison with experiment, it is important to determine which solution will be observed in the system when we begin at a certain initial state and chosen parameters vary slowly (quasi-stationary behavior). In the evolution diagram in Fig. 4.13, the results of simulation are

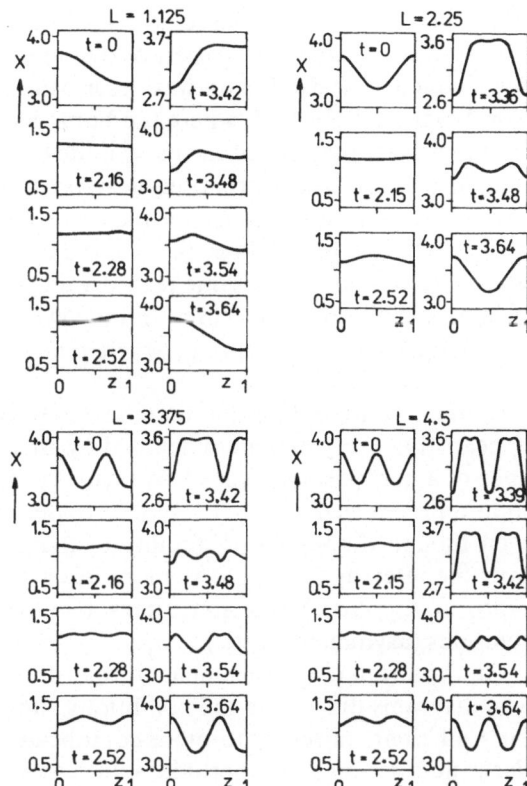

Figure 4.12 Time periodic spatial
concentration profiles of the
component x—Brusselator,
Example 6, Chapter 1, BC 2,
$A = 2$, $B = 5.45$, $D_x = 0.008$,
$D_y = 0.004$, $T_p = 3.64$.
(a) $L = 1.125$, (b) $L = 2.25$,
(c) $L = 3.375$, (d) $L = 4.5$.

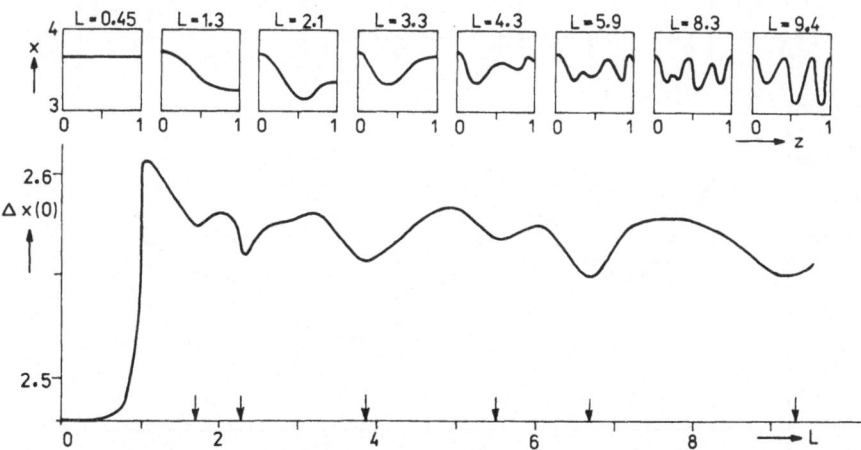

Figure 4.13 Evolution diagram of stable time periodic regimes with increasing characteristic
dimension, Example 6, Chapter 1, BC 2, $A = 2$, $B = 5.45$, $D_x = 0.008$, $D_y = 0.004$, $L_0 = 0.45$,
$a = 0.0055$; constant period $T_p = 3.64$.

illustrated, where the consecutive evolution of periodic wave-like regimes was studied through simulation of dynamic equations. The slow increase of the characteristic dimension was described as (4.12), $a = 0.0055$. The simulation began at $L_0 = 0.45$, where the homogeneous periodic solution is stable. Random perturbations of the magnitude 0.01 were applied for $L \in (0.45, 2.2)$, and, further on, no perturbations were used. In the upper part of Fig. 4.13, the characteristic spatial profiles of the eight individual stable periodic regimes are shown. We may follow the consecutive development of more complex wave-like solutions. The same period value (within computational error) was observed in all calculated cases. In the lower part of Fig. 4.13, the dependence

$$\Delta x(0) = \max_{T_p}[x(0, t)] - \min_{T_p}[x(0, t)] \tag{4.14}$$

(maximum amplitude at the boundary during the period) on L is shown. It appears that every minimum denoted by an arrow corresponds to the appearance of a new periodic regime. We can see that a nonhomogeneous periodic solution is realized at $L \doteq L^+ = 0.52$ under the effect of random fluctuations even though the region of stability of the homogeneous limit cycle extends over the interval $L \in (0, 2)$. As more extensive studies have shown [4.4], the ratios $\Delta x(z_i)/\Delta x(z_j)$ can sometimes be used for identifying properties of solutions (asymmetries, etc.).

An evolution diagram similar to that in Fig. 4.13 is presented in Fig. 4.14. The diagrams differ in initial conditions chosen at $t = 0$. The system evolves on a different trajectory to more complicated periodic dissipative structures (large numbers of stable periodic solutions coexist in this case [4.4]).

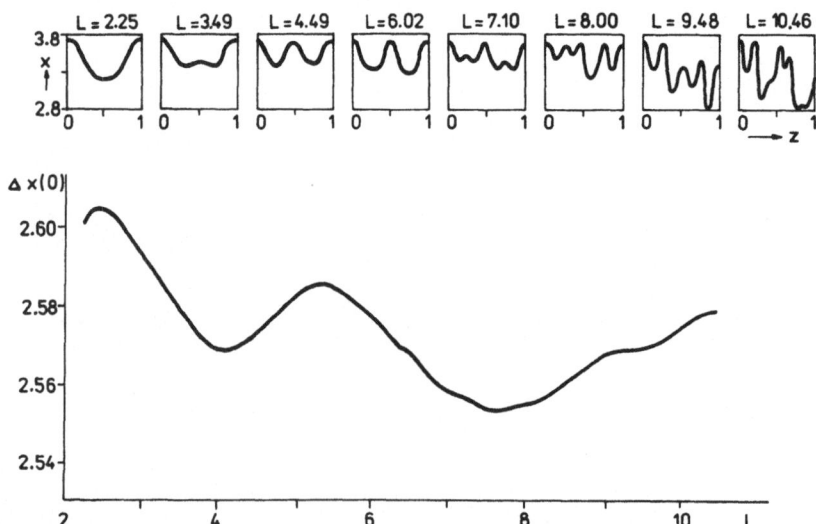

Figure 4.14 Evolution diagram of stable time periodic regimes with increasing characteristic dimension, Examples 6, Chapter 1, BC 2, $A = 2$, $B = 5.45$, $D_x = 0.008$, $D_y = 0.004$, $L_0 = 2.25$, $a = 0.0055$; constant period $T_p = 3.64$.

CHAPTER 5

Perspectives

Various computational methods of analysis of nonlinear dynamic systems have been described in Chapters 1–4. These methods have been developed and used mainly over the past 10 years by our research group. Relations among specific procedures are schematically shown in Fig. 5.1. Every procedure is denoted by its corresponding section number (or numbers when both LPS and DPS are involved). An increasing number of papers on the numerical methods discussed in this text have recently appeared in journals specializing in numerical methods and nonlinear analysis. In this chapter, we shall attempt to identify connections among the various numerical methods described here and the methods recently published. We shall also discuss the most common problems encountered in the application of particular numerical methods. The numbers of blocks in brackets in Fig. 5.1 will be used in the discussion.

(1) The computation of stationary solutions in LPS is more or less a routine process. Problems arise when we seek to determine all solutions of the problem, including the isolated branches of solutions. If we cannot make use of physical considerations, we can use a generator of random numbers to choose initial estimates for Newton's method. The situation is more complicated in DPS because nonlinear boundary-value problems can sometimes be difficult to solve. Nevertheless, when we consider problems in one spatial coordinate, we can say that the development of methods of analysis has, in principle, been completed (cf., the references in Section 3.1). Procedures for computing stationary solutions in more than one spatial coordinate are still under development, even though a number of iterative solution methods already exist. New vector computers enable us to directly solve large systems of nonlinear differential equations with a special structure by the global Newton method. It appears that there is great potential for the development of more universal software based on finite-element methods.

(2) The stability of steady-state solutions can be determined relatively easily for the case of LPS of lower dimensions (approximately up to dimension 20). Coefficients of the characteristic polynomial of the Jacobian matrix are computed, and the proper criterion is applied. For systems of higher dimensions, the effects of rounding errors may appear. These errors must be carefully controlled, e.g., by using double- or multiple-precision arithmetic. No generally applicable methods have been developed to directly determine the stability

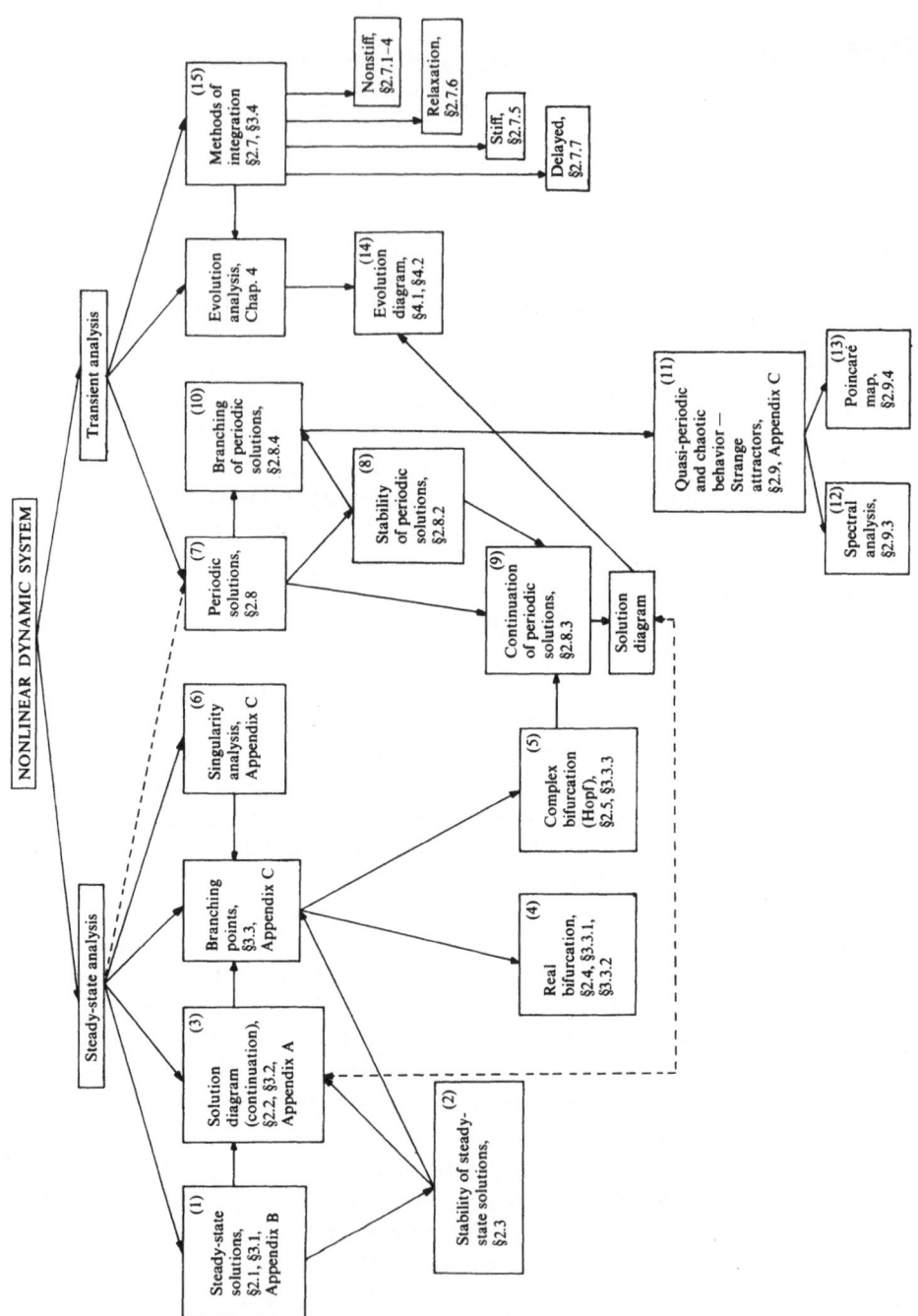

Figure 5.1 Relations among specific procedures of analysis of nonlinear dynamical system.

of stationary solutions in DPS. It seems that the most effective method is still the dynamic simulation of the originally considered DPS. The initial condition is chosen in the form of the slightly perturbed steady-state solution.

The determination of stability of delayed differential equations (where an infinite number of eigenvalues exist) causes problems in itself. The task can be difficult, particularly for systems of higher dimensions.

(3) A great deal of attention has recently been paid to the development of continuation algorithms in LPS. Thus, Keller and co-workers [5.1–5.4, 3.12, 3.28] have built a theoretical background of several continuation algorithms. They have worked out several modifications of the method of continuing along the arc (or "pseudo-arc") of solutions. They have also studied in detail the problems of stability and convergence of these proposed continuation methods theoretically. In particular, they have studied the methods of passing through bifurcation points. The developed numerical procedures have been applied to the solution of the problem of the flow between two rotating disks [5.4, 3.28] and to other problems as well.

In the method described in this text, an arc-length parameter is used during continuation. The dependence of the solution on a parameter is assumed to be continuously smooth in the $(n + 1)$-dimensional space (the dimension of the system plus the parameter). When crossing a bifurcation point, the algorithm usually continues in the direction of the original branch of solutions. The bifurcation point is not located. Rheinboldt suggested a similar continuation algorithm including step-length determination and location of critical points [5.5, 5.6]. A Fortran code of the algorithm has been published in [5.7]. Keller's methods [5.1, 5.2, 3.12] can deal with singularities occurring at bifurcation points. Keller adds "normalization" equations and a new parameter to the system to avoid the problem of noninvertibility of the Jacobian matrix in the neighborhood of the bifurcation point.

Keller claims that better numerical properties exist near bifurcation points, and he proves the nonsingularity of the augmented Jacobian matrix J. For dealing with limit points alone, Keller's method seems to be quite complicated because of the normalization choice for the arc-length relationship (in comparison with the standard arc-length relationship used in this text). However, at bifurcation points, his method is more comprehensive.

Menzel and Schwetlick [5.8, 5.9] and Riks [5.10] also deal with an extension of continuation methods to the case of the singular Jacobian matrix, as it is at bifurcation or limit points.

At a bifurcation point, as already mentioned, two principal steps must be accomplished: skipping over the bifurcation point (any predictor-corrector method can be used for this purpose) and finding the other bifurcation branches at the bifurcation point (multiple bifurcation points may have more than two branches intersecting at the bifurcation point). At a simple bifurcation point, the two branches leaving the bifurcation point lie tangent to a plane defined by two eigenvectors associated with the zero eigenvalues of $J^T J$. Here J is an

$n \times (n + 1)$ Jacobian matrix. Once J is found, the plane can be determined and all branches can be located by searching for equilibrium solutions.

The algorithm described in Section 2.4.2 is, on the contrary, direct (non-iterative) and requires only that the second partial derivatives be evaluated. During the continuation of steady-state solutions in DPS, it is necessary to form first a finite-dimensional approximation of the original problem and then use the procedures for LPS. Brown *et al.* [5.11] have also used a computational algorithm based, in principle, on the continuation technique described in this text. The chosen nonlinear hydrodynamic problem described by a set of partial differential equations is discretized by using a finite-element method. The coefficients in the corresponding expansions are derived from the Galerkin weighted residuals scheme. The Newton–Raphson method is then used for solving the resulting sets of nonlinear algebraic equations for the coefficients. The obtained Jacobian matrix of the solution is used not only for continuation, but also for the analysis of the stability of solutions, the detection of bifurcations of new families of solutions, and for computing asymptotic estimates of the effects of small changes in the parameter values, boundary conditions, and boundary shapes on the solution. The problems of describing the shape and stability of a rotating captive drop, and certain problems of the viscous film flow with contact lines, have been solved by this technique. Brown *et al.* [5.11] have used finite elements to discretize balance equations for several transport-phenomena problems, and they have applied a continuation algorithm to develop the dependence of solutions on a parameter.

Mehra and Carroll [5.12, 5.13, 5.14] have used the continuation technique, based on procedures described in Chapter 2, for the analysis of stability and control of an aircraft flying at high angles of attack. The system is described by a set of five or eight ordinary differential equations. Both equilibrium surfaces depending on parameters (bifurcation surfaces) and time-dependent solutions as functions of parameters have been studied. The authors have also studied the possibilities of using numerical differentiation routines to avoid the necessity of an analytical definition of the Jacobian matrix required in the continuation algorithm.

What kind of further developments can be expected in this area? Of course, the situation is simpler for LPS, where we can already reasonably well predict the form of proper software for the numerical treatment of the problem. The continuation algorithm described in this text will likely be improved when dealing with bifurcation points, particularly with respect to the location of newly bifurcating branches of solutions, according to procedures given in Section 2.4.2. This improved algorithm will have complicated logic because we need ensure that each branch of the solution is continued only once. Nevertheless, we believe that a random generator of initial estimates will still be necessary in cases where isolated families of solutions exist. The improved continuation algorithm should also include location of the points of complex (Hopf) bifurcation to enable connections with the continuation algorithm for computing periodic solutions.

(4) The procedure for evaluating branch points given in Section 2.4.1 is just one possible method. A good survey of various approaches for the case of limit points is given by Schwetlick [5.9]. A method analogous to the method of evaluating complex bifurcation points given in Section 2.5.2 is described for the case of real bifurcations in a paper by Seydel [2.10]. Simpson [5.15], Abbot [5.16], and Rheinboldt [5.17] have discussed numerical methods for a class of finite-dimensional bifurcation problems. Simpson [5.15] has described a numerical technique for determining a simple turning (limit) point on a branch of solutions of an algebraic system of equations depending on a scalar parameter. An inverse interpolation technique is used and the method is tested on several mildly nonlinear problems.

Moore and Spence [5.18, 5.19, 5.20] have discussed numerical methods for determining branch points for a general operator equation. It can be assumed that the general continuation routine mentioned above will contain, in the form of a subroutine, some effective method for determining bifurcation points (not limit points). Only afterwards will the method for determining directions of bifurcating branches be used. If the shooting method cannot be used in DPS, the finite-difference method becomes necessary, and differentiation between a limit and a bifurcation point becomes more complicated. The same conclusion holds for DPS problems in two and three spatial dimensions. A number of unsolved problems remain in this area.

(5) The methods of iterative computation of the points of the Hopf bifurcation have rarely been discussed in the literature. Hassard et al. [2.17, 5.21] have developed techniques for determining the Hopf (complex) bifurcation points based on a "trial-and-error technique" and a procedure for constructing asymptotic expansions of periodic solutions valid in the neighborhood of such points (cf., Section 2.5.3). The only existing direct iterative methods are described in Sections 2.5.1, 2.5.2 and 3.3.3.

(6) In the future, singularity theory [1.30, 5.22] will likely be used for predicting the form of solution diagrams, particularly in cases involving the existence of isolated branches of solutions. Individual possible solution diagrams will be classified. When singularity theory is combined with simple numerical calculations, the parameter values for the branch points and the points of isola formation can be determined in certain situations [2.14].

(7) The area of development of numerical methods for determining periodic solutions of nonlinear systems is one of great interest because it is connected with a number of technically important problems (e.g., stability problems). Stable periodic solutions are usually obtained as stationary solutions of dynamic simulations of the original systems. New algorithms had to be derived for computing unstable periodic solutions. Two of them, the difference algorithm and the algorithm based on the shooting method, are described in Section 2.8 for the case of LPS. No direct algorithms have been published for DPS. One algorithm based on a method analogous to the difference method for LPS is currently being prepared [5.23]. Hadeler [5.24] has recently published results of computation of periodic solutions for a system of ordinary

differential equations with a time delay using the sequential continuation method.

(8) The problem of determining the stability change of a periodic solution under study is closely connected with the solution method used for calculating the periodic solution under consideration. The simplest way to determine stability is to follow the behavior of the periodic solution through direct dynamic simulation. This procedure is used in DPS (it was not possible to obtain unstable periodic solutions in this case). Dynamic simulation is unnecessary in LPS if the shooting method has been used in the continuation algorithm and the monodromy matrix has been computed, cf., Sections 2.8.2 and 2.8.3.

(9) Recently the continuation of periodic solutions in LPS has been intensively studied. Mainly, sequential procedures have been used. The first algorithm is described in Section 2.8.3 [2.35] and is based on a predictor-corrector technique derived from the procedure used for continuing steady-state solutions. Mehra and co-workers [5.12–5.14] have began to develop a continuation algorithm for a limit-cycle type of solution. It seems that the derived algorithm passes well through limit points on the curve of periodic solutions in the solution diagram.

No methods for continuing periodic solutions in DPS have been discovered as yet. The sequential procedure have produced parts of solution and evolution diagrams in reaction-diffusion problems [4.4].

(10) Numerical methods for computing branch points on the branches of parametric curves of periodic solutions are still being developed. Points of bifurcation of new branches of periodic solutions can be detected over the course of the continuation algorithm for periodic solutions. Several numerical techniques for locating bifurcation points have been recently developed [2.36].

(11) Strange attractors connected with the chaotic behavior of dynamic solutions of the LPS and DPS systems have been intensively studied (numerically) over the last 5 or 6 years, even though the first report on a numerical study of the chaotic attractor is nearly 20 years old [2.37]. Two of the most often described ways in which chaotic solution in LPS appear are (1) by an infinite sequence of period-doubling bifurcations (Feigenbaum), cf., Section 2.9, and (2) by a sequence of bifurcations where the strange attractor bifurcates from a quasi-periodic solution in the form of a torus which has bifurcated from a periodic solution arising through Hopf (periodic) bifurcation. (This behavior can also be observed in the model of two coupled reaction-diffusion cells [2.57].) Since chaotic solutions are becoming a subject of great interest both from theoretical and practical points of view, we believe that numerical methods which characterize the chaotic attractors (e.g., computation of the values of the Liapunov exponents) and other characteristics of chaotic solutions will become a common tool in the near future. Up until now, spectral analysis and the computation of various Poincaré maps are the only methods commonly used [2.76].

(12) The computation of power spectra is based on standard computer routines for the Fourier transformation and the fast Fourier transformation. These routines are now available in the form of EPROM's. Thus, dedicated microcomputers can be also used for the studies of specific examples of strange attractors.

(13) The Poincaré map can yield (particularly in low-dimensional systems) condensed information on the form of the attractor in the phase space. In certain cases, we can use a sequence of Poincaré maps to predict the behavior of the attractor for intermediate values of the parameter in question.

(14) Constructing evolution diagrams seems to be a useful way to describe how the dissipative system develops over time in cases when one or more parameters vary relatively slowly in comparison with the characteristic transient processes in the system. This construction also appears to be useful when studying the system response upon sudden (small) parameter variations and when studying possibilities of control of the system development. Knowledge of corresponding solution diagrams can help to increase efficiency in construction evolution diagrams and aid us in interpreting the obtained results. Since the transient simulations necessary for construction of the evolution diagram can be time consuming, the use of efficient software for dynamic simulations is mandatory.

(15) Methods of dynamic simulation of both LPS and DPS are under continuous development. In LPS, efficient algorithms exist even for stiff problems. The development of proper algorithms for the simulation of relaxation oscillations appears necessary (adaptive algorithms would probably be very useful). The situation is more complicated in DPS, where standard software based on finite-difference techniques is available for quasi-linear parabolic equations in one spatial dimension. For problems in two or three spatial dimensions, the situation is far worse. The choice can be made between finite-difference methods (ADI) and finite-element approximations. The development of standard software with adaptive regulation of spatial approximations and automatic control of the time step size are problems which need to be solved in the near future. Perhaps the development of vector computers would be helpful.

We have not attempted to give a complete list of available references in this chapter, but we have tried to foresee the development of this very interesting field of numerical description of evolution in nonlinear problems. Personally, we believe that further development of efficient software will be helpful to natural scientists and engineers in their studies of nonlinear systems of both theoretical and practical interest.

DERPAR—A Continuation Algorithm

```
 1.       SUBROUTINE DERPAR(N,X,XLOW,XUPP,EPS,W,INITAL,ITIN,HH,HMAX,PREF,     DRPR    1
         C                 NDIR,E,MXADMS,NCORR,NCRAD,NOUT,OUT,MAXOUT,NPRNT)DRPR    2
 2.       DIMENSION X(11),XLOW(11),XUPP(11),W(11),HMAX(11),PREF(11),NDIR(11)DRPR    3
         C        ,OUT(100,12),F(11),G(10,11),BETA(11),MARK(11),DXDT(11)     DRPR    4
         C                                                                  DRPR    5
         C OBTAINING OF DEPENDENCE OF SOLUTION X(ALFA) OF EQUATION           DRPR    6
         C F ( X , ALFA ) = 0                                               DRPR    7
         C ON PARAMETER ALFA BY MODIFIED METHOD OF DIFFERENTIATION           DRPR    8
         C WITH RESPECT TO PARAMETER                                        DRPR    9
         C                                                                  DRPR   10
         C N: NUMBER OF UNKNOWNS X(I)                                       DRPR   11
         C X(1),....,X(N): INITIAL VALUES OF X(I), AFTER RETURN FINAL VALUES OF X(DRPR   12
         C X(N+1): INITIAL VALUE OF PARAMETER ALFA, AFTER RETURN FINAL VALUE OF ADRPR   13
         C XLOW(1),....,XLOW(N): LOWER BOUNDS FOR X(I)                       DRPR   14
         C XUPP(1),....,XUPP(N): UPPER BOUNDS FOR X(I)                       DRPR   15
         C XLOW(N+1),XUPP(N+1): LOWER AND UPPER BOUNDS FOR ALFA              DRPR   16
         C     IF XLOW OR XUPP IS EXCEEDED, THEN END OF DERPAR               DRPR   17
         C     AND MAXOUT=-2 AFTER RETURN                                   DRPR   18
         C EPS: ACCURACY DESIRED IN NEWTON ITERATION FOR                    DRPR   19
         C     SUM OF (W(I)*ABS(XNEW(I)-XOLD(I))), I=1,....,N+1              DRPR   20
         C W(1),....,W(N+1): WEIGHTS USED IN TERMINATION CRITERION OF NEWTON PROCEDRPR   21
         C INITAL: IF (INITAL.NE.0) THEN SEVERAL STEPS IN NEWTON ITERATION   DRPR   22
         C         ARE MADE BEFORE COMPUTATION IN ORDER TO INCREASE         DRPR   23
         C         ACCURACY OF INITIAL POINT:                               DRPR   24
         C     IF (INITAL.EQ.1.AND.EPS-ACCURACY IS NOT FULFILLED IN ITIN ITERATIODRPR   25
         C     THEN RETURN.   IF (INITAL.EQ.2) THEN ALWAYS RETURN            DRPR   26
         C     AFTER INITIAL NEWTON ITERATION, RESULTS ARE IN X.            DRPR   27
         C     IF (INITAL.EQ.3) THEN CONTINUE IN DERPAR AFTER INITIAL NEWTON. DRPR   28
         C     IF (INITAL.EQ.0) THEN NO INITIAL NEWTON ITERATION IS USED.    DRPR   29
         C ITIN: MAXIMAL NUMBER OF INITIAL NEWTON ITERATIONS, IF EPS-ACCURACY DRPR   30
         C       IS NOT FULFILLED IN ITIN ITERATIONS THEN ITIN=-1 AFTER RETURN. DRPR   31
         C HH: INTEGRATION STEP ALONG ARC LENGTH OF SOLUTION LOCUS           DRPR   32
         C HMAX(1),....,HMAX(N+1): UPPER BOUNDS FOR INCREMENTS OF X(I) IN    DRPR   33
         C                     ONE INTEGRATION STEP (APPROXIMATION ONLY)     DRPR   34
         C                                                                  DRPR   35
         C PREF(1),....,PREF(N+1): PREFERENCE NUMBERS (EXPLANATION SEE IN SUBR.GAUDRPR   36
         C NDIR(1),....,NDIR(N+1): INITIAL CHANGE OF X(I) IS POSITIVE ALONG   DRPR   37
         C     SOLUTION LOCUS (CURVE) IF NDIR(I)=1 AND NEGATIVE IF NDIR(I)=-1. DRPR   38
         C E: CRITERION FOR TEST ON CLOSED CURVE, IF                        DRPR   39
         C     (SUM OF (W(I)*ABS(X(I)-XINITIAL(I))),I=1,....,N+1).LE.E)      DRPR   40
         C     THEN CLOSED CURVE MAY BE EXPECTED                            DRPR   41
         C MXADMS: MAXIMAL ORDER OF ADAMS-BASHFORTH FORMULA, 1.LE.MXADMS.LE.4 DRPR   42
         C NCORR: MAXIMAL NUMBER OF NEWTON CORRECTIONS AFTER PREDICTION      DRPR   43
         C     BY ADAMS-BASHFORTH METHOD.                                   DRPR   44
         C NCRAD: IF (NCRAD.NE.0) THEN ADDITIONAL NEWTON CORRECTION          DRPR   45
         C NOUT: AFTER RETURN NUMBER OF CALCULATED POINTS ON THE CURVE X(ALFA), DRPR   47
         C     NOUT.LE.MAXOUT.                                              DRPR   48
         C NCRAD: IF (NCRAD.NE.0) THEN ADDITIONAL NEWTON CORRECTION          DRPR   45
         C OUT(J,1),OUT(J,2),....,OUT(J,N+1): J-TH POINT X(1),....,X(N),ALFA DRPR   49
         C     ON CURVE X(ALFA)                                            DRPR   50
         C OUT(J,N+2): VALUE OF SQRT(SUM OF SQUARES OF F)                    DRPR   51
         C     IF (OUT(J,N+2).LT.0.0) THEN  ABS(OUT(J,N+2)) CORRESPONDS TO X AND ADRPR   52
         C     NOT EXACTLY, BECAUSE ADDITIONAL NEWTON CORRECTION WAS USED (NCRAD.NDRPR   53
         C     VALUES F(I) ARE NOT COMPUTED FOR X AND ALFA PRINTED AND/OR STORED ADRPR   54
         C     THEREFORE LAST TIME COMPUTED VALUE OF SQRT(SUM OF SQUARES OF F) IS DRPR   55
         C     ON OUR DISPOSAL ONLY.                                        DRPR   56
         C MAXOUT: MAXIMAL NUMBER OF CALCULATED POINTS ON CURVE X(ALFA).     DRPR   57
         C     IF MAXOUT AFTER RETURN EQUALS TO:                           DRPR   58
         C         -1: THEN CLOSED CURVE X(ALFA) MAY BE EXPECTED           DRPR   59
         C         -2: THEN BOUND XLOW OR XUPP WAS EXCEEDED                DRPR   60
         C         -3: THEN SINGULAR JACOBIAN MATRIX OCCURED, ITS RANK IS.LT.N DRPR   61
         C NPRNT: IF (NPRNT.EQ.3) THEN RESULTING POINTS ON CURVE X(ALFA)     DRPR   62
```

```
     C    ARE IN ARRAY OUT( , ) AFTER RETURN. IF (NPRNT.EQ.1) THEN THESE POINDRPR   63
     C    ARE PRINTED ONLY. IF (NPRNT.EQ.2) THEN THESE POINTS ARE BOTH PRINTEDRPR   64
     C    AND IN ARRAY OUT.                                               DRPR   65
     C SUBROUTINE FCTN MUST BE PROGRAMMED IN FOLLOWING FORM:              DRPR   66
     C    SUBROUTINE FCTN(N,X,F,G)                                        DRPR   67
     C    DIMENSION X(11),F(10),G(10,11)                                  DRPR   68
     C    F(I)= FI (X(1),X(2),...,X(N),ALFA) FOR I=1,...,N                DRPR   69
     C    G(I,J)= D FI (X(1),...,X(N),ALFA)/ D X(J) FOR  I,J=1,...,N       DRPR   70
     C    G(I,N+1)= D FI (X(1),...,X(N),ALFA)/ D ALFA  FOR I=1,...,N       DRPR   71
     C    RETURN                                                          DRPR   72
     C    END                                                             DRPR   73
     C LW IS PRINTER DEVICE NUMBER                                        DRPR   74
 3.       N1=N+1                                                          DRPR   75
 4.       LW=3                                                            DRPR   76
 5.       IF (INITAL) 10,50,10                                            DRPR   77
     C INITIAL NEWTON ITERATIONS                                         DRPR   78
 6.    10 DO 30 L=1,ITIN                                                  DRPR   79
 7.       CALL FCTN(N,X,F,G)                                             DRPR   80
 8.       SQUAR=0.0                                                      DRPR   81
 9.       DO 15 I=1,N                                                    DRPR   82
10.    15    SQUAR=SQUAR+F(I)**2                                          DRPR   83
11.       LL=L-1                                                         DRPR   84
12.       SQUAR=SQRT(SQUAR)                                              DRPR   85
13.       IF (NPRNT.NE.3) WRITE (LW,20) LL,(X(I),I=1,N1),SQUAR            DRPR   86
14.    20 FORMAT (3X,7H DERPAR,I3,25H,INITIAL NEWTON ITERATION/22X,       DRPR   87
     C          11HX,ALFA,SQF=,5F15.7/(33X,5F15.7))                       DRPR   88
15.       CALL GAUSE(N,G,F,M,10,11,PREF,BETA,K)                          DRPR   89
16.       IF (M.EQ.0) GO TO 220                                          DRPR   90
17.       P=0.0                                                          DRPR   91
18.       DO 25 J=1,N1                                                   DRPR   92
19.          X(J)=X(J)-F(J)                                              DRPR   93
20.    25    P=P+ABS(F(J))*W(J)                                          DRPR   94
21.       IF (P.LE.EPS) GO TO 40                                         DRPR   95
22.    30 CONTINUE                                                       DRPR   96
23.       WRITE (LW,35) ITIN                                             DRPR   97
24.    35 FORMAT (/16H DERPAR OVERFLOW,I5,2X,18HINITIAL ITERATIONS//)     DRPR   98
25.       ITIN=-1                                                        DRPR   99
26.       IF (INITAL.EQ.1) RETURN                                        DRPR  100
27.    40 IF (NPRNT.NE.3) WRITE (LW,45) (X(I),I=1,N1)                     DRPR  101
28.    45 FORMAT (3X,39H DERPAR AFTER INITIAL NEWTON ITERATIONS/22X,      DRPR  102
     C          7HX,ALFA=,4X,5F15.7/(33X,5F15.7))                        DRPR  103
29.       IF (INITAL.EQ.2) RETURN                                        DRPR  104
     C AFTER INITIAL NEWTON ITERATIONS                                   DRPR  105
30.    50 INDIC=4H****                                                    DRPR  106
31.       INDSP=4H                                                       DRPR  107
32.       IF (NPRNT.NE.3) WRITE (LW,55)                                   DRPR  108
33.    55 FORMAT (/6X,63H DERPAR RESULTS (VARIABLE CHOSEN AS INDEPENDENT IS DRPR 109
     C    CMARKED BY *)/)                                                DRPR  110
34.       KOUT=0                                                         DRPR  111
35.       NOUT=0                                                         DRPR  112
36.       MADMS=0                                                        DRPR  113
37.       NC=1                                                           DRPR  114
38.       K1=0                                                           DRPR  115
39.    60 CALL FCTN(N,X,F,G)                                             DRPR  116
40.       SQUAR=0.0                                                      DRPR  117
41.       DO 65 I=1,N                                                    DRPR  118
42.    65    SQUAR=SQUAR+F(I)**2                                          DRPR  119
43.       CALL GAUSE(N,G,F,M,10,11,PREF,BETA,K)                          DRPR  120
44.       IF (M.EQ.0) GO TO 220                                          DRPR  121
45.       IF (K1.EQ.K) GO TO 70                                          DRPR  122
     C CHANGE OF INDEPENDENT VARIABLE (ITS INDEX = K NOW)                DRPR  123
46.       MADMS=0                                                        DRPR  124
47.       K1=K                                                           DRPR  125
48.    70 SQUAR=SQRT(SQUAR)                                              DRPR  126
49.       IF (NCRAD.EQ.1) SQUAR=-SQUAR                                    DRPR  127
50.       P=0.0                                                          DRPR  128
51.       DO 75 I=1,N1                                                   DRPR  129
52.    75    P=P+W(I)*ABS(F(I))                                          DRPR  130
53.       IF (P.LE.EPS) GO TO 95                                         DRPR  131
54.       IF (NC.GE.NCORR) GO TO 85                                       DRPR  132
55.       DO 80 I=1,N1                                                   DRPR  133
56.    80    X(I)=X(I)-F(I)                                              DRPR  134
     C ONE ITERATION IN NEWTON METHOD                                    DRPR  135
57.       NC=NC+1                                                        DRPR  136
58.       GO TO 60                                                       DRPR  137
59.    85 IF (NCORR.EQ.0) GO TO 95                                        DRPR  138
60.       WRITE (LW,90) NCORR,P                                           DRPR  139
61.    90 FORMAT (/37H DERPAR NUMBER OF NEWTON CORRECTIONS=,I5,31H IS NOT SDRPR 140
     C    CUFFICIENT,ERROR OF X=,F15.7/)                                 DRPR  141
62.    95 NC=1                                                           DRPR  142
63.       IF (NCRAD.EQ.0) GO TO 105                                       DRPR  143
```

```
          C ADDITIONAL NEWTON CORRECTION                                        DRPR 144
 64.          DO 100 I=1,N1                                                      DRPR 145
 65.   100    X(I)=X(I)-F(I)                                                     DRPR 146
 66.   105 NOUT=NOUT+1                                                           DRPR 147
 67.          DO 110 I=1,N1                                                      DRPR 148
 68.   110    MARK(I)=INDSP                                                      DRPR 149
 69.          MARK(K)=INDIC                                                      DRPR 150
 70.          IF (NPRNT.NE.3) WRITE (LW,115) (X(I),MARK(I),I=1,N1),SQUAR         DRPR 151
 71.   115 FORMAT (6X,27H DERPAR RESULTS X,ALFA,SQF=,5(F15.7,A1)/                DRPR 152
          C         (33X,5(F15.7,A1)))                                           DRPR 153
 72.          IF (NPRNT.EQ.1) GO TO 135                                          DRPR 154
 73.          IF (NOUT.LE.100) GO TO 125                                         DRPR 155
 74.          WRITE (LW,120)                                                     DRPR 156
 75.   120 FORMAT (/28H DERPAR OUTPUT ARRAY IS FULL//)                           DRPR 157
 76.          RETURN                                                             DRPR 158
 77.   125 DO 130 I=1,N1                                                         DRPR 159
 78.   130    OUT(NOUT,I)=X(I)                                                   DRPR 160
 79.          OUT(NOUT,N+2)=SQUAR                                                DRPR 161
 80.          GO TO 140                                                          DRPR 162
 81.   135 IF (NOUT.EQ.1) GO TO 125                                              DRPR 163
 82.   140 IF (NOUT.GE.MAXOUT) RETURN                                            DRPR 164
 83.          DO 145 I=1,N1                                                      DRPR 165
 84.          IF (X(I).LT.XLOW(I).OR.X(I).GT.XUPP(I)) GO TO 210                  DRPR 166
 85.   145    CONTINUE                                                           DRPR 167
 86.          IF (NOUT.LE.3) GO TO 155                                           DRPR 168
 87.          P=0.0                                                              DRPR 169
 88.          DO 150 I=1,N1                                                      DRPR 170
 89.   150    P=P+W(I)*ABS(X(I)-OUT(1,I))                                        DRPR 171
 90.          IF (P.LE.E) GO TO 200                                              DRPR 172
          C CLOSED CURVE MAY BE EXPECTED                                        DRPR 173
 91.   155 DXK2=1.0                                                              DRPR 174
 92.          DO 160 I=1,N1                                                      DRPR 175
 93.   160 DXK2=DXK2+BETA(I)**2                                                  DRPR 176
 94.          DXDT(K)=1.0/SQRT(DXK2)*FLOAT(NDIR(K))                             DRPR 177
          C DERIVATIVE OF INDEPENDENT VARIABLE X(K) WITH RESPECT TO ARC LENGTH   DRPR 178
          C OF SOLUTION LOCUS IS COMPUTED HERE                                  DRPR 179
 95.          H=HH                                                               DRPR 180
 96.          DO 170 I=1,N1                                                      DRPR 181
 97.          NDIR(I)=1                                                          DRPR 182
 98.          IF (I.EQ.K) GO TO 165                                             DRPR 183
 99.          DXDT(I)=BETA(I)*DXDT(K)                                           DRPR 184
100.   165    IF (DXDT(I).LT.0.0) NDIR(I)=-1                                    DRPR 185
101.          IF (H*ABS(DXDT(I)).LE.HMAX(I)) GO TO 170                          DRPR 186
102.          MADMS=0                                                            DRPR 187
103.          H=HMAX(I)/ABS(DXDT(I))                                            DRPR 188
104.   170    CONTINUE                                                           DRPR 189
105.          IF (NOUT.LE.KOUT+3) GO TO 175                                     DRPR 190
106.          IF (H*ABS(DXDT(K)).LE.0.8*ABS(X(K)-OUT(1,K))) GO TO 175           DRPR 191
107.          IF ((OUT(1,K)-X(K))*FLOAT(NDIR(K)).LE.0.0) GO TO 175              DRPR 192
108.          MADMS=0                                                            DRPR 193
109.          IF (H*ABS(DXDT(K)).LE.ABS(X(K)-OUT(1,K))) GO TO 175               DRPR 194
110.          H=ABS(X(K)-OUT(1,K))/ABS(DXDT(K))                                 DRPR 195
111.          KOUT=NOUT                                                          DRPR 196
112.   175 CALL ADAMS(N,DXDT,MADMS,H,X,MXADMS)                                   DRPR 197
113.          GO TO 60                                                           DRPR 198
114.   200 WRITE (LW,205)                                                        DRPR 199
115.   205 FORMAT (/36H DERPAR CLOSED CURVE MAY BE EXPECTED/)                    DRPR 200
116.          MAXOUT=-1                                                          DRPR 201
117.          RETURN                                                             DRPR 202
118.   210 MAXOUT=-2                                                             DRPR 203
119.          RETURN                                                             DRPR 204
120.   220 WRITE (LW,225) (X(I),I=1,N1)                                          DRPR 205
121.   225 FORMAT (/48H DERPAR SINGULAR JACOBIAN MATRIX FOR X AND ALFA=,         DRPR 206
          C         5F12.6/(48X,5F12.6))                                         DRPR 207
122.          MAXOUT=-3                                                          DRPR 208
123.          RETURN                                                             DRPR 209
124.          END                                                               DRPR 210
  1.          SUBROUTINE GAUSE(N,A,B,M,NN,MM,PREF,BETA,K)                        DRPR 211
  2.          DIMENSION A(NN,MM),B(MM),PREF(MM),BETA(MM),Y(11),X(11),IRK(11),    DRPR 212
          C         IRK(11)                                                      DRPR 213
          C SOLUTION OF N LINEAR EQUATIONS FOR N+1 UNKNOWNS                      DRPR 214
          C N: NUMBER OF EQUATIONS                                              DRPR 216
          C A: N X (N+1) MATRIX OF SYSTEM                                       DRPR 217
          C SOLUTION OF N LINEAR EQUATIONS FOR N+1 UNKNOWNS                      DRPR 214
          C B: RIGHT-HAND SIDES                                                 DRPR 218
          C M: IF (M.EQ.0) AFTER RETURN THEN RANK(A).LT.N                        DRPR 219
          C PREF(I): PREFERENCE NUMBER FOR X(I) TO BE INDEPENDENT VARIABLE,      DRPR 220
          C 0.0.LE.PREF(I).LE.1.0, THE LOWER IS PREF(I) THE HIGHER IS PREFERENCE DRPR 221
          C BETA(I): COEFFICIENTS IN EXPLICIT DEPENDENCES OBTAINED IN FORM:      DRPR 222
          C         X(I)=B(I)+BETA(I)*X(K), I.NE.K.                              DRPR 223
          C K: RESULTING INDEX OF INDEPENDENT VARIABLE                           DRPR 224
```

```
3.          N1=N+1                                                      DRPR 225
4.          ID=1                                                        DRPR 226
5.          M=1                                                         DRPR 227
6.          DO 1 I=1,N1                                                 DRPR 228
7.           IRK(I)=0                                                   DRPR 229
8.      1    IRR(I)=0                                                   DRPR 230
9.      2 IR=1                                                          DRPR 231
10.         IS=1                                                        DRPR 232
11.         AMAX=0.0                                                    DRPR 233
12.         DO 6 I=1,N                                                  DRPR 234
13.          IF (IRR(I)) 6,3,6                                          DRPR 235
14.     3    DO 5 J=1,N1                                                DRPR 236
15.          P=PREF(J)*ABS(A(I,J))                                      DRPR 237
16.          IF (P-AMAX) 5,5,4                                          DRPR 238
17.     4    IR=I                                                       DRPR 239
18.          IS=J                                                       DRPR 240
19.          AMAX=P                                                     DRPR 241
20.     5    CONTINUE                                                   DRPR 242
21.     6 CONTINUE                                                      DRPR 243
22.         IF (AMAX.NE.0.0) GO TO 7                                    DRPR 244
    C THIS CONDITION FOR SINGULARITY OF MATRIX MUST BE SPECIFIED        DRPR 245
    C MORE EXACTLY WITH RESPECT TO COMPUTER ACTUALLY USED               DRPR 246
23.         M=0                                                         DRPR 247
24.         GO TO 15                                                    DRPR 248
25.     7 IRR(IR)=IS                                                    DRPR 249
26.         DO 9 I=1,N                                                  DRPR 250
27.          IF (I.EQ.IR.OR.A(I,IS).EQ.0.0) GO TO 9                     DRPR 251
28.          P=A(I,IS)/A(IR,IS)                                         DRPR 252
29.          DO 8 J=1,N1                                                DRPR 253
30.     8    A(I,J)=A(I,J)-P*A(IR,J)                                    DRPR 254
31.          A(I,IS)=0.0                                                DRPR 255
32.          B(I)=B(I)-P*B(IR)                                          DRPR 256
33.     9 CONTINUE                                                      DRPR 257
34.         ID=ID+1                                                     DRPR 258
35.         IF (ID.LE.N) GO TO 2                                        DRPR 259
36.         DO 10 I=1,N                                                 DRPR 260
37.          IR=IRR(I)                                                  DRPR 261
38.          X(IR)=B(I)/A(I,IR)                                         DRPR 262
39.     10   IRK(IR)=1                                                  DRPR 263
40.         DO 11 K=1,N1                                                DRPR 264
41.          IF(IRK(K).EQ.0) GO TO 12                                   DRPR 265
42.     11 CONTINUE                                                     DRPR 266
43.     12 DO 13 I=1,N                                                  DRPR 267
44.          IR=IRR(I)                                                  DRPR 268
45.     13   Y(IR)=-A(I,K)/A(I,IR)                                      DRPR 269
46.         DO 14 I=1,N1                                                DRPR 270
47.          B(I)=X(I)                                                  DRPR 271
48.     14   BETA(I)=Y(I)                                               DRPR 272
49.         B(K)=0.0                                                    DRPR 273
50.         BETA(K)=0.0                                                 DRPR 274
51.     15 RETURN                                                       DRPR 275
52.         END                                                        DRPR 276
1.          SUBROUTINE ADAMS(N,D,MADMS,H,X,MXADMS)                     DRPR 277
2.          DIMENSION DER(4,11),X(11),D(11)                            DRPR 278
    C ADAMS-BASHFORTH METHODS                                          DRPR 279
3.          N1=N+1                                                      DRPR 280
4.          DO 10 K=1,3                                                 DRPR 281
5.          I=4-K                                                       DRPR 281
6.          DO 10 J=1,N1                                                DRPR 282
7.     10    DER(I+1,J)=DER(I,J)                                        DRPR 283
8.          MADMS=MADMS+1                                               DRPR 284
9.          IF (MADMS.GT.MXADMS) MADMS=MXADMS                          DRPR 285
10.         IF (MADMS.GT.4) MADMS=4                                     DRPR 286
11.         DO 60 I=1,N1                                                DRPR 287
12.          DER(1,I)=D(I)                                             DRPR 288
13.          GO TO (20,30,40,50),MADMS                                 DRPR 289
14.     20   X(I)=X(I)+H*DER(1,I)                                      DRPR 290
15.          GO TO 60                                                  DRPR 291
16.     30   X(I)=X(I)+0.5*H*(3.0*DER(1,I)-DER(2,I))                   DRPR 292
17.          GO TO 60                                                  DRPR 293
18.     40   X(I)=X(I)+H*(23.0*DER(1,I)-16.0*DER(2,I)+5.0*DER(3,I))/12.0  DRPR 294
19.          GO TO 60                                                  DRPR 295
20.     50   X(I)=X(I)+H*(55.0*DER(1,I)-59.0*DER(2,I)+37.0*DER(3,I)-   DRPR 296
    C                 9.0*DER(4,I))/24.0                               DRPR 297
21.     60   CONTINUE                                                  DRPR 298
22.         RETURN                                                     DRPR 299
23.         END                                                        DRPR 300
```

SHOOT—An Algorithm for Solving Nonlinear Boundary-Value Problems by the Shooting Method

Consider a set of nonlinear first-order differential equations

$$\frac{dy_i}{dx} = F_i(x, y_1, \ldots, y_N), \quad i = 1, 2, \ldots, N, \tag{B1}$$

subject to linear two-point boundary conditions

$$\sum_{j=1}^{N} a_{ij} y_j(a) = c_i, \quad i = 1, 2, \ldots, N_1, \tag{B2a}$$

$$\sum_{j=1}^{N} b_{ij} y_j(b) = d_i, \quad i = 1, 2, \ldots, N_2, \tag{B2b}$$

$$\sum_{j=1}^{N} (e_{ij} y_j(a) + f_{ij} y_j(b)) = s_i, \quad i = 1, 2, \ldots, N_3, \tag{B2c}$$

where the rank of the $N_3 \times N$ matrix $F = \{f_{ij}\}$ is equal to N_3.

The condition $N_1 + N_2 + N_3 = N$ is evidently valid for a correctly formulated boundary-value problem, and the ranks of the matrices $A = \{a_{ij}\}$ or $B = \{b_{ij}\}$ are N_1 or N_2, respectively. The functions F_i are assumed to be differentiable.

The shooting method, i.e., transformation of the boundary-value problem into a relevant initial-value problem, is used for solving the nonlinear boundary-value problem (B1), (B2). Let us select $N - N_1$ initial values of y at $x = a$:

$$y_{j_i}(a) = \eta_i, \quad i = 1, 2, \ldots, N_4. \tag{B3}$$

Here $j_i, i = 1, 2, \ldots, N_4 = N - N_1$, are indices chosen almost arbitrarily.

Denote the $N_1 \times N_1$ matrix

$$A_1 = \{a_{ij}\}, \quad i = 1, 2, \ldots, N_1, j = 1, 2, \ldots, N, j \neq j_k, k = 1, 2, \ldots, N_4, \tag{B4}$$

and the $N_1 \times N_4$ matrix

$$A_2 = \{a_{ij}\}, \quad i = 1, 2, \ldots, N_1, j = j_1, j_2, \ldots, j_{N_4}. \tag{B5}$$

A complete set of initial conditions can be obtained after evaluating of (B6):

$$y_i(a) = \sum_{k=1}^{N_1} B_{ik}\left[c_k - \sum_{m=1}^{N_4} a_{kj_m}\eta_m\right],$$

$$i = 1, 2, \ldots, N, i \neq j_k, k = 1, 2, \ldots, N_4. \qquad \text{(B6)}$$

Here $\{B_{ik}\} = A_1^{-1}$. Relation (B6) can also be written in matrix notation:

$$\boldsymbol{\gamma} = A_1^{-1}\mathbf{c} - A_1^{-1}A_2\boldsymbol{\eta}, \qquad \text{(B7)}$$

where $\boldsymbol{\gamma}$ is a vector of $y_i(a)$, $i \neq j_k$, $k = 1, 2, \ldots, N_4$, and $\mathbf{c} = (c_1, \ldots, c_{N_1})^T$.

The solution of the initial-value problem (B1) subject to initial conditions (B3) and (B6) depends on the choice of the initial values of $\mathbf{y}$, i.e.,

$$y_i = y_i(x, \eta_1, \ldots, \eta_{N_4}). \qquad \text{(B8)}$$

The solution (B8) obeys the original boundary-value problem (B1), (B2) if boundary conditions (B2b) and (B2c) are satisfied:

$$\sum_{j=1}^{N} b_{ij}y_j(b, \boldsymbol{\eta}) - d_i = 0, \quad i = 1, 2, \ldots, N_2, \qquad \text{(B9a)}$$

$$\sum_{j=1}^{N} (e_{ij}y_j(a, \boldsymbol{\eta}) + f_{ij}y_j(b, \boldsymbol{\eta})) - s_i = 0, \quad i = 1, 2, \ldots, N_3. \qquad \text{(B9b)}$$

Relations (B6) assure that the boundary conditions (B2a) are satisfied automatically. Equations (B9a) and (B9b) form a set of N_4 equations for N_4 unknowns $\eta_1, \ldots, \eta_{N_4}$:

$$R_i(\boldsymbol{\eta}) = 0, \quad i = 1, 2, \ldots, N_4. \qquad \text{(B10)}$$

For solving system (B10), the Newton–Raphson method is used here:

$$\Gamma(\boldsymbol{\eta}^k)\Delta\boldsymbol{\eta}^k = -R(\boldsymbol{\eta}^k) \qquad \text{(B11)}$$

and

$$\boldsymbol{\eta}^{k+1} = \boldsymbol{\eta}^k + \alpha\Delta\boldsymbol{\eta}^k, \qquad \text{(B12)}$$

where $\alpha \in (0, 1\rangle$ is chosen so that the sum of squares of the residuals is reduced, i.e., the condition

$$\sum_{i=1}^{N_4} R_i^2(\boldsymbol{\eta}^{k+1}) < \sum_{i=1}^{N_4} R_i^2(\boldsymbol{\eta}^k) \qquad \text{(B13)}$$

is satisfied. The Jacobian matrix $\{\partial R_i/\partial\eta_j\}$ can be written as

$$\Gamma(\boldsymbol{\eta}) = \begin{pmatrix} \dfrac{\partial R_1}{\partial\eta_1}, \dfrac{\partial R_1}{\partial\eta_2}, \ldots, \dfrac{\partial R_1}{\partial\eta_{N_4}} \\[2ex] \dfrac{\partial R_2}{\partial\eta_1}, \ldots, \dfrac{\partial R_2}{\partial\eta_{N_4}} \\[2ex] \dfrac{\partial R_{N_4}}{\partial\eta_1}, \ldots, \dfrac{\partial R_{N_4}}{\partial\eta_{N_4}} \end{pmatrix},$$

where

$$\frac{\partial R_i}{\partial \eta_k} = \sum_{j=1}^{N} b_{ij} p_{jk}(b), \quad i = 1, 2, \ldots, N_2 \tag{B14a}$$

and

$$\frac{\partial R_{i+N_2}}{\partial \eta_k} = \sum_{j=1}^{N} [e_{ij} p_{jk}(a) + f_{ij} p_{jk}(b)], \quad i = 1, 2, \ldots, N_3. \tag{B14b}$$

Here, for the variational variables

$$p_{jk}(x) = \frac{\partial y_j(x)}{\partial \eta_k}, \tag{B15}$$

a set of differential equations can be developed:

$$\frac{dp_{ik}(x)}{dx} = \sum_{j=1}^{N} \frac{\partial F_i(x, y_1, \ldots, y_N)}{\partial y_j} p_{jk}(x),$$

$$i = 1, 2, \ldots, N; k = 1, 2, \ldots, N_4. \tag{B16}$$

The initial conditions for (B16) take the form

$$p_{j_k i}(a) = 0 \quad \text{for} \quad i \neq k; \quad p_{j_k k}(a) = 1; \quad k = 1, 2, \ldots, N_4. \tag{B17a}$$

The missing initial conditions can be evaluated after differentiating of (B6) or (B7), which leads to

$$p_{im}(a) = -\sum_{k=1}^{N_1} B_{ik} a_{kj_m},$$

$$i = 1, 2, \ldots, N, i \neq j_k, \quad k = 1, 2, \ldots, N_4, \quad m = 1, 2, \ldots, N_4. \tag{B17b}$$

Integration of the coupled system of $N(1 + N_4)$ differential equations (B1) and (B16) is performed by the Runge–Kutta–Merson method with automatic step-size control.

The special form of the right-hand sides F_i of Eq. (B1) must be programmed in the form of a subroutine RHS. The evaluation of the partial derivatives $\partial F_i/\partial y_j$ must also be included in this subroutine. The parameter values (e.g., of a physical nature) in the right-hand sides F_i can be read in a subroutine READIN and transferred by using the COMMON statement. All further explanations are included as comments in the programm.

```
      C SHOOTING METHOD FOR NONLINEAR BOUNDARY VALUE PROBLEM FOR A SET OF
      C FIRST-ORDER ORDINARY DIFFERENTIAL EQUATIONS (NEWTON-FOX METHOD)
1.          DIMENSION A(10,10),B(10,10),E(10,10),F(10,10),C(10),D(10),S(10),
            CY(100),ETA(10),A1(10,10),B1(10,10),R(10,1),YF(100),ETAS(10),
            CDETA(10),JVAR(10)
2.          COMMON /NUMA/ N,N4
3.          COMMON /PRNT/ IW
4.          COMMON /SMALLH/ KK
5.          IR=1
6.          IW=3
      C IR IS FEADER DEVICE NUMBER
      C IW IS PRINTER DEVICE NUMBER
```

```
7.     10 WRITE (IW,20)
8.     20 FORMAT (1H1////74H SHOOTING METHOD FOR NONLINEAR BOUNDARY VALUE PR
          COBLEMS (NEWTON-FOX METHOD)////)
9.        CALL READIN
10.       READ (IR,30) N,N1,N2,N3,AA,BB
11.    30 FORMAT (4I5,2F10.5)
12.       IF (N.LE.0) STOP
   C N: NUMBER OF DIFFERENTIAL EQUATIONS
   C N1: NUMBER OF LINEAR BOUNDARY CONDITIONS AT X=AA
   C N2: NUMBER OF LINEAR BOUNDARY CONDITIONS AT X=BB
   C N3: NUMBER OF LINEAR MIXED BOUNDARY CONDITIONS AT X=AA AND X=BB
   C AA,BB: LEFT AND RIGHT ENDS OF INTERVAL (AA,BB)
13.       WRITE (IW,40) N,AA,BB
14.    40 FORMAT (//9H NBVP FOR,I5,19H EQS ON INTERVAL (,F12.6,2H ,,F12.6,
          C2H )//27H LINEAR BOUNDARY CONDITIONS/)
15.       WRITE (IW,50) (I,I=1,N)
16.    50 FORMAT (6X,9(3H*Y(,I2,1H),6X))
17.       IF (N1.EQ.0) GO TO 100
18.       WRITE (IW,60)
19.    60 FORMAT (/17H BOUNDARY AT X=A:/)
20.       DO 80 I=1,N1
21.          READ (IR,70) (A(I,J),J=1,N),C(I)
22.    70    FORMAT (8F10.5)
23.          WRITE (IW,90) (A(I,J),J=1,N),C(I)
24.    80 CONTINUE
25.    90 FORMAT (1X,9F12.5)
   C A(I,J),C(I): COEFFICIENTS AND RIGHT-HAND SIDES IN BOUNDARY CONDITIONS
   C                 AT  X=A
26.   100 IF (N2.EQ.0) GO TO 130
27.       WRITE (IW,110)
28.   110 FORMAT (/17H BOUNDARY AT X=B:/)
29.       DO 120 I=1,N2
30.          READ (IR,70)  (B(I,J),J=1,N),D(I)
31.          WRITE (IW,90) (B(I,J),J=1,N),D(I)
32.   120 CONTINUE
   C B(I,J),D(I): COEFFICIENTS AND RIGHT-HAND SIDES IN BOUNDARY CONDITIONS
   C                 AT  X=B
33.   130 IF (N3.EQ.0) GO TO 160
34.       WRITE (IW,140)
35.   140 FORMAT (/15H MIXED BOUNDARY/)
36.       DO 150 I=1,N3
37.          READ (IR,70) (E(I,J),J=1,N)
38.          WRITE (IW,90) (E(I,J),J=1,N)
39.          READ (IR,70) (F(I,J),J=1,N),S(I)
40.          WRITE (IW,90) (F(I,J),J=1,N),S(I)
41.   150 CONTINUE
   C E(I,J),F(I,J),S(I): COEFFICIENTS AND RIGHT-HAND SIDES IN MIXED
   C                       BOUNDARY CONDITIONS
42.   160 N4=N-N1
43.       WRITE (IW,170) N4
44.   170 FORMAT (//28H AT X=A ARE CHOSEN FOLLOWING,I4,19H INITIAL CONDITION
          CS/)
45.       READ (IR,180) (JVAR(I),ETA(I),I=1,N4)
46.   180 FORMAT (I5,F10.5)
   C JVAR(I): INDEX OF Y WHICH WILL BE CHOSEN AT X=A AS MISSING
   C               INITIAL CONDITION
   C ETA(I): INITIAL CHOICE OF THIS Y AT X=A. (N-N1) SUCH VALUES MUST BE CHOSEN
47.       WRITE (IW,190) (JVAR(I),ETA(I),I=1,N4)
48.   190 FORMAT (4H Y( ,I2,2H)=,F12.6)
49.       IF (N1.EQ.0) GO TO 260
50.       M=0
51.       J1=0
   C OBTAINING OF EXPLICIT DEPENDENCES OF INITIAL VALUES Y(A) ON CHOSEN ETA(I)
52.       DO 240 J=1,N
53.          DO 200 JJ=1,N4
54.             IF (JVAR(JJ).EQ.J) GO TO 220
55.   200    CONTINUE
56.          J1=J1+1
57.          DO 210 I=1,N1
58.   210    A1(I,J1)=A(I,J)
59.          GO TO 240
60.   220    M=M+1
61.          DO 230 I=1,N1
62.   230    B1(I,M+1)=-A(I,J)
63.   240 CONTINUE
64.       DO 250 I=1,N1
65.   250 B1(I,1)=C(I)
66.       M1=M+1
67.       CALL GAUSS(N1,A1,B1,M1,10,10)
68.       IF (M1.EQ.0) GO TO 770
69.   260 READ (IR,270) EMERS,HMAX,HMIN,KERR
```

```
 70.     270 FORMAT (3E10.2,2I5)
         C EMERS: ACCURACY OF INTEGRATION BY RUNGE-KUTTA-MERSON ROUTINE
         C HMAX,HMIN: MAXIMAL AND MINIMAL PERMISSIBLE STEP LENGTH
         C KERR: CONTROL OF STEP LENGTH IS PERFORMED WITH RESPECT TO ERROR FOR
         C FIRST KERR DIFFERENTIAL EQUATIONS ONLY
 71.         READ (IR,280) EPS,MAXIT,NPRINT,NHX
 72.     280 FORMAT (E10.2,3I5)
         C EPS: DESIRED ACCURACY IN BOUNDARY CONDITIONS AT X=B
         C MAXIT: MAXIMAL NUMBER OF ITERATIONS
         C NPRINT: IF NPRINT=0 THEN PRINTING NO ITERATIONS AND PROFILES
         C         IF NPRINT=1 THEN PRINTING ITERATIONS ONLY
         C         IF NPRINT=2 THEN PRINTING ITERATIONS AND PROFILES
         C NHX: NUMBER OF EQUIDISTANT MESH POINTS IN INTERVAL (A,B)
         C         WHERE PROFILES Y(X) ARE PRINTED
 73.         IF (NHX.LE.0) NHX=1
 74.         IT=0
 75.         KEND=0
 76.         WRITE (IW,290) EMERS,HMAX,HMIN,KERR,EPS,MAXIT
 77.     290 FORMAT (///25H ACCURACY OF INTEGRATION=,E12.2,5X,33H MAXIMAL AND M
            CINIMAL STEP LENGTH=,2E12.2,5X,5HKERR=,I5//65H PRESCRIBED ACCURACY
            CIN MISSING INITIAL CONDITIONS CHOSEN AT X=A:,E10.2,5X,29HMAXIMAL N
            CUMBER OF ITERATIONS=,I5//)
         C
         C INITIAL CONDITIONS FOR VARIATIONAL VARIABLES
 78.         J1=0
 79.         DO 340 J=1,N
 80.             DO 300 JJ=1,N4
 81.             IF (JVAR(JJ).EQ.J) GO TO 320
 82.     300     CONTINUE
 83.             J1=J1+1
 84.             DO 310 I=1,N4
 85.             K1=(J-1)*N4+I+N
 86.             YP(K1)=B1(J1,I+1)
 87.     310     CONTINUE
 88.             GO TO 340
 89.     320     DO 330 I=1,N4
 90.             K1=(J-1)*N4+I+N
 91.             YP(K1)=0.0
 92.     330     CONTINUE
 93.             K1=(J-1)*N4+JJ+N
 94.             YP(K1)=1.0
 95.     340 CONTINUE
         C
 96.     350 ALFA=1.0
         C SET OF INITIAL CONDITIONS AT X=A
         C
 97.     360 J1=0
 98.         DO 400 J=1,N
 99.             DO 370 JJ=1,N4
100.             IF (JVAR(JJ).EQ.J) GO TO 390
101.     370     CONTINUE
102.             J1=J1+1
103.             Y(J)=B1(J1,1)
104.             DO 380 I=1,N4
105.     380     Y(J)=Y(J)+B1(J1,I+1)*ETA(I)
106.             GO TO 400
107.     390     Y(J)=ETA(JJ)
108.     400 CONTINUE
109.         IF (KEND.EQ.1) WRITE (IW,410)
110.     410 FORMAT (///47H ***** SOLUTION OF BOUNDARY VALUE PROBLEM *****)
111.         IF (NPRINT.EQ.0.AND.KEND.EQ.0) GO TO 430
112.         WRITE (IW,420) IT,(Y(J),J=1,N)
113.     420 FORMAT (//27H INITIAL CONDITIONS Y(A) IN,I3,11H. ITERATION/(5X,
            C8E14.6))
114.     430 NEQS=N*(1+N4)
115.         DO 440 J=1,N
116.     440 YP(J)=Y(J)
117.         DO 450 J=N,NEQS
118.     450 Y(J)=YP(J)
         C INTEGRATION OF INITIAL VALUE PROBLEM
119.         H=HMAX
120.         KK=0
121.         IF (NPRINT.NE.2.AND.KEND.EQ.0) GO TO 490
122.         IF (KEND.EQ.0) WRITE (IW,460)
123.     460 FORMAT (//34H SOLUTION OF INITIAL VALUE PROBLEM/)
124.         DO 480 K=1,NHX
125.         ZAC=AA+FLOAT(K-1)*(BB-AA)/FLOAT(NHX)
126.         AKON=AA+FLOAT(K)*(BB-AA)/FLOAT(NHX)
127.         CALL MERSON (NEQS,Y,ZAC,AKON,EMERS,H,HMAX,HMIN,KERR)
128.         WRITE (IW,470) AKON,(Y(J),J=1,N)
129.     470 FORMAT (3H X=,F10.4,3H Y=,8F12.6/(19X,8F12.6))
```

```
130.    480 CONTINUE
131.        GO TO 500
132.    490 CALL MERSON (NEQS,Y,AA,BB,EMERS,H,HMAX,HMIN,KERR)
133.    500 IF (NPRINT.EQ.0.AND.KEND.EQ.0) GO TO 520
134.        WRITE (IW,510) IT,(Y(J),J=1,N)
135.    510 FORMAT (//24H TERMINAL VALUES Y(B) IN,I3,11H. ITERATION/(5X,8E14.6
        C))
136.    520 IF (N2.EQ.0) GO TO 560
    C EVALUATION OF RESIDUALS IN BOUNDARY CONDITIONS AT X=B
    C EVALUATION OF JACOBI MATRIX
137.        DO 550 I=1,N2
138.        P(I,1)=D(I)
139.        DO 530 J=1,N4
140.    530 A1(I,J)=0.0
141.        DO 550 J=1,N
142.        R(I,1)=R(I,1)-B(I,J)*Y(J)
143.        DO 540 K=1,N4
144.        K1=(J-1)*N4+K+N
145.        A1(I,K)=A1(I,K)+B(I,J)*Y(K1)
146.    540 CONTINUE
147.    550 CONTINUE
148.    560 IF (N3.EQ.0) GO TO 600
149.        DO 590 I=1,N3
150.        I1=I+N2
151.        R(I1,1)=S(I)
152.        DO 570 J=1,N4
153.    570 A1(I1,J)=0.0
154.        J1=0
155.        DO 590 J=1,N
156.        R(I1,1)=R(I1,1)-E(I,J)*YP(J)-F(I,J)*Y(J)
157.        DO 580 K=1,N4
158.        K1=(J-1)*N4+K+N
159.        A1(I1,K)=A1(I1,K)+F(I,J)*Y(K1)+E(I,J)*YP(K1)
160.    580 CONTINUE
161.    590 CONTINUE
    C EVALUATION OF SUM OF SQUARES OF RESIDUALS
162.    600 SUMSQ=0.0
163.        DO 610 I=1,N4
164.    610 SUMSQ=SUMSQ+R(I,1)**2
165.        IF (KEND.EQ.1) GO TO 730
166.        IF (IT.EQ.0) GO TO 640
167.        IF (SUMSQ.LE.SUMSQ1*1.0001) GO TO 640
    C NEWTON INCREMENT MUST BE HALVED
168.        ALFA=ALFA/2.0
169.        IF (NPRINT.NE.0) WRITE (IW,660) (R(I,1),I=1,N4)
170.        IF (NPRINT.NE.0) WRITE (IW,615) SUMSQ
171.    615 FORMAT (/31H SUM OF SQUARES OF RESIDUALS IS,E14.5,
        C33H. NEWTON INCREMENT MUST BE HALVED/)
172.        DO 620 I=1,N4
173.    620 ETA (I)=ETAS(I)+ALFA*DETA(I)
174.        IF (ALFA*SQ.GE.EPS/2.0) GO TO 360
175.        WRITE (IW,630) IT
176.    630 FORMAT (///71H SUM OF SQUARES OF RESIDUALS HAS NOT DECREASED IN NEW
        CTON'S DIRECTION IN,I3,11H. ITERATION//)
177.        KEND=1
178.        GO TO 360
179.    640 SUMSQ1=SUMSQ
    C SUM OF SQUARES OF RESIDUALS HAS BEEN REDUCED
180.        DO 650 I=1,N4
181.    650 ETAS(I)=ETA(I)
182.        IF (NPRINT.EQ.0) GO TO 670
183.        WRITE (IW,660) (R(I,1),I=1,N4)
184.    660 FORMAT (/84H BOUNDARY CONDITIONS AT X=B AND MIXED BOUNDARY CONDITI
        CONS ARE FULFILLED WITH ERRORS:/(10X,8E15.5))
185.        IF (IT.EQ.0) WRITE (IW,665) SUMSQ
186.    665 FORMAT (/39H INITIAL SUM OF SQUARES OF RESIDUALS IS,E14.5)
187.        IF (IT.NE.0) WRITE (IW,666) ALFA,SUMSQ
188.    666 FORMAT (/42H NEWTON INCREMENT WAS MULTIPLIED BY FACTOR,F9.5,
        C42H. RESULTING SUM OF SQUARES OF RESIDUALS IS, E14.5)
189.    670 M=1
190.        CALL GAUSS(N4,A1,R,M,10,1)
191.        IF (M.EQ.0) GO TO 710
192.        SQ=0.0
193.        DO 680 I=1,N4
194.        DETA(I)=R(I,1)
195.        ETA(I)=ETAS(I)+DETA(I)
196.        SQ=SQ+DETA(I)**2
197.    680 CONTINUE
198.        SQ=SQRT(SQ)
199.        IT=IT+1
200.        IF (IT.GT.MAXIT) GO TO 690
```

```
201.         IF (SQ.GT.EPS) GO TO 350
      C ACCURACY FOR SOLUTION HAS BEEN FULFILLED
202.         KEND=1
203.         GO TO 360
      C
204.   690 WRITE (IW,700) IT
205.   700 FORMAT (///50H MAXIMAL NUMBER OF ITERATIONS HAS BEEN EXCEEDED IN,
             CI3,11H. ITERATION/26H RESULTS MAY BE INACCURATE//)
206.         KEND=1
207.         GO TO 360
      C
208.   710 WRITE (IW,720)
209.   720 FORMAT (//48H SINGULAR MATRIX FOR INVERSION - END OF SHOOTING//)
210.         KEND=1
211.         GO TO 360
      C
212.   730 WRITE (IW,740) (R(I,1),I=1,N4)
213.   740 FORMAT (//95H THE SOLUTION SATISFIES BOUNDARY CONDITIONS AT X=B AN
             CD MIXED BOUNDARY CONDITIONS WITH RESIDUALS/(5X,6E15.6))
214.         WRITE (IW,750)
215.   750 FORMAT (///15H END OF PROGRAM//)
      C
216.   760 READ (IR,30) JUMP
      C SWITCHING OF ALGORITHM FOR NEXT RUN
      C ACCORDING TO VALUE OF JUMP
217.         IF (JUMP.LT.1.OR.JUMP.GT.3) STOP
218.         GO TO (10,160,260),JUMP
      C
219.   770 WRITE (IW,780)
220.   780 FORMAT (/113H INDICES FOR MISSING INITIAL CONDITIONS AT X=A ARE IN
             CACCEPTABLE WITH RESPECT TO BOUNDARY CONDITIONS AT THIS POINT//)
221.         READ (IR,270)
222.         READ (IR,280)
223.         GO TO 760
224.         END
  1.         SUBROUTINE PRSTR (NN,X,Y,F)
  2.         DIMENSION Y(100),F(100),G(10,10)
  3.         COMMON /NUMB/ N,N4
      C SUBPROGRAM FOR RIGHT-HAND-SIDES OF VARIATIONAL SYSTEM
  4.         CALL RHS(N,X,Y,F,G)
  5.         DO 2 I=1,N
  6.           DO 2 K=1,N4
  7.           KN=K+N
  8.           P=0.0
  9.           DO 1 J=1,N
 10.             K1=(J-1)*N4+KN
 11.             P=P+G(I,J)*Y(K1)
 12.     1     CONTINUE
 13.           K1=(I-1)*N4+KN
 14.           F(K1)=P
 15.     2   CONTINUE
 16.         RETURN
 17.         END
  1.         SUBROUTINE RHS(N,X,Y,F,G)
  2.         DIMENSION Y(10),F(10),G(10,10)
      C THIS SUBROUTINE IS PROGRAMMED BY USER
      C X: INDEPENDENT VARIABLE
      C Y(1),....,Y(N): DEPENDENT VARIABLES IN DIFFERENTIAL EQUATIONS
      C               DY(J)/DX=FJ(X,Y(1),....,Y(N))
      C F(1),....,F(N): RIGHT-HAND-SIDES OF DIFFERENTIAL EQUATIONS, I.E.,
      C               F(J)=FJ(X,Y(1),....,Y(N))
      C G(J,K): PARTIAL DERIVATIVES D FJ(X,Y(1),....,Y(N))/D Y(K)
      C               FOR  J,K=1,2,....,N
      C VALUES OF F(1),....,F(N),G(1,1),....,G(N,N) MUST BE EVALUATED HERE
  3.         COMMON /PAR/ BB,BET,TC
  4.         RETURN
  5.         END
  1.         SUBROUTINE GAUSS(N,A,B,M,NN,MM)
  2.         DIMENSION A(NN,NN),B(NN,MM),IRR(10),X(10)
      C SOLUTION OF SYSTEM OF LINEAR EQUATIONS FOR M RIGHT SIDES
      C N: NUMBER OF EQUATIONS
      C A: MATRIX OF SYSTEM
      C B: RIGHT HAND SIDES. AFTER RETURN RESULTING SOLUTION.
      C M: IF M.EQ.0 AFTER RETURN THEN MATRIX A WAS SINGULAR
  3.         ID=1
  4.         DO 10 I=1,N
  5.           IRR(I)=0
  6.     10 CONTINUE
  7.     20 ID=1
  8.         IS=1
  9.         AMAX=0.0
```

```
10.        DO 60 I=1,N
11.          IF (IRR(I)) 60,30,60
12.    30    DO 50 J=1,N
13.          P=ABS(A(I,J))
14.          IF (P-AMAX) 50,50,40
15.    40    IR=I
16.          IS=J
17.          AMAX=P
18.    50    CONTINUE
19.    60 CONTINUE
20.        IF (AMAX.NE.0.0) GO TO 70
     C     THIS CONDITION MUST BE SPECIFIED MORE EXACTLY WITH RESPECT TO THE
     C     COMPUTER ACTUALLY USED
21.        M=0
22.        GO TO 120
23.    70 IRR(IR)=IS
24.        DO 90 I=1,N
25.          IF (I.EQ.IR.OR.A(I,IS).EQ.0.0) GO TO 90
26.          P=A(I,IS)/A(IR,IS)
27.          DO 80 J=1,N
28.            IF (A(IR,J).NE.0.0) A(I,J)=A(I,J)-P*A(IR,J)
29.    80    CONTINUE
30.          A(I,IS)=0.0
31.          DO 85 J=1,M
32.            B(I,J)=B(I,J)-P*B(IR,J)
33.    85    CONTINUE
34.    90 CONTINUE
35.        ID=ID+1
36.        IF (ID.LE.N) GO TO 20
37.        DO 115 J=1,M
38.          DO 100 I=1,N
39.          IR=IRR(I)
40.          X(IR)=B(I,J)/A(I,IR)
41.   100    CONTINUE
42.          DO 110 I=1,N
43.          B(I,J)=X(I)
44.   110    CONTINUE
45.   115 CONTINUE
46.   120 RETURN
47.        END
 1.        SUBROUTINE MERSON (N,F,BGNX,ENDX,ACCUR,HH,HMAX,HMIN,KERR)
 2.        DIMENSION  F(100),Y(100),S(100),YK(100),YK2(100),YK3(100)
 3.        COMMON /PRNT/ IW
 4.        COMMON /SMALLH/ KK
     C SUBROUTINE FOR INTEGRATION OF SYSTEM OF N FIRST-ORDER EQUATIONS
     C WITH AUTOMATIC CONTROL OF STEP-LENGTH
     C F(1),....,F(N): INITIAL CONDITIONS AND AFTER RETURN RESULTS
     C (BGNX,ENDX): INTERVAL OF INTEGRATION
     C ACCUR: DESIRED ACCURACY FOR SOLUTION OF FIRST KERR EQUATIONS
     C IF KERR=0 THEN CONSTANT STEP IS USED
     C HH,HMAX,HMIN: INITIAL,MAXIMAL AND MINIMAL STEP LENGTH
     C RIGHT-HAND SIDES ARE TO BE PROGRAMMED IN THE FORM OF SUBROUTINE PRSTR
 5.        X=BGNX
 6.        KKK=0
 7.        PP=(ENDX-BGNX)/ABS(ENDX-BGNX)
 8.        HH=ABS(HH)*PP
 9.        IF (ABS(HH).LT.ABS(HMIN)) HH=ABS(HMIN)*PP
10.        IF (ABS(ENDX-BGNX)-ABS(HH)) 10,10,20
11.    10 H=ENDX-BGNX
12.        GO TO 30
13.    20 H=HH
14.    30 IF (ABS(H).GT.ABS(HMAX)) H=ABS(HMAX)*PP
15.        IF (ABS(H).GE.ABS(HMIN)*1.001) GO TO 60
16.        IF (KKK.EQ.1) GO TO 80
17.        IF (KK.EQ.1) GO TO 50
18.        KK=1
19.        WRITE (IW,40) H,HMIN,X
20.    40 FORMAT (10H MERSON H=,E12.3,29H IS TOO SMALL, REPLACED BY H=,E12.3
     C,6H AT X=,F12.5)
21.    50 H=ABS(HMIN)*PP
22.        IF (ABS(X-ENDX).LT.ABS(H)) H=ENDX-X
23.        GO TO 80
24.    60 IF (KK.EQ.0) GO TO 80
25.        KK=0
26.        WRITE (IW,70) H,X
27.    70 FORMAT (10H MERSON H=,E12.3,41H ALREADY SATISFIES DESIRED ACCURACY
     C AT X=,F12.5)
28.    80 DO 90 I=1,N
29.    90 Y(I)=F(I)
30.        CALL PRSTR(N,X,Y,S)
31.        DO 100 I=1,N
```

```
32.            YK(I)=H*S(I)/3.0
33.            Y(I)=F(I)+YK(I)
34.  100 CONTINUE
35.            CALL PRSTR(N,X+H/3.0,Y,S)
36.            YK2(I)=H*S(I)/3.0
37.            DO 110 I=1,N
38.            Y(I)=F(I)+0.5*(YK(I)+YK2(I))
39.  110 CONTINUE
40.            CALL PRSTR(N,X+H/3.0,Y,S)
41.            DO 120 I=1,N
42.            YK2(1)=H*S(I)/3.0
43.            Y(I)=F(I)+0.375*YK(I)+1.125*YK2(I)
44.  120 CONTINUE
45.            CALL PRSTR(N,X+H/2.0,Y,S)
46.            DO 130 I=1,N
47.            YK3(I)=H*S(I)/3.0
48.            Y(I)=F(I)+1.5*YK(I)+6.0*YK3(I)-4.5*YK2(I)
49.  130 CONTINUE
50.            CALL PRSTR(N,X+H,Y,S)
51.            E=0.00001*ACCUR
52.            DO 140 I=1,N
53.            IF (I.GT.KERR) GO TO 140
54.            P=ABS(0.2*(YK(I)+4.0*YK3(I)-4.5*YK2(I)-0.5*S(I)*H/3.0))
55.            IF (P.GT.E) E=P
56.  140 CONTINUE
57.            IF (E-ACCUR) 160,150,150
58.  150 IF (KK.EQ.1) GO TO 160
59.            KKK=0
60.            H=0.8*H*(ACCUR/E)**0.2
61.            GO TO 30
62.  160 DO 170 I=1,N
63.  170 F(I)=F(I)+0.5*(YK3(I)*4.0+YK(I)+H*S(I)/3.0)
64.            X=X+H
65.            H=0.8*H*(ACCUR/E)**0.2
66.            IF (KKK.EQ.1) GO TO 180
67.            HH=H
68.            IF (ABS(X-ENDX).GE.ABS(H)) GO TO 30
69.            H=ENDX-X
70.            IF (ABS(H).GT.ABS(HMAX)) GO TO 30
71.            KKK=1
72.            GO TO 30
73.  180 RETURN
74.            END
```

Bifurcation and Stability Theory

The classical bifurcation and stability theory, as it has been developed over the last 80 years in the works of Poincaré, Liapunov, Andronov, Pontryagin, Hopf, and others, is well documented in a number of mathematical texts [2.33, C.1–C.6]. The techniques based on the notions of linearized and structural stability of solutions have been applied mainly to the description of systems of lower dimensionality. A complete description of behavior of two-dimensional dynamical systems in a phase plane is, for example, available in [C.1, C.2]. We shall assume that the reader is familiar with the description of the bifurcation and stability theory to the extent contained in one of the above-cited texts. Later we shall briefly discuss the basic notions of the modern theoretical background of investigations of bifurcation and stability of non-linear systems. Even though most computational algorithms described earlier in the text were developed without specifically using modern bifurcation theory, a proper interpretation of solution diagrams and further development of computational algorithms, particularly in situations with higher codimension (to be discussed later) often require an application of the concepts discussed in the following.

The center manifold and the normal-form apparatus is usually used in current investigations of bifurcations and the stability of bifurcated solutions. A number of recent texts deal with the above-mentioned subjects [1.3, 1.4, 1.22, 1.82, 2.16, C.7–C.10].

We shall briefly describe the basic ideas involved in applying the theory of invariant manifolds (center-manifold) and normal forms.

Let us consider a system of nonlinear ordinary differential equations [C.11]:

$$\dot{\mathbf{x}} = A\mathbf{x} + \mathbf{f}(\mathbf{x}), \tag{C1}$$

where $\mathbf{x} = [x_1, \ldots, x_n] \in R^n$, A is an $n \times n$ matrix (with constant terms), and every component $f_k(\mathbf{x})$, $k = 1, 2, \ldots, n$ of $\mathbf{f}(\mathbf{x})$ is a power series in the variables $x_1, x_2, \ldots, x_n$. The power series does not contain linear terms. In this case, $\mathbf{f}$ vanishes along with its Jacobian matrix at the origin. The matrix A depends on the parameter $\mathbf{\mu}$ (i.e., $A = A(\mathbf{\mu})$).

Then $\mathbf{x} = \mathbf{0}$ is a fixed point (i.e., stationary solution) for all values of $\mathbf{\mu}$. We shall study the stability of $\mathbf{x} = \mathbf{0}$ and bifurcations to new nontrivial solutions which occur from **0** as **μ** varies. The function **f** might also depend upon **μ**.

The local stability of the solution **x** = **0** is determined by the eigenvalues of A: when parameters are varied, one or more eigenvalues may cross the imaginary axis and the stability may change. When a single parameter (scalar) is varied, two generic cases may occur: either a single real eigenvalue passes through the origin or two complex-conjugate eigenvalues cross the imaginary axis. When more than one parameter is varied, several eigenvalues may cross the imaginary axis at the same time.

C.1 Invariant Manifolds and the Center-Manifold Theorem (Reduction of Dimension)

The solution structure of the linear system

$$\dot{\mathbf{x}} = A\mathbf{x}, \tag{C2}$$

is determined by the eigenvalues and the corresponding eigenspaces of the matrix A, i.e., the solutions with initial conditions on the eigenspace remain in that space. Each of the eigenspaces is invariant under the flow of (C2). In nonlinear systems (C1), which possess a fixed point of (C1) at **0**, invariant sets are retained, but they are no longer linear, but "deformed" and form smooth invariant manifolds composed of solution curves.

To separate those solutions which locally decay and those which grow, we define stable, unstable, and center eigenspaces E^s, E^u, and E^c of the matrix A (cf., Fig. C.1) as those spaces spanned by eigenvectors belonging to eigenvalues of A which have negative, positive, and zero real parts, respectively. The stable-unstable-manifold theorem [1.22, C.9, C.12] and the center-manifold theorem [2.16] show that in the nonlinear case we have invariant stable, unstable, and center manifolds W^s, W^u, and W^c (cf. Fig. C.1(b)) which are tangent to E^s, E^u, and E^c at **x**(0) = **0**.

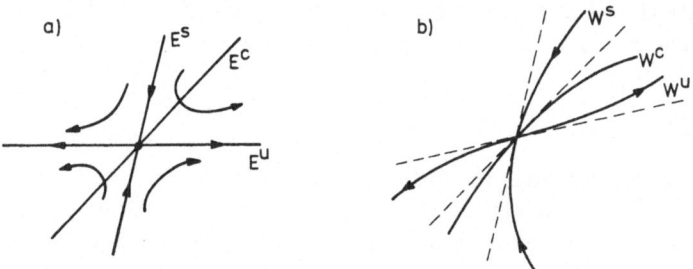

Figure C.1 Eigenspaces and invariant manifolds, (a) linear, (b) nonlinear.

Hence, using the center manifold, we can follow what happens as the parameter μ is varied, i.e., when the eigenvalues which have a zero real part for some critical value μ_0 depart from the imaginary axis. When one or more eigenvalues cross the imaginary axis from left to right, (and, when originally no eigenvalue has been located on the right) a loss of stability occurs.

Let us assume an n-dimensional system (C1). As $\mu = (\mu_1, \mu_2, \ldots, \mu_m)^T$ varies, d eigenvalues cross the imaginary axis simultaneously, all other $(n - d)$ eigenvalues remaining in the left-hand half-plane. The center-manifold theorem implies that there is a μ-dependent local nonlinear change of co-ordinates $\psi : \mathbf{x} \to \mathbf{y}$ or $\mathbf{y} = \psi(\mathbf{x})$ such that in $\mathbf{y}$ coordinates, Eq. (C1) can be studied in the form [C.22]

$$\dot{\mathbf{y}}_c = \mathbf{B}_\mu \mathbf{y}_c + \mathbf{g}_\mu^1(\mathbf{y}_c), \quad \mathbf{y}_c \in R^d \quad (d\text{-dimensional}), \tag{C3a}$$

$$\dot{\mathbf{y}}_s = \mathbf{C}_\mu \mathbf{y}_s + \mathbf{g}_\mu^2(\mathbf{y}_s), \quad \mathbf{y}_s \in R^{n-d} \quad ((n - d)\text{-dimensional}), \tag{C3b}$$

where $\mathbf{y} = (\mathbf{y}_c, \mathbf{y}_s)$, and Eq. (C3a) and Eq. (C3b) are restrictions to center and stable manifolds, respectively.

The linear part of ψ is the linear transformation to Jordan canonical form. The solutions $\mathbf{y}_s(t)$ in the stable manifold decay ($\mathbf{y}_s(t) \to \mathbf{0}$ as $t \to \infty$), since their behavior is dominated by the eigenvalues of $\mathbf{C}_\mu$ which have negative real parts. The bifurcation phenomena may thus be studied by restricting ourselves to the d-dimensional equation (C3a).

The matrix $\mathbf{B}_\mu$ may be obtained from $\mathbf{A}_\mu$ directly by using the Jordan decomposition theorem of linear algebra [2.5]. The nonlinear function $\mathbf{g}_\mu^1$ generally cannot be calculated explicitly, since to do so would require a knowledge of exact solutions of the equation. However, it can be approximated locally in the form of a Taylor series expansion [5.18, C.13]. Hassard has developed computer programs available for automated calculations of the expansions [2.17].

C.2 Normal Forms

Another method which can be used for simplifying the system (C1) study in the neighborhood of $\mathbf{x} = \mathbf{0}$ is based on the theory of normal forms of differential equations. Let us assume a system of nonlinear ordinary differential equations (C1a),

$$\dot{\mathbf{y}} = \mathbf{B}\mathbf{y} + \mathbf{g}(\mathbf{y}). \tag{C1a}$$

Let $\mathbf{B}$ be an $n \times n$ matrix in Jordan canonical form and let every component $g_k(\mathbf{y})$, $k = 1, 2, \ldots, n$ of $\mathbf{g}(\mathbf{y})$ be a power series in the variables $y_1, y_2, \ldots, y_n$.

Results of Brjuno, Takens and Arnold [C.14, C.15, 1.3] show that in certain cases the theory of normal forms may give results similar to those of the center-manifold theorem.

Let us apply a transformation to (C1a):

$$\mathbf{y} = \mathbf{z} + \boldsymbol{\varphi}(\mathbf{z}) \tag{C4}$$

(where $\boldsymbol{\varphi}(\mathbf{z})$ is a formal power series, with $\boldsymbol{\varphi}(0) = \boldsymbol{\varphi}'(0) = 0$, and obtain, from (C1a), the system

$$\dot{\mathbf{z}} = \mathbf{B}\mathbf{z} + \mathbf{h}(\mathbf{z}), \tag{C5}$$

where the linear part $\mathbf{B}\mathbf{z}$ is unchanged.

We are attempting to obtain the system (C5) in the simplest form. The systems (C1a) and (C5) coupled by transformation (C4) are called formally equivalent. If $\mathbf{g}(\mathbf{y})$, $\mathbf{h}(\mathbf{z})$, and $\boldsymbol{\varphi}(\mathbf{z})$ are convergent power series, the systems (C1a) and (C5) are called analytically equivalent.

We shall discuss the questions of formal equivalence. The problems of convergence are treated in [1.3, C.14–C.16].

Formal power series $\mathbf{g}(\mathbf{y})$, $\boldsymbol{\varphi}(\mathbf{z})$ can be written in the following form:

$$\mathbf{g}(\mathbf{y}) = \mathbf{w}_2(\mathbf{y}) + \mathbf{w}_3(\mathbf{y}) + \cdots = \sum_{s=2}^{\infty} \mathbf{w}_s(\mathbf{y}), \tag{C6}$$

$$\boldsymbol{\varphi}(\mathbf{z}) = \boldsymbol{\varphi}_2(\mathbf{z}) + \boldsymbol{\varphi}_3(\mathbf{z}) + \cdots = \sum_{s=2}^{\infty} \boldsymbol{\varphi}_s(\mathbf{z}), \tag{C7}$$

where $\mathbf{w}_s(\mathbf{y})$, $\boldsymbol{\varphi}_s(\mathbf{z})$ are homogeneous vector polynomials of the degree s. For example, the general form of a homogeneous vector polynomial of the second (third) degree is ($d = 2$)

$$\mathbf{w}_2(\mathbf{y}) = \mathbf{w}_2(y_1, y_2) = \begin{bmatrix} a_1 y_1^2 + a_2 y_1 y_2 + a_3 y_2^2 \\ b_1 y_1^2 + b_2 y_1 y_2 + b_3 y_2^2 \end{bmatrix},$$

$$\boldsymbol{\varphi}_3(\mathbf{z}) = \boldsymbol{\varphi}_3(z_1, z_2) = \begin{bmatrix} a_1 z_1^3 + a_2 z_1^2 z_2 + a_3 z_1 z_2^2 + a_4 z_2^3 \\ b_1 z_1^3 + b_2 z_1^2 z_2 + b_3 z_1 z_2^2 + b_4 z_2^3 \end{bmatrix}.$$

Let us denote H^s the linear vector space of all homogeneous vector polynomials of degree s defined on R^n and set $H = \bigcup_{s=0}^{\infty} H^s$. Then H is a direct sum of the spaces H^s and forms an infinite-dimensional linear vector space. Let us introduce new variables into (C1a) by the relation

$$\mathbf{y} = \mathbf{z} + \boldsymbol{\varphi}_s(\mathbf{z}). \tag{C8}$$

Now we shall derive the form of the system (C5). We shall differentiate (C8) with respect to t:

$$\dot{\mathbf{y}} = \dot{\mathbf{z}} + \left[\frac{\partial \boldsymbol{\varphi}_s}{\partial \mathbf{z}} \right] \dot{\mathbf{z}}. \tag{C9}$$

Here $\partial \boldsymbol{\varphi}_s / \partial \mathbf{z} = [\partial(\varphi_s)_i / \partial z_j]$ is the Jacobian matrix.

We shall substitute the relation obtained from (C1a) for $\dot{\mathbf{y}}$ (where $\mathbf{g}(\mathbf{y})$ is written in the form (C6)), and at the same time we substitute the relation for $\mathbf{y}$.

$$B[\mathbf{z} + \varphi_s(\mathbf{z})] + \mathbf{w}_2[\mathbf{z} + \varphi_s(\mathbf{z})] + \cdots + \mathbf{w}_s[\mathbf{z} + \varphi_s(\mathbf{z})] + \cdots$$

$$= \dot{\mathbf{z}} + \left[\frac{\partial \varphi_s}{\partial \mathbf{z}}\right]\dot{\mathbf{z}}. \tag{C10}$$

Since $\mathbf{w}_i[\mathbf{z} + \varphi_s(\mathbf{z})] = \mathbf{w}_i(\mathbf{z}) +$ the terms of higher degrees than s (denoted A), we have

$$B\mathbf{z} + B\varphi_s(\mathbf{z}) + \mathbf{w}_2(\mathbf{z}) + \mathbf{w}_3(\mathbf{z}) + \cdots + \mathbf{w}_s(\mathbf{z}) + A = \dot{\mathbf{z}} + \left[\frac{\partial \varphi_s}{\partial \mathbf{z}}\right]\dot{\mathbf{z}}.$$

If I is a unit matrix of order n we have

$$\left[I + \frac{\partial \varphi_s}{\partial \mathbf{z}}\right]\dot{\mathbf{z}} = B\mathbf{z} + \mathbf{w}_2(\mathbf{z}) + \cdots + \mathbf{w}_s(\mathbf{z}) + B\varphi_s(\mathbf{z}) + \cdots, \tag{C11}$$

i.e.,

$$\dot{\mathbf{z}} = \left[I + \frac{\partial \varphi_s}{\partial \mathbf{z}}\right]^{-1}(B\mathbf{z} + \mathbf{w}_2(\mathbf{z}) + \cdots + \mathbf{w}_s(\mathbf{z}) + B\varphi_s(\mathbf{z}) + \cdots) \tag{C12}$$

as

$$\left[I + \frac{\partial \varphi_s}{\partial \mathbf{z}}\right]^{-1} = I - \left[\frac{\partial \varphi_s}{\partial \mathbf{z}}\right] + \left[\frac{\partial \varphi_s}{\partial \mathbf{z}}\right]^2 - \cdots$$

and from (C12) we obtain

$$\dot{\mathbf{z}} = B\mathbf{z} + \sum_{i=2}^{s-1}\mathbf{w}_i(\mathbf{z}) + \left\{\mathbf{w}_s(\mathbf{z}) - \left[\left[\frac{\partial \varphi_s}{\partial \mathbf{z}}\right]B\mathbf{z} - B\varphi_s(\mathbf{z})\right]\right\} + \cdots. \tag{C13}$$

Let us denote the expression in the brackets $\mathscr{L}$, i.e.,

$$\mathscr{L}\varphi_s(\mathbf{z}) = \left[\frac{\partial \varphi_s}{\partial \mathbf{z}}\right]B\mathbf{z} - B\varphi_s(\mathbf{z}). \tag{C14}$$

The relation (C14) defines the linear mapping $\mathscr{L}: H^s \to H^s$; hence, $\mathscr{L}$ is a linear operator on H^s. The expression (C14) can be interpreted as the Poisson bracket of vector fields $B\mathbf{z}$ and $\varphi_s(\mathbf{z})$ (see [1.3, C.15]), i.e.,

$$\mathscr{L}\varphi_s(\mathbf{z}) = [B\mathbf{z}, \varphi_s(\mathbf{z})] = \left[\frac{\partial \varphi_s}{\partial \mathbf{z}}\right]B\mathbf{z} - B\varphi_s(\mathbf{z}). \tag{C15}$$

The Poisson bracket is defined in the following way: let $\mathbf{a} = (a_1, \ldots, a_n)$, $\mathbf{b} = (b_1, \ldots, b_n)$ be two vector fields on R^n. Then the Poisson bracket $[\mathbf{a}, \mathbf{b}]$ of the vector fields $\mathbf{a}, \mathbf{b}$ also is a vector field on R^n with the ith component

$$[\mathbf{a}, \mathbf{b}]_i = \sum_{j=1}^{n} a_j \frac{\partial b_i}{\partial x_j} - b_j \frac{\partial a_i}{\partial x_j}. \tag{C16}$$

From (C16), it follows that

$$[\mathbf{a}, \mathbf{b}] = \left[\frac{\partial \mathbf{b}}{\partial \mathbf{x}}\right] \cdot \mathbf{a} - \left[\frac{\partial \mathbf{a}}{\partial \mathbf{x}}\right] \mathbf{b}, \tag{C17}$$

where $[\partial \mathbf{a}/\partial \mathbf{x}]$, $[\partial \mathbf{b}/\partial \mathbf{x}]$ denote the Jacobian matrix.

Let us return to relation (C13). The transformation (C8) did not change the linear term and the terms up to the $(s - 1)$th degree. The term of the sth degree takes the form

$$\mathbf{w}_s(\mathbf{z}) - \mathcal{L}\boldsymbol{\varphi}_s(\mathbf{z}).$$

If we can choose $\boldsymbol{\varphi}_s(\mathbf{z})$ in the transformation (C8) so that

$$\mathcal{L}\boldsymbol{\varphi}_s(\mathbf{z}) = \mathbf{w}_s(\mathbf{z}), \tag{C18}$$

then the terms of the sth degree will be removed from (C1a), and (C5) will be simpler. Equation (C18) is called a homologous equation, and its solvability is connected with the possibility of performing the simplification of (C3a). If the homologous equations

$$\mathcal{L}\boldsymbol{\varphi}_i(\mathbf{z}) = \mathbf{w}_i(\mathbf{z}), \quad i = 1, 2, \ldots, k \tag{C19}$$

are solvable (i.e., to a given $\mathbf{w}_i(\mathbf{z})$, we can find a homogeneous vector polynomial $\boldsymbol{\varphi}_i(\mathbf{z})$ of the degree i such that (C19) holds), (C1a) can be transformed into the form (C5), where $\mathbf{h}(\mathbf{z})$ contains the terms of order higher than k only. If (C19) is solvable for all $i = 1, 2, \ldots$, (C5) takes the form

$$\dot{\mathbf{z}} = \mathbf{B}\mathbf{z},$$

i.e., it is fully linearized. If $R^s = \mathcal{L}(H^s)$ denotes the range of values of the linear operator $\mathcal{L}$ (R^s is a linear subspace of H^s), the homologous equation (C19) is solvable if $\mathbf{w}_s(\mathbf{z}) \in R^s$. The following theorem holds [C.11, C.15].

Theorem. *Let $\mathbf{X}$ be a vector field on R^n with $\mathbf{X}(0) = 0$ and $A = [\partial \mathbf{X}/\partial \mathbf{y}]_{y=0} \neq 0$. Let us denote $\mathbf{X}^s$ the part of the Taylor series of the field $\mathbf{X}$ containing the terms up to order $s \geq 2$. Then if*

$$\mathbf{Y}^s - \mathbf{X}^s \in R^s = \mathcal{L}_A(H^s), \tag{C20}$$

a local diffeomorphism Φ exists such that

$$\mathbf{Y}^s = \Phi_* \mathbf{X}^s$$

(i.e., Φ_* transforms the field $\mathbf{X}^s$ into $\mathbf{Y}^s$) and, hence, the equations $\dot{\mathbf{y}} = \mathbf{X}^s(\mathbf{y})$, $\dot{\mathbf{z}} = \mathbf{Y}^s(\mathbf{z})$ are analytically equivalent.

Let us illustrate the consequences of this theorem on an example. Let us have an equation

$$\dot{\mathbf{y}} = \mathbf{B}\mathbf{y} + \mathbf{w}_2(\mathbf{y}) + \cdots + \mathbf{w}_s(\mathbf{y}), \tag{C21}$$

i.e., $\mathbf{X}^s(\mathbf{y}) = \mathbf{B}\mathbf{y} + \sum_{i=2}^{s} \mathbf{w}_i(\mathbf{y})$. If $\mathbf{w}_s(\mathbf{y}) \in R^s$, then the homologous equation has a solution and a transformation of coordinates exists which transforms (C21) into the form

$$\dot{z} = \mathbf{B}\mathbf{z} + \sum_{i=1}^{s-1} \mathbf{w}_i(\mathbf{z}), \tag{C22}$$

i.e., we have removed the term $\mathbf{w}_s(\mathbf{z})$. If, however, $\mathbf{w}_s(\mathbf{y}) \notin R^s$, then $\mathbf{w}_s(\mathbf{y})$ can be decomposed into two parts

$$\mathbf{w}_s(\mathbf{y}) = \mathbf{W}_s^1(\mathbf{y}) + \mathbf{W}_s^2(\mathbf{y}),$$

where $\mathbf{W}_s^1 \in R^s$ and $\mathbf{W}_s^2 \in H^s \backslash R^s$. If the theorem is applied to the field

$$\mathbf{X}^s = \mathbf{B}\mathbf{y} + \sum_{i=1}^{s} \mathbf{w}_i(\mathbf{y}) \quad \text{and} \quad \mathbf{Y}^s = \mathbf{B}\mathbf{y} + \sum_{i=1}^{s-1} \mathbf{w}_i(\mathbf{y}) + \mathbf{W}_s^2(\mathbf{y}),$$

then $\mathbf{Y}^s - \mathbf{X}^s = \mathbf{W}_s^1 \in R^s$, and (C22) is equivalent to

$$\dot{z} = \mathbf{B}\mathbf{z} + \sum_{i=1}^{s-1} \mathbf{w}_i(\mathbf{z}) + \mathbf{W}_s^2(\mathbf{z}). \tag{C23}$$

Hence. all the terms of sth order in (C22) which lie in H^s can be deleted. Thus, for every $s \geq 2$, we can compute H^s in such a way that we can choose a base in H^s, $\mathbf{h}^1, \ldots, \mathbf{h}^N$, $N = \dim H^s$, to determine $\mathscr{L} \mathbf{h}^i$; thus, H^s is determined.

Let us consider a specific example of a two-dimensional system, the linear part of which has purely imaginary eigenvalues ($\pm i\omega$, $\omega > 0$), which might correspond to the restriction of a higher-dimensional problem to the center manifold ($\mathbf{y} \sim \mathbf{y}_c$) [C.11]). The system can be written in the form (Jordan theorem)

$$\begin{pmatrix} \dot{y}_1 \\ \dot{y}_2 \end{pmatrix} = \begin{pmatrix} 0 & -\omega \\ \omega & 0 \end{pmatrix} \begin{pmatrix} y_1 \\ y_2 \end{pmatrix} + \begin{pmatrix} f_1(y_1, y_2) \\ f_2(y_1, y_2) \end{pmatrix} = \mathbf{A}\mathbf{y} + \mathbf{f}(\mathbf{y}), \tag{C24}$$

where f_1, f_2 contain the terms of order greater than or equal to 2. Let us start with $s = 2$. H^2 is a six-dimensional space generated, for example, by vectors

$$\left\{ \begin{pmatrix} y_1^2 \\ 0 \end{pmatrix}, \begin{pmatrix} y_1 y_2 \\ 0 \end{pmatrix}, \begin{pmatrix} y_2^2 \\ 0 \end{pmatrix}, \begin{pmatrix} 0 \\ y_1^2 \end{pmatrix}, \begin{pmatrix} 0 \\ y_1 y_2 \end{pmatrix}, \begin{pmatrix} 0 \\ y_2^2 \end{pmatrix} \right\}$$

which will be denoted $\mathbf{h}^1, \ldots, \mathbf{h}^6$, respectively. We shall determine $\mathscr{L}\mathbf{h}^i(\mathbf{y}) = [\mathbf{A}\mathbf{y}, \mathbf{h}^i(\mathbf{y})]$ (for $i = 1, 2, \ldots, 6$). From (C16) and (C18) it follows that $\mathbf{h}^i(\mathbf{y}) = (h_1^i(\mathbf{y}), h_2^i(\mathbf{y}))$:

$$\mathscr{L}\mathbf{h}^i(\mathbf{y}) = [\mathbf{A}\mathbf{y}, \mathbf{h}^i(\mathbf{y})] = \left[\frac{\partial \mathbf{h}^i}{\partial \mathbf{y}} \right] \mathbf{A}\mathbf{y} - \mathbf{A}\mathbf{h}^i(\mathbf{y})$$

$$= \begin{pmatrix} \dfrac{\partial h_1^i}{\partial y_1} & \dfrac{\partial h_1^i}{\partial y_2} \\ \dfrac{\partial h_2^i}{\partial y_1} & \dfrac{\partial h_2^i}{\partial y_2} \end{pmatrix} \begin{pmatrix} -\omega y_2 \\ \omega y_1 \end{pmatrix} - \begin{pmatrix} 0 & -\omega \\ \omega & 0 \end{pmatrix} \begin{pmatrix} h_1^i(\mathbf{y}) \\ h_2^i(\mathbf{y}) \end{pmatrix},$$

$$i = 1, 2, \ldots, 6.$$

We obtain the following six vectors:

$$\left\{ \begin{pmatrix} 2y_1y_2 \\ y_1^2 \end{pmatrix}, \begin{pmatrix} y_2^2 - y_1^2 \\ y_1y_2 \end{pmatrix}, \begin{pmatrix} 2y_1y_2 \\ -y_2^2 \end{pmatrix}, \begin{pmatrix} y_1^2 \\ -2y_1y_2 \end{pmatrix}, \begin{pmatrix} y_1y_2 \\ y_1^2 - y_2^2 \end{pmatrix}, \begin{pmatrix} y_2^2 \\ 2y_1y_2 \end{pmatrix} \right\}$$

which generate R^2 and it can be shown that these vectors also generate H^2 and, therefore, $R^2 = H^2$. Hence (C24) is equivalent to

$$\begin{pmatrix} \dot{z}_1 \\ \dot{z}_2 \end{pmatrix} = \begin{pmatrix} 0 & -\omega \\ \omega & 0 \end{pmatrix} \begin{pmatrix} z_1 \\ z_2 \end{pmatrix} + \begin{pmatrix} g_1(z_1, z_2) \\ g_2(z_1, z_2) \end{pmatrix}, \tag{C24a}$$

where g_1, g_2 contain terms of third and higher orders because all terms of the second order can be deleted. In a similar way, we may continue for $s = 3$, $4, \ldots$.

The above procedure is, however, very laborious even for small s and is practically unmanageable for higher s (the manipulations have to be performed on a computer). Now we shall present another procedure which is often easier to perform. It can be applied in cases where B in (C1a) does not have multiple eigenvalues. We shall introduce the following definition:

Let the eigenvalues $\lambda_1, \ldots, \lambda_n$ of the matrix B be different. This set of eigenvalues is called resonant of the s-th order if such non-negative integers $q_1, \ldots, q_n$ exist that $\sum_{i=1}^n q_i = s$ and

$$q_1 \lambda_1 + q_2 \lambda_2 + \cdots + q_n \lambda_n = \lambda_i \tag{C25}$$

for some $i \in \{1, 2, \ldots, n\}$. We shall denote eigenvectors of the matrix B as $\mathbf{e}_1, \ldots, \mathbf{e}_n$ and $\mathbf{q} = [q_1, \ldots, q_n]$. Let $\mathbf{y} = [y_1, \ldots, y_n]$ be coordinates in this basis. Then monomials of the form $\mathbf{y}^q \cdot \mathbf{e}_i$, where $\mathbf{y}^q = y_1^{q_1} \cdot y_2^{q_2} \cdot \ldots \cdot y_n^{q_n}$ and $q_1, \ldots, q_n$ satisfy (C25), are called resonance monomials. The following theorem can be formulated.

Theorem. *For the eigenvalues and eigenvectors of the operator $\mathscr{L}$ holds*

$$\mathscr{L} \mathbf{y}^q \mathbf{e}_i = [(q_1 \lambda_1 + \cdots + q_n \lambda_n) - \lambda_i] \cdot \mathbf{y}^q \mathbf{e}_i; \tag{C26}$$

i.e., the eigenvalues of the operator $\mathscr{L}$ are given by relation

$$\Lambda(\mathbf{q}) = \sum_{j=1}^n q_j \lambda_j - \lambda_i. \tag{C27}$$

From (C27) it follows that the operator $\mathscr{L}$ maps all resonance monomials (and only those monomials) into a zero vector. The images of nonresonance monomials generate R^s. Hence, from (C23), we can delete all nonresonance monomials by an appropriate transformation of coordinates and transform the equation into the form

$$\dot{\mathbf{z}} = \mathbf{B}\mathbf{z} + \mathbf{w}(\mathbf{z}), \tag{C28}$$

where $\mathbf{w}(\mathbf{z})$ contains only resonance monomials of the order 2, 3, $\ldots$.

Let us return to the above example. The eigenvalues of the matrix $A = \begin{pmatrix} 0 & -\omega \\ \omega & 0 \end{pmatrix}$ are $\lambda_{1,2} = \pm\omega i$, consequently different. As a consequence of $\lambda_1 + \lambda_2 = 0$,

$$2\lambda_1 + \lambda_2 = \lambda_1$$

or

$$\lambda_1 + 2\lambda_2 = \lambda_2,$$

and generally

$$(k+1)\lambda_1 + k\lambda_2 = \lambda_1 \quad \text{or} \quad k\lambda_1 + (k+1)\lambda_2 = \lambda_2,$$

respectively; hence, the resonance monomials take the form

$$z_1^2 z_2 \mathbf{e}_1, z_1 z_2^2 \mathbf{e}_2, \ldots, z_1^{k+1} z_2^k \mathbf{e}_1, z_1^k z_2^{k+1} \mathbf{e}_2. \tag{C29}$$

The coordinates z_1, z_2 are considered with respect to the base $\mathbf{e}_1$, $\mathbf{e}_2$ of the eigenvectors of the matrix A. Since there are no second-order monomials in (C29), we can delete all terms of second order. So we have obtained the same result as in the first procedure, but now all terms which cannot be deleted from (C23) are determined by (C29). The normal form of Eq. (C24) is then where

$$\begin{pmatrix} \dot{z}_1 \\ \dot{z}_2 \end{pmatrix} = \begin{pmatrix} i\omega & 0 \\ 0 & -i\omega \end{pmatrix} \begin{pmatrix} z_1 \\ z_2 \end{pmatrix} + \begin{pmatrix} a_1 z_1^2 z_2 + a_2 z_1^3 z_2^2 + a_3 z_1^4 z_2^3 + a_4 z_1^5 z_2^4 + \cdots \\ b_1 z_1 z_2^2 + b_2 z_1^2 z_2^3 + b_3 z_1^3 z_2^4 + b_4 z_1^4 z_2^5 + \cdots \end{pmatrix}. \tag{C30}$$

The matrix in (C30) is in Jordan canonical form because we work with the basis formed by the eigenvectors of A. Relation (C30) can be transformed back into the coordinates with respect to the canonical basis. The normal form (C30) or an analogous polar coordinate form is used in most of the bifurcation studies in the case where a pair of complex-conjugate eigenvalues crosses the imaginary axis (Hopf or Poincaré–Andronov bifurcation).

Let us discuss several illustrative examples. We have a system

$$\begin{pmatrix} \dot{x}_1 \\ \dot{x}_2 \end{pmatrix} = A \begin{pmatrix} x_1 \\ x_2 \end{pmatrix} + \begin{pmatrix} f_1(x_1, x_2) \\ f_2(x_1, x_2) \end{pmatrix}, \quad A = \begin{pmatrix} 1 & 0 \\ 1 & 2 \end{pmatrix}. \tag{C31}$$

The eigenvalues and the eigenvectors of the matrix A are $\lambda_1 = 1$, $\lambda_2 = 2$, $\mathbf{e}_1 = \begin{pmatrix} 1 \\ -1 \end{pmatrix}$, $\mathbf{e}_2 = \begin{pmatrix} 0 \\ 1 \end{pmatrix}$, respectively.

If we transform the coordinates

$$y_1 = x_1,$$

$$y_2 = x_1 + x_2, \tag{C32}$$

then (C31) will take the form

$$\begin{pmatrix} \dot{y}_1 \\ \dot{y}_2 \end{pmatrix} = \begin{pmatrix} 1 & 0 \\ 0 & 2 \end{pmatrix} \begin{pmatrix} y_1 \\ y_2 \end{pmatrix} + \begin{pmatrix} g_1(y_1, y_2) \\ g_2(y_1, y_2) \end{pmatrix}, \tag{C33}$$

where

$$g_1(y_1, y_2) = f_1(y_1, y_2 - y_1),$$

$$g_2(y_1, y_2) = f_1(y_1, y_2 - y_1) + f_2(y_1, y_2 - y_1).$$

The matrix A has been transformed into the Jordan form, and the coordinates y_1, y_2 are the coordinates in the basis of eigenvectors e_1, e_2 of the matrix A. For the eigenvalues of the matrix A, the only resonance relation is

$$\lambda_2 = 2\lambda_1 + 0\lambda_2.$$

Therefore, only a single resonance monomial $y_1^2 \cdot y_2^0 \cdot e_2 = \begin{pmatrix} 0 \\ y_1^2 \end{pmatrix}$ exists. The transformation of the coordinates $y = z + \varphi(z)$ transforms (C33) into

$$\begin{pmatrix} \dot{z}_1 \\ \dot{z}_2 \end{pmatrix} = \begin{pmatrix} 1 & 0 \\ 0 & 2 \end{pmatrix} \cdot \begin{pmatrix} z_1 \\ z_2 \end{pmatrix} + \begin{pmatrix} 0 \\ cz_1^2 \end{pmatrix}$$

or

$$\dot{z}_1 = z_1,$$

$$\dot{z}_2 = 2z_2 + cz_1^2. \tag{C34}$$

If the system (C34) is transformed back into the coordinates with respect to the canonical basis, we obtain

$$\dot{x}_1 = x_1,$$

$$\dot{x}_2 = x_1 + 2x_2 + cx_1^2. \tag{C35}$$

Let us consider another system,

$$\begin{pmatrix} \dot{x}_1 \\ \dot{x}_2 \end{pmatrix} = \begin{pmatrix} 0 & \omega \\ \omega & 0 \end{pmatrix} \begin{pmatrix} x_1 \\ x_2 \end{pmatrix} + \begin{pmatrix} f_1(x_1, x_2) \\ f_2(x_1, x_2) \end{pmatrix}. \tag{C36}$$

The eigenvalues are $\lambda_1 = \omega$, $\lambda_2 = -\omega$. The corresponding eigenvectors are $e_1 = \begin{pmatrix} 1 \\ 1 \end{pmatrix}$, $e_2 = \begin{pmatrix} 1 \\ -1 \end{pmatrix}$. The system (C36) can be rewritten in the coordinates with respect to the basis e_1, e_2. We obtain

$$\begin{pmatrix} \dot{y}_1 \\ \dot{y}_2 \end{pmatrix} = \begin{pmatrix} \omega & 0 \\ 0 & -\omega \end{pmatrix} \begin{pmatrix} y_1 \\ y_2 \end{pmatrix} + \begin{pmatrix} g_1(y_1, y_2) \\ g_2(y_1, y_2) \end{pmatrix}. \tag{C37}$$

The eigenvalues satisfy resonance relations

$$(k + 1)\lambda_1 + k\lambda_2 = \lambda_1,$$

$$k\lambda_1 + (k + 1)\lambda_2 = \lambda_2.$$

Resonance monomials then take the form

$$\left. \begin{matrix} y_1^{k+1} y_2^k e_1 \\ y_1^k y_2^{k+1} e_2 \end{matrix} \right\}, \quad k = 1, 2, \ldots,$$

and the system (C37) can be rewritten in the following form by means of the transformation $\mathbf{y} = \mathbf{z} + \boldsymbol{\varphi}(\mathbf{z})$:

$$\dot{z}_1 = \omega z_1 + a_1 z_1^2 z_2 + a_2 z_1^3 z_2^2 + \cdots,$$

$$\dot{z}_2 = -\omega z_2 + b_1 z_1 z_2^2 + b_2 z_1^2 z_2^3 + \cdots. \tag{C38}$$

By means of the transformation $x_1 = z_1 + z_2$ and $x_2 = z_1 - z_2$, (C38) can be rewritten in the coordinates x_1, x_2 with respect to the canonical basis:

$$\dot{x}_1 = \omega x_2 + (x_1^2 - x_2^2)(c_1 x_1 + d_1 x_2) + (x_1^2 - x_2^2)(c_2 x_1 + d_2 x_2) + \cdots,$$

$$\dot{x}_2 = \omega x_1 + (x_1^2 - x_2^2)(\tilde{c}_1 x_1 + \tilde{d}_1 x_2) + (x_1^2 - x_2^2)(\tilde{c}_2 x_1 + \tilde{d}_2 x_2)$$

The transformation into normal form can help to determine the stability of the trivial solution of the system (C1) in cases where the eigenvalues of the matrix A have zero real parts and we cannot evaluate the stability from the linearized terms alone.

Let us consider the system (C24), which can be rewritten in the form (see (C30))

$$\dot{z}_1 = i\omega z_1 + a_1 z_1^2 z_2 + a_2 z_1^3 z_2^2 + \cdots,$$

$$\dot{z}_2 = -i\omega z_2 + b_1 z_1 z_2^2 + b_2 z_1^2 z_2^3 + \cdots, \tag{C39}$$

where the variables z_1, z_2 are related to the basis of eigenvectors of the matrix $\begin{pmatrix} 0 & -\omega \\ \omega & 0 \end{pmatrix}$, which are $\mathbf{e}_1 = \begin{pmatrix} i \\ 1 \end{pmatrix}$, $\mathbf{e}_2 = \begin{pmatrix} -i \\ 1 \end{pmatrix}$, $\mathbf{e}_1$ and $\mathbf{e}_2$ are complex conjugate. This is why the variables (coordinates) z_1, z_2 are complex conjugate as well and can be written as $z_1 = z$ and $z_2 = \bar{z}$. Then the first equation in (C39) can be rewritten in the form

$$\dot{z} = i\omega z + a_1 |z|^2 z + a_2 |z|^4 z + \cdots, \tag{C40}$$

and the second equation can be deleted because it is complex conjugate to the first equation. Let us set $r^2 = |z|^2$. Then $\dot{r}^2 = \dot{z}\bar{z} + z\dot{\bar{z}}$, and from (C40), after substitution, we obtain

$$\dot{r}^2 = 2\,\mathrm{Re}\{a_1\}r^4 + O(r^6), \quad \left(\dot{r}^2 = \frac{d(r^2)}{dt} \right). \tag{C41}$$

If $\mathrm{Re}\{a_1\} > 0$, i.e., $\dot{r}^2 > 0$, r^2 is increasing; r^2 is decreasing for $\mathrm{Re}\{a_1\} < 0$. The phase portrait of Eq. (C40) in the neighborhood of zero is shown in Fig. C.2.

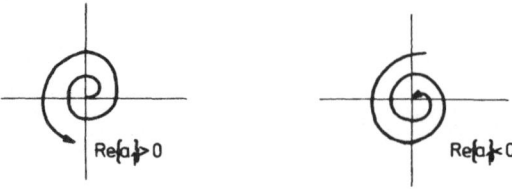

Re$\{a_1\} > 0$ Re$\{a_1\} < 0$

Figure C.2 Phase portrait of (C40).

C.3 Bifurcation of Singular Points of Vector Fields

We shall illustrate the use of normal forms in the current theory of bifurcations. First, we shall discuss several notions and definitions beginning with the concept of structural stability.

Let M be C^r-manifold and $\Gamma(M)$ the set of all C^r vector fields on M. We can provide $\Gamma(M)$ with a topology [C.12]. In this case, $\Gamma(M)$ is a topological space.

A differential equation describing a certain physical process,

$$\dot{\mathbf{x}} = \mathbf{v}(\mathbf{x}), \quad \mathbf{v} \in \Gamma(M), \tag{C42}$$

will be defined on the manifold M. The variables of the physical process are subjected to inherent fluctuations, causing inherent fluctuations of the vector field $\mathbf{v}(\mathbf{x})$. If these fluctuations do not cause qualitative changes in the behavior of physical variables, the process is structurally stable. If the model is adequate, the property of structural stability has to be found in the mathematical model as well, i.e., in the system (C42), in this case. This means that if we change the right-hand side of Eq. (C42) so that we take the vector field $\mathbf{w}(\mathbf{x})$ from a certain small neighborhood of the field $\mathbf{v} \in \Gamma(M)$, then the phase portrait of the equation

$$\dot{\mathbf{x}} = \mathbf{w}(\mathbf{x}) \tag{C43}$$

will be qualitatively equivalent to the phase portrait of (C42).

We shall present more exact definitions of the notions used above.

Definition. Two vector fields $\mathbf{v}(\mathbf{x})$ and $\mathbf{w}(\mathbf{x})$ on M are called *topologically orbitally equivalent* (TOE) if a homeomorphism $h: M \to M$ exists such that it transforms the orbits of the field $\mathbf{w}$ on the orbits of the field $\mathbf{v}$ while preserving the direction of the course of the trajectory (though parametrization need not be preserved).

Phase portraits of TOE fields are homeomorphic.

Definition. The field $\mathbf{v}(\mathbf{x})$ is called *structurally stable* if a neighborhood U of the field $\mathbf{v}$ in $\Gamma(M)$ exists such that all vector fields $\mathbf{w} \in U$ are TOE with $\mathbf{v}$.

From this definition follows that the set of all structurally stable vector fields belonging to $\Gamma(M)$ is open in $\Gamma(M)$. Let us denote this set $\varphi(M)$. Its complement $\Gamma(M) \backslash \varphi(M) = B(M)$ is called a bifurcation set. The elements of the set $B(M)$ are vector fields, which are in a certain sense degenerate. If we have (C42) with the right-hand side $\mathbf{v}(x) \in \varphi(M)$, than a small perturbation of the field $\mathbf{v}$ will not qualitatively change its phase portrait. If the right-hand side of (C42) is an element of $B(M)$, a small perturbation of the right-hand side of (C42) may change the phase portrait qualitatively. This phenomenon is called bifurcation, and it generally denotes the qualitative change of the phase portrait of (C42).

When we study systems (C42) with the right-hand side $v \in B(M)$, we are mainly interested in the question of how many topologically different phase portraits can arise from the original phase portrait through bifurcation. It would be very convenient if the set $\varphi(M)$ were dense in $\Gamma(M)$. In this case, we would obtain a structurally stable field with any small change of the field $w \in B(M)$. However, the results of Smale [C.7] have shown that in the neighborhood of certain vector fields, no structurally stable fields exist. In the following, we shall study only such vector fields v from which we can obtain a structurally stable field by a small perturbation (in other words, the field v is not an inner point of the set $B(M)$).

C.4 Codimension of a Vector Field. Unfolding of a Vector Field

Let $v \in B(M)$, and let a curve $c_1 : [0, 1] \to \Gamma(M)$ exist such that $c_1(\lambda_0) = v$, $\lambda_0 \in (0,1)$, and $c_1(\lambda) \subset \varphi(M)$ for $\forall \lambda \neq \lambda_0$. Let us denote $v_0 = c_1(0)$ and $v_1 = c_1(1)$ (see Fig. C.3). Furthermore, let the curve c_1 cross $B(M)$ transversally, i.e., in a schematic way, the tangent vector to the curve c_1 at the point $v = c_1(\lambda_0)$ does not lie in the tangent plane to $B(M)$ passing through the point v [1.22].

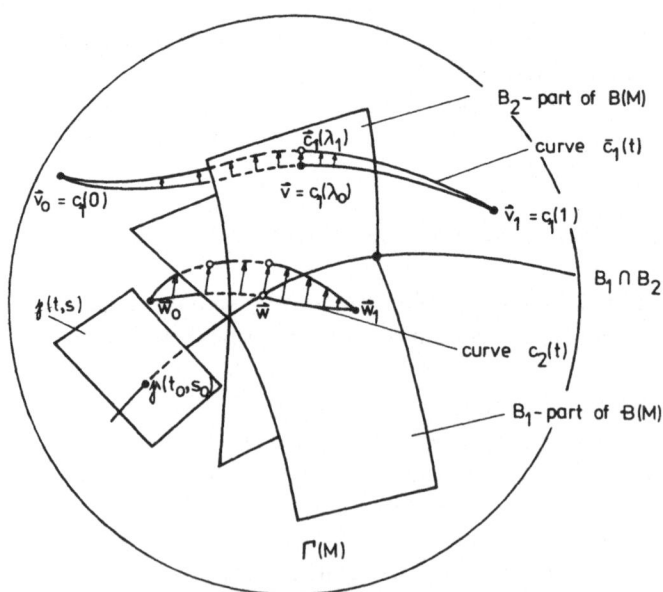

Figure C.3 The space $\Gamma(M)$.

The curve c_1 can be interpreted as a one-parametric system of vector fields. If the curve c_1 is perturbed slightly, we obtain the curve $\bar{c}_1$ which again crosses the surface $B(M)$. Hence, the perturbed one-parametric system $\bar{c}_1$ (for a certain λ_1) again contains a structurally unstable vector field $\bar{c}_1(\lambda_1) \in B(M)$. As can be seen from Fig. C.3, we cannot avoid an intersection with $B(M)$ by a small perturbation of the curve c_1. Such singularities (i.e., vector fields from $B(M)$ which cannot be removed by a small perturbation of a one-parametric system of vector fields) are called singularities of codimension 1.

On the contrary, by a small perturbation of the curve c_2 (see Fig. C.3), the singularity on $B_1 \cap B_2$ can be removed.

Similarly, if we take a two-parametric system of vector fields $\gamma(s, t)$ so that it intersects $B_1 \cap B_2$ transversally, we cannot remove the singularity from $B_1 \cap B_2$ by a small perturbation of γ. Such singular vector fields are of co-dimension 2. The singularities which cannot be removed by a small perturbation of a k-parametric system of vector fields are of codimension k. The co-dimension is a measure of instability of the vector field: the higher the codimension, the higher the instability.

Definition. A k-parametric systems $\mathbf{w}(\mathbf{x}, \boldsymbol{\mu})$, $\boldsymbol{\mu} \in R^k$ of vector fields such that $\mathbf{w}(\mathbf{x}, \mathbf{0}) = \mathbf{v}(\mathbf{x})$ is called a k-parametric unfolding of the vector field $\mathbf{v}(\mathbf{x})$.

The study of stability of a trivial solution $\mathbf{x} = \mathbf{0}$ of (C1) depending on the parameter $\boldsymbol{\mu}$ is a problem which has a local character. Let us discuss the application of the above notions.

Consider a vector field $\mathbf{v}(\mathbf{x})$ which has an isolated singular point $(\mathbf{v}(\mathbf{x}_0) = \mathbf{0})$ at the point $\mathbf{x}_0 \in R^n$. We can choose a neighborhood $U \in R^n$ of the point $\mathbf{x}_0$ which does not contain any other singular point of the field $\mathbf{v}$. Let $\boldsymbol{\varepsilon}_0 \in R^k$ be a point of a k-dimensional space R^k and let V be a certain neighborhood of the point $\boldsymbol{\varepsilon}_0$ in R^k. A smooth map

$$\mathbf{v}: U \times V \to R^n \tag{C44}$$

can be interpreted as a vector field on $U \times V$, tangent to the verticals of the form

$$U \times \{\boldsymbol{\varepsilon}\}, \quad \boldsymbol{\varepsilon} \in V. \tag{C45}$$

Definition. The map (C44) is called *a local system of vector fields* and denoted $(\mathbf{v}; \mathbf{x}_0, \boldsymbol{\varepsilon}_0)$.

The local system $(\mathbf{v}; \mathbf{x}_0, \boldsymbol{\varepsilon}_0)$ is also called a k-parametric unfolding of the vector field $\mathbf{v}(\cdot, \boldsymbol{\varepsilon}_0)$.

Let us mention that $(\mathbf{v}; \mathbf{x}_0, \boldsymbol{\varepsilon}_0)$ is a k-parametric system of vector fields on U, smoothly depending on $\boldsymbol{\varepsilon} \in V$. Individual elements of the system $(\mathbf{v}; \mathbf{x}_0, \boldsymbol{\varepsilon}_0)$ will be denoted $\mathbf{v}(\mathbf{x}, \boldsymbol{\varepsilon}) = \mathbf{v}_\varepsilon(\mathbf{x})$. The local system $(\mathbf{v}; \mathbf{x}_0, \boldsymbol{\varepsilon}_0)$ can also be considered a smooth map of $V \in R^k$ into the set of vector fields on U, $\Gamma(U)$.

Definition. Local systems

$$(\mathbf{v}; \mathbf{x}_0, \boldsymbol{\varepsilon}_0) \tag{C46}$$

and

$$(\mathbf{w}; \mathbf{y}_0, \boldsymbol{\varepsilon}_0) \tag{C47}$$

are called equivalent if a continuous map $h: U_1 \times V \to U_2$, $h(\mathbf{x}, \boldsymbol{\varepsilon}) = \mathbf{y}$, $h(\mathbf{x}_0, \boldsymbol{\varepsilon}_0) = \mathbf{y}_0$ exists such that for every fixed $\boldsymbol{\varepsilon} \in V$, $h(\cdot, \boldsymbol{\varepsilon})$ is a homeomorphism of U_1 on U_2. The homeomorphism transforms the orbits of the equation $\dot{\mathbf{x}} = \mathbf{v}(\mathbf{x}, \boldsymbol{\varepsilon})$ on the orbits of the system $\dot{\mathbf{y}} = \mathbf{w}(\mathbf{y}, \boldsymbol{\varepsilon})$ with preserved orientation.

Definition. The local system

$$(\mathbf{u}; \mathbf{x}_0, \boldsymbol{\mu}_0) \tag{C48}$$

is induced from the system (C46) if a continuous map $\boldsymbol{\psi}: V_1 \to V_2$, $\boldsymbol{\varepsilon} = \boldsymbol{\psi}(\boldsymbol{\mu})$, $\boldsymbol{\varepsilon}_0 = \boldsymbol{\psi}(\boldsymbol{\mu}_0)$ exists and

$$\mathbf{u}(\mathbf{x}, \boldsymbol{\mu}) = \mathbf{v}(\mathbf{x}, \boldsymbol{\psi}(\boldsymbol{\mu})). \tag{C49}$$

Definition. The local system $(\mathbf{v}; \mathbf{x}_0, \boldsymbol{\varepsilon}_0)$ is called a *versal deformation* of the vector field $\mathbf{v}(\mathbf{x}, \boldsymbol{\varepsilon}_0)$ if every local system containing the vector field $\mathbf{v}(\mathbf{x}, \boldsymbol{\varepsilon}_0)$ is equivalent to a local system induced from $(\mathbf{v}; \mathbf{x}_0, \boldsymbol{\varepsilon}_0)$.

When we study the local behavior of vector fields in a neighborhood of a singular point, we can also define a codimension of the vector field at its singular point in the following way: a singular point $\mathbf{x}_0 \in R^n$ of the vector field $\mathbf{v} \in \Gamma(U)$, $U \subset R^n$ has a codimension c if the k-jet of the field $\mathbf{v}$ at the point $\mathbf{x}_0$ is an element of some subvariety of a codimension $n + c$ in the space of all k-jets of the vector fields on U (see Arnold [1.4]).

EXAMPLE A: Let us have a differential equation

$$\dot{x} = x^2, \quad x \in R^1. \tag{C50}$$

Let us denote the right-hand side of (C50) as $v_0(x) = x^2$. This vector field has an isolated singular point at $x_0 = 0$, and its linearization at this point has zero eigenvalue.

The field $v_0(x)$ has a codimension 1 at the point $x_0 = 0$ [1.4]. Let us take the one-parametric unfolding of the field $v_0(x) = x^2$ of the form

$$w(x, \mu) = \mu x + x^2, \tag{C51}$$

where $\mu \in R^1$, $w(x, 0) = v_0(x)$. The phase portraits of the equation

$$\dot{x} = w(x, \mu) = \mu x + x^2 \tag{C52}$$

are shown schematically in Fig. C.4(a)–(c).

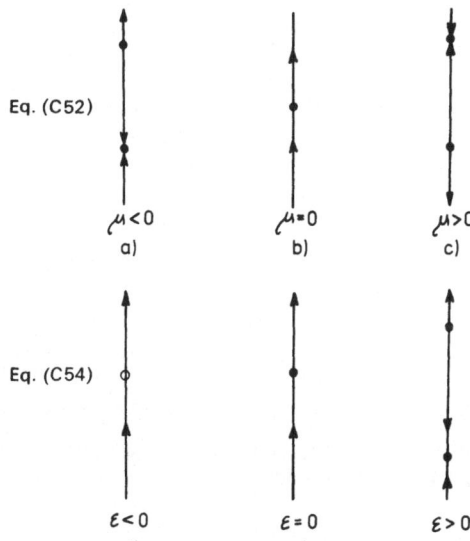

Figure C.4 Phase portraits of (C52) and (C54).

Let us consider an unfolding

$$v(x, \varepsilon) = x^2 - \varepsilon, \quad \varepsilon \in R^1, \tag{C53}$$

$v(x, 0) = v_0(x) = x^2$, instead of the unfolding (C51). The phase portraits of the equation

$$\dot{x} = v(x, \varepsilon) = x^2 - \varepsilon \tag{C54}$$

are shown in Fig. C.4(d)–(f).

When comparing phase portraits in Fig. C.4(a)–(c) with those in Fig. C.4(d)–(f), we observe that the phase portraits C.4(a) and C.4(c) are homeomorphic (topologically equivalent), but that the phase portraits in Fig. C.4(d) and Fig. C.4(f) are not. If we sketch the individual unfoldings $w(x, \mu)$ and $v(x, \varepsilon)$ as curves in the space $\Gamma(U)$ (as in Fig. C.3), we obtain Fig. C.5(a) and (b). Figure C.5(a) corresponds to Figs. C.4(a)–(c), and Fig. C.5(b) to Figs. C.4(d)–(f).

The local system of vector fields (C53) is a versal deformation of the field $v_0(x) = x^2$ [1.4]. The local system (C52) is equivalent to the system induced from (C53), as can be shown in the following way: let $w(y, \mu) = \mu y + y^2$ and $v(x, \varepsilon) = x^2 - \varepsilon$. Let us set $x = h(y, \mu) = y + \mu/2$ and $\varepsilon = \varphi(\mu) = \mu^2/4$ and substitute:

$$v(x, \varepsilon) = v(h(y, \mu), \varphi(\mu))$$

$$= v\left(y + \frac{\mu}{2}, \frac{\mu^2}{4}\right)$$

$$= \left(y + \frac{\mu}{2}\right)^2 - \frac{\mu^2}{4} = \mu y + y^2 = w(y, \mu),$$

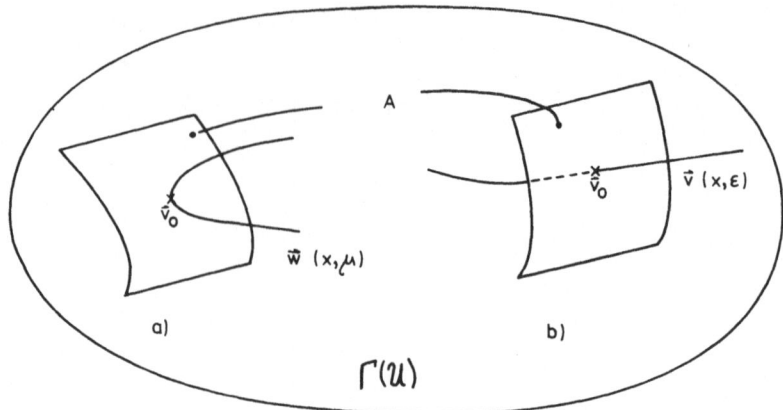

Figure C.5 The space $\Gamma(U)$ with $\mathbf{v}(\mathbf{x}, \varepsilon)$, $\mathbf{w}(\mathbf{x}, \mu)$. A—hypersurface of codimension 1, where the point $\mathbf{v}_0(x) = x^2$ is located.

In the above expressions, we can take any μ but only ε such that $\varepsilon = \mu^2/4 \geq 0$. This means that all terms of the system $w(x, \mu)$ are "mutually" topologically orbitally equivalent (TOE) with only those elements of the system $v(x, \varepsilon)$ which correspond to the nonnegative values of the parameter ε. It follows immediately that the system $w(x, \mu)$ is not a versal deformation of the field $v_0(x) = x^2$.

EXAMPLE B: Let us have an equation

$$\dot{x} = x^3 \quad (x \in R^1). \tag{C55}$$

The vector field $v(x) = x^3$ on the right-hand side of (C55) has an isolated singular point of codimension 2 at the point $x_0 = 0$.

Versal deformation [1.4] of the field $v_0(x)$ is the local system

$$v(x, \varepsilon) = x^3 - \varepsilon_1 x + \varepsilon_2; \quad \varepsilon = (\varepsilon_1, \varepsilon_2) \in R^2. \tag{C56}$$

The set of singular points of the field $v(x, \varepsilon_1, \varepsilon_2)$ depending on the parameters $\varepsilon_1, \varepsilon_2$ is shown in Fig. C.6 (it corresponds to the well-known "cusp catastrophe" surface).

The parametric plane $(\varepsilon_1, \varepsilon_2)$ is divided into two parts by the curve $4\varepsilon_1^3 = 27\varepsilon_2^2$. There are three stationary states for the parameter values belonging to the arrow-like cross-hatched part of the plane (the field $v(x, \varepsilon)$ has three singular points here), and only one stationary solution for the other parts of the plane. For parameter values corresponding to points of the curve (except

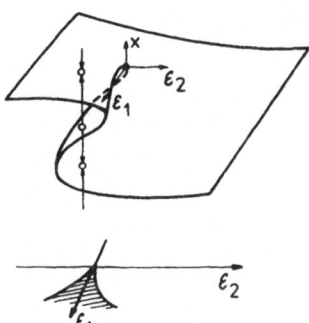

Figure C.6 The set of singular points of (C56).

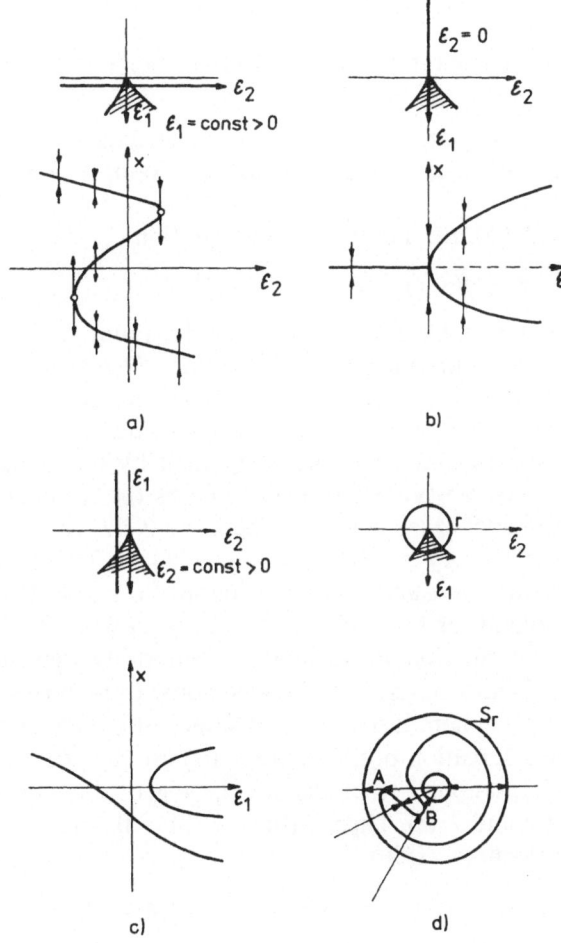

Figure C.7 Dependence of
stationary solutions on a
parameter, (C56).

the tip of the arrow), the equation has two stationary states, one of them having codimension 1. The tip of the arrow corresponds to the stationary point of codimension 2.

The dependence of the stationary states on one of the parameters $(\varepsilon_1, \varepsilon_2)$ (with the other parameter fixed) is obtained as an intersection of the surface from Fig. C.6 with the plane perpendicular to the parametric plane with the line of intersection parallel to one of the coordinate axis ($\varepsilon_1 = 0$ or $\varepsilon_2 = 0$) (see Figs. C.7(a)–(c)).

If the parameters $\varepsilon_1, \varepsilon_2$ are varied so that the point $(\varepsilon_1, \varepsilon_2)$ moves on a circle S_r^1, $\varepsilon_1^2 + \varepsilon_2^2 = r^2$, the dependence of phase portraits of (C56) on $(\varepsilon_1, \varepsilon_2) \in S_r^1$ can be illustrated by Fig. C.7(d), where half-lines emanating from the center of S_r^1 depict the phase space of the variable x corresponding to the point on the circle S_r^1 of the parametric space.

C.5 Construction of a Versal Deformation

Normal forms are used in the construction of versal deformations. We shall discuss the main ideas through an example [1.3].

EXAMPLE: Let us have an equation

$$\dot{x} = v(x, \varepsilon); \quad x \in R^2, \quad \varepsilon \in R^1. \tag{C57}$$

Let $v(x, 0)$ have an isolated singular point $x_0 = 0$. Let the matrix $[\partial v/\partial x]_{\varepsilon=0}^{x=0}$ has two pure imaginary eigenvalues. Hence, for the zero value of the parameters, the eigenvalues $\lambda_{1,2} = \pm i\omega$ are in resonance, and for $\varepsilon \neq 0$, no resonance takes place. Then, for $\varepsilon = 0$, resonance terms will be contained in (C57), which will take the form (obtained by the transformation $x = y + h(y, 0)$) $\dot{y} = Ay + w(y)$, where $w(y)$ contains the resonant terms only. For $\varepsilon \neq 0$, no resonance takes place and therefore (C57) can be transformed into the linear form $\dot{y} = Ay$ by setting $x = y + h(y, \varepsilon)$. This transformation of coordinates depends on ε and is not continuous for $\varepsilon = 0$. Therefore, the terms which are resonant for $\varepsilon = 0$ cannot be removed from (C57) (for $\varepsilon \neq 0$). The resulting transformation of variables is smoothly dependent on ε and transforms (C57) into an equation in which only those terms which become resonant for $\varepsilon = 0$ remain. If the terms of sufficiently high degree (which do not influence the bifurcation of phase portraits) are deleted, we obtain the required versal deformation. Let us denote the eigenvalues of the matrix $[\partial v(x, \varepsilon)/\partial x]$ as $\lambda_1(\varepsilon)$ and $\lambda_2(\varepsilon)$. Then $\lambda_1(0) + \lambda_2(0) = 0$, and (C57) can be transformed into the form (cf., C30)

$$\dot{z}_1 = \lambda_1(\varepsilon)z_1 + a_1(\varepsilon)z_1^2 z_2 + a_2(\varepsilon)z_1^3 z_2^2 + \cdots,$$

$$\dot{z}_2 = \lambda_2(\varepsilon)z_2 + b_1(\varepsilon)z_1 z_2^2 + b_2(\varepsilon)z_1^2 z_2^3 + \cdots, \tag{C58}$$

Figure C.8 Phase portrait of (C62). $\varepsilon > 0$ $\varepsilon < 0$

where $z_2 = \bar{z}_1$, and thus we can consider a single equation instead of (C58), i.e., the equation

$$\dot{z} = \lambda(\varepsilon)z + a_1(\varepsilon)z^2\bar{z} + a_2(\varepsilon)z^3\bar{z}^2 + \cdots. \tag{C59}$$

We shall limit ourselves to the first two terms on the right-hand side and obtain

$$\dot{z} = \lambda(\varepsilon)z + a_1(\varepsilon)z^2\bar{z}. \tag{C60}$$

We shall use the function

$$\rho(z) = |z|^2$$

to study (C59) and (C60). The equations can be rewritten in the forms

$$\dot{\rho} = 2\rho(\text{Re}\{\lambda_1(\varepsilon)\} + \text{Re}\{a_1(\varepsilon)\} \cdot \rho + O(\rho^2)), \tag{C61}$$

$$\dot{\rho} = 2\rho(\text{Re}\{\lambda_1(\varepsilon)\} + \text{Re}\{a_1(\varepsilon)\}\rho). \tag{C62}$$

If $\lambda_1(0) \neq 0$, $d\,\text{Re}\{\lambda_1(\varepsilon)\}/d\varepsilon|_{\varepsilon=0} \neq 0$ and $\text{Re}\{a_1(0)\} \neq 0$, then the term $O(\rho^2)$ will not affect bifurcation of phase portraits in a certain neighborhood of the singular point 0 (see [1.4]).

The family of equations for ρ can be followed easily. If $\text{Re}\{\lambda_1(\varepsilon)\}$ and $\text{Re}\{a_1(\varepsilon)\}$ have different signs, in addition to the singular point $\rho = 0$, another singular point of (C62) $\rho = -\text{Re}\{\lambda_1(\varepsilon)\}/\text{Re}\{a_1(\varepsilon)\}$ exists. When $\text{Re}\{a_1(\varepsilon)\} > 0$, depending on the sign of $\text{Re}\{\lambda_1(\varepsilon)\}$, the vector field has one of the two phase portraits shown in Fig. C.8.

An origin of coordinates corresponds to the point $\rho = 0$ on the plane. A limit cycle corresponds to the point $\rho = -\text{Re}\{\lambda_1(\varepsilon)\}/\text{Re}\{a_1(\varepsilon)\}$.

The behavior of the limit cycle can be followed on the diagram shown in Fig. C.9. Here ε is plotted on one axis and $|z|$ on both sides of the other axis.

First, let us consider the case $\text{Re}\{a_1(0)\} < 0$. When ε crosses zero, the focus at the origin of coordinates loses its stability. At $\text{Re}\{\lambda_1(\varepsilon)\} = 0$, the phase trajectories approach zero nonexponentially. This behavior is schematically

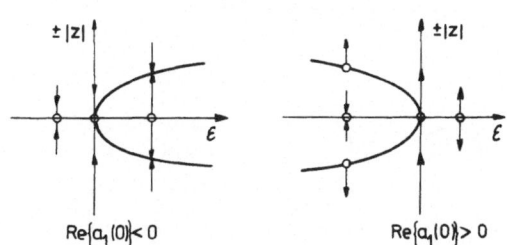

Figure C.9 The behavior of a limit cycle, see (C62).

$\text{Re}\{a_1(0)\} < 0$ $\text{Re}\{a_1(0)\} > 0$

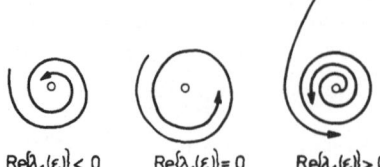

Figure C.10 Creation of a stable limit cycle,
$Re\{a_1(0)\} < 0$.

shown in Fig. C.10. When $Re\{\lambda_1(\varepsilon)\} > 0$, the phase curves approach the stable limit cycle. Hence, the stationary state loses its stability, and a stable periodic solution appears. In physical and engineering problems, this phenomenon is called soft excitation. If $Re\{a_1(0)\} > 0$, the limit cycle exists for $Re\{\lambda_1(\varepsilon)\} < 0$ and is unstable, cf., Fig. C.11. Here, at $Re\{\lambda_1(\varepsilon)\} < 0$, a stable focus and an unstable limit cycle exist; for $Re\{\lambda_1(\varepsilon)\} = 0$ the limit cycle coincides with the focus, which is unstable for $Re\{\lambda_1(\varepsilon)\} > 0$. This case of the loss of stability is called hard excitation.

Another aid for constructing versal deformations can be obtained from the Shoshitaishvili Theorem [1.3] which we shall present here in a restricted form useful for our purposes.

Theorem. *Let*

$$\dot{x} = v(x, \varepsilon)$$

be a k-parametric system of differential equations (determined by a k-parametric local system $(v; x_0, \varepsilon_0)$), where, for the sake of simplicity, $x_0 = 0 \in R^n$, $\varepsilon_0 = 0 \in R^k$, and furthermore, $v(0, 0) = 0$.

For $\varepsilon = 0$ and $x = 0$, let the matrix $[\partial v/\partial x]$ have k^0 eigenvalues with zero real part, k^+ eigenvalues with positive real part, and k^- eigenvalues with negative real part ($k^0 + k^+ + k^- = n$). Then the original system is equivalent (TOE) to the system

$$\dot{y}^0 = w(y^0, \varepsilon), \quad y^0 \in R^{k^0},$$

$$\dot{y}^+ = y^+, \quad y^+ \in R^{k^+},$$

$$\dot{y}^- = -y^-, \quad y^- \in R^{k^-}.$$

If $v(x, \varepsilon)$ is a versal deformation of the field $v(x, 0)$, $w(y^0, \varepsilon)$ is a versal deformation of the field $w(y^0, 0)$.

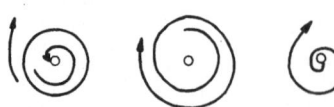

Figure C.11 Creation of an unstable limit cycle,
$Re\{a_1(0)\} > 0$.

This theorem helps us construct the versal deformation of a vector field on n-dimensional phase space because it is necessary only to construct the versal deformation on the corresponding k^0-dimensional phase space.

Thus, for example, when we have constructed the versal deformation of the vector field $v(x) = x^2$; $x \in R^1$ (see the previous example), $v(x, \varepsilon) = x^2 + \varepsilon$, we have in fact constructed the versal deformation for any vector field $\mathbf{w}(\mathbf{y})$, $\mathbf{y} \in R^n$, where $\mathbf{w}(\mathbf{0}) = \mathbf{0}$, and $[(\partial \mathbf{w}/\partial \mathbf{y})_{\mathbf{y}=0}$ has only a unique eigenvalue with zero real part. The versal deformation then takes the form

$$\dot{y} = y^2 + \varepsilon, \quad y \in R^1,$$

$$\dot{\mathbf{y}}^+ = \mathbf{y}^+, \quad \mathbf{y}^+ \in R^{k^+},$$

$$\dot{\mathbf{y}}^- = -\mathbf{y}^-, \quad \mathbf{y}^- \in R^{k^-}, k^+ + k^- = n - 1.$$

C.6 Bifurcations of Codimension 2

The above cases of creation and annihilation of a pair of singular points (stationary states) and creation or disappearance of limit cycles correspond to the only available bifurcations of phase portraits in the neighborhood of a singular point of typical one-parametric families of vector fields (C1).

In two-parameter families, these bifurcations (singularities) occur on the curves in parametric planes. Moreover, we observe more complex singularities at certain points of the parametric plane; Arnold [1.3] has listed five types of such singularities which we encounter in two-parametric families of vector fields. Let us consider two real parameters $\varepsilon_1, \varepsilon_2$. The following five cases can be studied:

1. One zero eigenvalue ($\lambda_1 = 0$) with an additional singularity. For example, for the system

$$\dot{x} = \pm x^3 + \varepsilon_1 x + \varepsilon_2, \quad x \in R^1 \tag{C63}$$

the solution diagram is shown in Fig. C.7a (the case $+ x^3$ is considered).

A half-cubic parabola divides the parametric plane into two parts. In the smaller part, the system has three stationary points in the neighborhood of $x = 0$; in the larger part, only one stationary point (cf., Fig. C.7(a)). The changes of the phase portrait with the parameter varying on a small-diameter circle ($\varepsilon_1^2 + \varepsilon_2^2 = r^2$) are shown in Fig. C.7(d).

2. One pair of pure imaginary complex-conjugate eigenvalues ($\lambda_{1,2} = \pm i\omega$) with an additional singularity. The system

$$\dot{z} = z(i\omega + \varepsilon_1 + \varepsilon_2 z\bar{z} \pm z^2\bar{z}^2), \quad z \in \mathbb{C} \tag{C64}$$

can be used as an example.

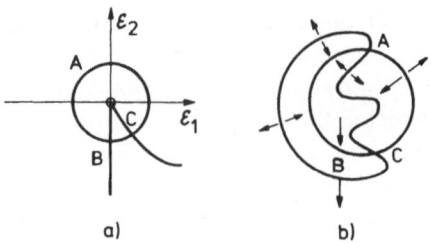

Figure C.12 Bifurcation diagram of the
codimension-2 bifurcation, case 2, see (C64).

a)   b)

A bifurcation diagram shown in Fig. C.12 consists of the straight line $\varepsilon_1 = 0$
and the tangent half of a parabola at zero, as shown in Fig. C.12(a) (here $+ z\bar{z}^2$
in considered). The changes of the phase portrait with the parameter varying
on a circle around zero are shown in Fig. C.12(b). The circle in Fig. C.12(b)
corresponds to the stationary point $z = 0$.

3. Two pure imaginary pairs of complex-conjugate eigenvalues,

$$\lambda_{1,2} = \pm i\omega_1, \lambda_{3,4} = \pm i\omega_2.$$

This case was treated by Iooss [C.17]. Arnold [1.3] states that it is not clear
whether in this case a two-parameter (or a finite-parameter) topologically
versal family exists (particularly in the case where ω_1/ω_2 is not a rational
number).

4. A pair of pure imaginary eigenvalues ($\lambda_{1,2} = \pm i\omega$) and one zero real
eigenvalue ($\lambda_3 = 0$). This case was studied by Guckenheimer [C.18]; Langford
[C.19] studied the unfolding of this case when nonvanishing quadratic terms
were present. Holmes [C.11] considered the symmetric case in which the first
nonvanishing terms were of order 3. The local model of the degenerate system
on the central manifold can be written in cylindrical coordinates as

$$((r^2 = y_1^2 + y_2^2, \theta = \arctan(y_2/y_1)),$$
$$\dot{r} = a_{11}r^3 + a_{12}rz^2 + O(|z|^m \cdot r^n), \quad m + n \geq 5,$$
$$\dot{\theta} = \omega + b_1 z^2 + b_2 r^2 + O(|z|^m \cdot r^n), \quad m + n \geq 4,$$
$$\dot{z} = a_{21}r^2z + a_{22}z^2 + O(|z|^m \cdot r^n), \quad m + n \geq 5. \tag{C65}$$

When the azimuthal component (θ) is decoupled, a two-dimensional system
in (r, z) is obtained. The study of a two-parameter unfolding of this system
has revealed the existence of a homoclinic loop and a continuous family of
periodic orbits, and has revealed the implication that the full problem may
exhibit chaotic solutions and a sensitive dependence on initial conditions near
the bifurcation point.

5. Two real zero eigenvalues, $\lambda_1 = \lambda_2 = 0$. As an example, we can consider
a system on a plane

$$\dot{x}_1 = x_2,$$
$$\dot{x}_2 = \varepsilon_1 + \varepsilon_2 x_1 + x_1^2 \pm x_1 x_2. \tag{C66}$$

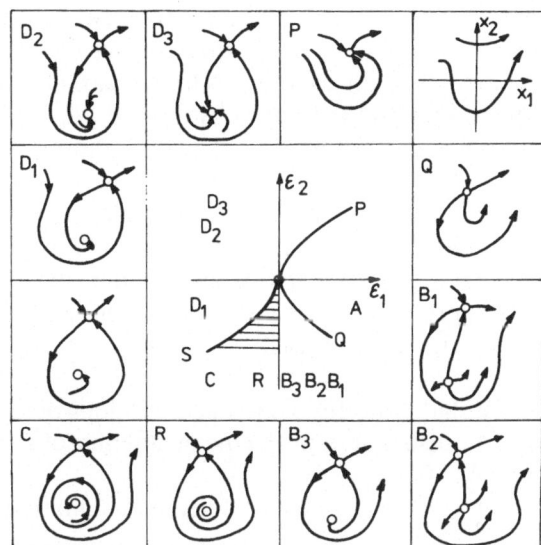

Figure C.13 Bifurcation diagram
of the codimension-2 bifurcation,
case 5, see (C66).

The bifurcation diagram divides the plane $\varepsilon = (\varepsilon_1, \varepsilon_2)$ into four parts, denoted A, B, C, and D in Fig. C.13.

The phase portraits are also given in Fig. C.13. The branches (P, Q, R, S) of the bifurcation diagram correspond to systems of codimension 1. The bifurcation on the branch S—creation of the orbit from the loop of the separatrix—does not belong to the discussed codimension-1 singularities because it is not a local (i.e., in the neighborhood of the singular point) but a global phenomenon. Hence, we can see that with the increasing number of parameters of the studied family of vector fields, we encounter global bifurcations of lower codimensions.

Case 5 was studied by Bogdanov [C.20].

C.7 Bifurcations from Limit Cycles

We shall discuss bifurcations from limit cycles in more detail. Let us consider a nonautonomous system

$$\dot{z} = F(\mu, t, z), \tag{C67}$$

where $z \in R^d$, $\mu \in R^k$ is a vector parameter and t is time. We shall assume that the system (C67) has a periodic solution $z = p_\mu(t)$ with the period $\omega(\mu)$, i.e., $p_\mu(t + \omega(\mu)) = p_\mu(t)$. Then the vector field $F(\mu, t, z)$ has a closed orbit $\gamma_\mu = \{z = p_\mu(t); t \in R\}$ in the phase space R^d.

We shall further discuss (a) the change of stability of a periodic solution when μ varies, (b) the creation or annihilation of a periodic solution when a parameter is varied, (c) the appearance of subharmonic solutions with a period $n\omega$, $n = 2, 3, \ldots$, from the given ω-periodic solution, and (d) the creation of an invariant torus from a closed orbit. A more detailed description can be found in the books by Arnold [1.3], Marsden and McCraken [2.16], and Iooss and Joseph [1.82], where additional references are cited.

Definition. The periodic solution $\mathbf{p}(t)$ of (C67) is orbitally stable (briefly, "stable") if an open neighborhood U of the closed orbit γ (in the phase space R^d) exists such that if $\mathbf{q} \in U$, then the trajectory of (C67) going through $\mathbf{q}$ approaches γ for $t \to \infty$ (i.e., the orbitally stable orbit "attracts" trajectories close to itself).

We shall now discuss the basic notions of the Floquet theory in order to be able to discuss the problems of stability of periodic solutions in more detail later (cf., [2.33]).

Let us have a system of differential equations,

$$\frac{d\mathbf{z}}{dt} = A(t)\mathbf{z}, \tag{C68}$$

$\mathbf{z} \in R^d$, where $A(t)$ is an $d \times d$ matrix in which the elements are continuous functions on R and for which ($\omega > 0$)

$$A(t + \omega) = A(t). \tag{C69}$$

In this case, the Floquet theorem states that the fundamental matrix $\boldsymbol{\Phi}(t)$ of the solution of the system (C69) is in the form

$$\boldsymbol{\Phi}(t) = \boldsymbol{P}(t)e^{t\boldsymbol{R}}, \tag{C70}$$

where the matrix $\boldsymbol{R}$ is constant and the elements of the regular matrix $\boldsymbol{P}(t)$ are continuously differentiable functions with a period ω, i.e.,

$$\boldsymbol{P}(t + \omega) = \boldsymbol{P}(t). \tag{C71}$$

A transformation of variables can be used to obtain $\boldsymbol{P}(0) = \boldsymbol{I}$, and, also, $\boldsymbol{\Phi}(0) = \boldsymbol{I}$. Furthermore, we assume that the fundamental matrix is always normed in this way. The matrix $\boldsymbol{\Phi}(\omega)$ is called a monodromy matrix. When we set $t = \omega$ in (C70) we obtain

$$\boldsymbol{\Phi}(\omega) = e^{\omega\boldsymbol{R}}, \tag{C72a}$$

and evidently,

$$\boldsymbol{R} = \frac{1}{\omega} \ln \boldsymbol{\Phi}(\omega). \tag{C72b}$$

The eigenvalues σ_j of the matrix $\boldsymbol{R}$ are called characteristic exponents (Floquet exponents) of the system (C68). The eigenvalues ρ_j of the monodromy matrix are called (Floquet) multipliers. For every multiplier ρ, a nontrivial solution $\mathbf{z}(t)$ of the system exists which satisfies the condition

$$\mathbf{z}(t + \omega) = \rho \mathbf{z}(t). \tag{C73}$$

Upon using multipliers, we can formulate conditions of stability of the zero solution of the system (C68). The zero solution of the system (C68) is asymptotically stable if $|\rho_j| < 1$ for $j = 1, 2, \ldots, d$, i.e., all multipliers lie inside the unit circle in the Gauss plane.

Let us discuss the bifurcation of $n\omega$-periodic solution of the system (C68) from the ω-periodic solution ($n \in \mathbb{N}$, $n > 1$). From the relations (C70), (C71), and (C72a), it follows that

$$\boldsymbol{\Phi}(t + \omega) = \mathbf{P}(t + \omega)e^{(t+\omega)\mathbf{R}} = \mathbf{P}(t)e^{t\mathbf{R}} \cdot e^{\omega\mathbf{R}}$$

$$= \boldsymbol{\Phi}(t) \cdot \boldsymbol{\Phi}(\omega),$$

i.e.,

$$\boldsymbol{\Phi}(t + \omega) = \boldsymbol{\Phi}(t) \cdot \boldsymbol{\Phi}(\omega). \tag{C74}$$

Then $\boldsymbol{\Phi}(2\omega) = \boldsymbol{\Phi}(\omega + \omega) = \boldsymbol{\Phi}^2(\omega)$, and generally

$$\boldsymbol{\Phi}(n \cdot \omega) = \boldsymbol{\Phi}^n(\omega). \tag{C75}$$

The eigenvalues $\rho(\omega)$ of the monodromy matrix $\boldsymbol{\Phi}(\omega)$ (the multipliers) satisfy the equation

$$\boldsymbol{\Phi}(\omega) \cdot \mathbf{x} = \rho(\omega) \cdot \mathbf{x}. \tag{C76}$$

The eigenvalues $\rho(n\omega)$ of the matrix $\boldsymbol{\Phi}(n\omega)$ satisfy the equation

$$\boldsymbol{\Phi}(n\omega) \cdot \mathbf{x} = \rho(n\omega)\mathbf{x},$$

which can be rewritten in the form

$$\boldsymbol{\Phi}^n(\omega)\mathbf{x} = \rho(n\omega)\mathbf{x} = \rho^n(\omega)\mathbf{x}, \tag{C77}$$

i.e., $\rho(n\omega) = \rho^n(\omega)$. The existence of a number $\sigma \in \mathbb{C}$ follows from (C77) such that

$$\rho(n\omega) = e^{\sigma n\omega}, \quad \rho(\omega) = e^{\sigma\omega}. \tag{C78}$$

The number σ is not uniquely defined. If σ satisfies the relations (C78), then so does $\sigma + 2k\pi i/\omega$, $k \in \mathbb{Z}$.

When the matrix $\mathbf{R}$ in the relation (C70) is chosen properly, σ is the Floquet exponent. Every solution $\mathbf{z}(t)$ of the system (C68) can be expressed by means of the fundamental matrix $\boldsymbol{\Phi}(t)$ in the form

$$\mathbf{z}(t) = \boldsymbol{\Phi}(t) \cdot \mathbf{x}_0, \tag{C79}$$

where $\mathbf{z}(0) = \mathbf{x}_0$ is an initial condition. Then

$$\mathbf{z}(t + \omega) = \boldsymbol{\Phi}(t + \omega)\mathbf{x}_0 = \boldsymbol{\Phi}(t)\boldsymbol{\Phi}(\omega)\mathbf{x}_0. \tag{C80}$$

If x_0 is an eigenvector of the matrix $\boldsymbol{\Phi}(\omega)$, i.e., $\boldsymbol{\Phi}(\omega)x_0 = \rho(\omega)x_0 = e^{\sigma\omega}x_0$, then from (C80), it follows that $z(t + \omega)\cdot = \boldsymbol{\Phi}(t)\cdot e^{\sigma\omega}x_0 = e^{\sigma\omega}\cdot\boldsymbol{\Phi}(t)x_0 = e^{\sigma\omega}\cdot z(t)$, i.e.,

$$z(t + \omega) = e^{\sigma\omega}\cdot z(t). \tag{C81}$$

Let us define

$$\xi(t) = e^{-\sigma t}\cdot z(t). \tag{C82}$$

$\xi(t)$ is ω-periodic since $\xi(t + \omega) = e^{-\sigma(t+\omega)}\cdot z(t + \omega) = e^{-\sigma t}\cdot e^{-\sigma\omega}\cdot e^{\sigma\omega}\cdot z(t) = e^{-\sigma t}z(t) = \xi(t)$ (we have used (C81)). We express $z(t)$ by means of $\xi(t)$:

$$z(t) = e^{\sigma t}\cdot\xi(t). \tag{C83}$$

Differentiating (C82), we obtain

$$\dot{\xi}(t) = -\sigma\cdot\xi(t) + e^{-\sigma t}\dot{z}(t) = -\sigma\xi(t) + e^{-\sigma t}A(t)z(t),$$

i.e.,

$$\sigma\cdot\xi(t) = -\frac{d\xi}{dt} + A(t)\cdot\xi(t). \tag{C84}$$

Let us denote $J\xi = -d\xi/dt + A(t)\cdot\xi$. Relation (C84) can be interpreted as an eigenvalue problem:

$$\sigma\xi = J\xi. \tag{C85}$$

The eigenvalues of the operator J (acting on a space of continuous ω-periodic vector functions) are Floquet exponents of the system (C68).

Let us discuss (C83). If $|\exp\sigma| = \exp(\text{Re}\{\sigma\}) < 1$, i.e., $\text{Re}\{\sigma\} < 0$, then $\lim_{t\to\infty}\exp(\sigma t) = 0$, and thus $\lim_{t\to\infty}z(t) = 0$, i.e., the zero solution of the system (C68) is asymptotically stable (cf., with (C73) and the following discussion).

Now let the matrix $A(t)$ in (C68) depend on the parameter μ continuously, i.e., we have

$$\dot{z} = A_\mu(t)z. \tag{C68a}$$

Then the Floquet exponents σ depend on μ. Let us write $\sigma(\mu) = \xi(\mu) + i\eta(\mu)$. Let us assume that all multipliers $\rho(\mu)$ lie inside the circle $|\rho| < 1$, i.e., $|\rho(\mu)| = \exp(\omega\,\text{Re}\{\sigma(\mu)\}) = \exp(\omega\xi(\mu))$ and $\xi(\mu) < 0$ for $\mu < 0$. For corresponding solutions $z_\mu(t) = \exp(\sigma(\mu)t)\xi(t)$, $\lim_{t\to\infty}z_\mu(t) = 0$.

Let us now discuss the cases where, for $\mu = 0$, one of the multipliers lies on a unit circle and the remaining multipliers lie inside it. There are three possibilities:

1. $\rho(0) = 1 \Rightarrow \sigma(0) = 0 + 2k\pi i$. From (C73) or (C81), it follows that

$$z(t + \omega) = z(t),$$

i.e., an ω-periodic solution exists.

2. $\rho(0) = -1 \Rightarrow \sigma(0) = i\pi + 2k\pi i$. From (C81), we obtain

$$z(t + \omega) = -z(t)$$

and

$$z(t + 2\omega) = z(t);$$

hence, a subharmonic 2ω-periodic solution arises.

3. $\rho(0) = e^{i\omega_0}$, $\qquad \sigma(0) = 0 + i\eta(0) = i\omega_0$. According to (C83), we obtain a solution in the form

$$z(t) = e^{i\omega_0 t}\xi(t), \tag{C86}$$

where $\exp(i\omega_0 t)$ is a periodic function with a period $2\pi/\omega_0$ and $\xi(t)$ has a period ω. Two cases can occur: either the numbers $2\pi/\omega_0$ and ω are rationally dependent, or they are not. In the first case, integers m, n exist such that $m2\pi/\omega_0 = n\omega$, i.e.,

$$\omega_0 = \frac{2\pi m}{n\omega}. \tag{C87}$$

The point ω_0 satisfying (C87) is called the point of resonance. For resonant ω_0, the solution (C86) is $n\omega$-periodic, since

$$z(t + n\omega) = e^{i\omega_0(t + m2\pi/\omega_0)} \cdot \xi(t + n\omega) = z(t).$$

If ω_0 is nonresonant, we obtain quasi-periodic $z(t)$.

As we have seen, the stability of the zero solution of (C68) depends on multipliers ρ or on Floquet exponents σ, and their dependence on the parameter μ decides the type of bifurcation of periodic (or quasi-periodic) solution for a critical value of the parameter.

Now we shall discuss the stability and bifurcation of periodic solutions of the nonlinear system (C67) where $F(\mu, t, z) = F(\mu, t + \omega, z)$, $\mu \in R^1$.

Let us assume that (C67) has a periodic solution $Z_\mu(t)$ with a period ω. Let us denote a closed trajectory in the phase space corresponding to the solution $Z_\mu(t)$ as $\gamma_\mu = \{z \in R^d; z = Z_\mu(t); t \in R^1\}$. The orbital stability of the solution $Z_\mu(t)$ can be evaluated so that we "add" a small perturbation $v_\mu(t)$ to the solution $Z_\mu(t)$ and evaluate the solution $U_\mu(t) = Z_\mu(t) + v_\mu(t)$ (see Fig. C.14).

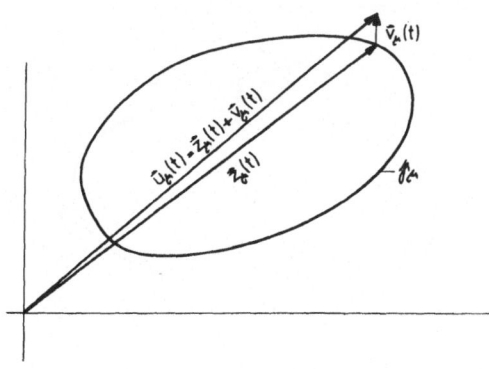

Figure C.14 Perturbation of a periodic solution $Z_\mu(t)$ by a small perturbation $v_\mu(t)$.

Let us substitute this expression into (C67):

$$\frac{dU_\mu(t)}{dt} = F(\mu, t, U_\mu(t))$$

$$\Rightarrow \frac{dZ_\mu(t)}{dt} + \frac{dv_\mu(t)}{dt} = F(\mu, t, Z_\mu(t) + v_\mu(t))$$

$$\Rightarrow \frac{dv_\mu(t)}{dt} = F(\mu, t, Z_\mu(t) + v_\mu(t)) - F(\mu, t, Z_\mu(t)), \tag{C88}$$

where

$$\frac{dZ_\mu(t)}{dt} = F(\mu, t, Z_\mu(t)).$$

If we substitute a linear term of the Taylor expansion of the right-hand side at the point $Z_\mu(t)$ ("with respect to powers $v_\mu(t)$") for the right-hand side of (C88), we obtain the equation in variations

$$\frac{dv_\mu}{dt} = F_z(\mu, t, Z_\mu(t))v_\mu(t). \tag{C89}$$

The investigation of orbital stability of a solution $Z_\mu(t)$ has been carried over into the study of stability (in the Liapunov sense) of the zero solution of the system (C89) (see Fig. C.15). Equation (C89) can be studied similarly to (C68). If (C88) has an ω-periodic solution, $U_\mu(t) = Z_\mu(t) + v_\mu(t)$ will also be ω-periodic (cf., Fig. C.16).

Similarly, the appearance of subharmonic $n\omega$-periodic solutions corresponds to the appearance of the subharmonic $n\omega$-periodic solution $U_\mu(t)$ (see Fig. C.17). In an analogous way, quasi-periodic solutions may arise.

The above procedures for evaluating periodic solutions of (C67) have been used by Iooss and Joseph [1.82]. In greater detail, we shall discuss another

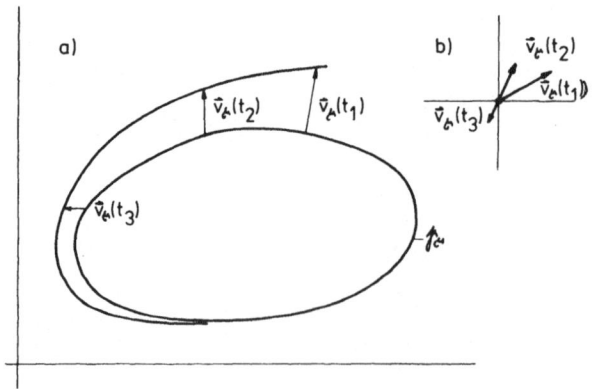

Figure C.15 Investigation of the stability of $Z_\mu(t)$ using variational equations.

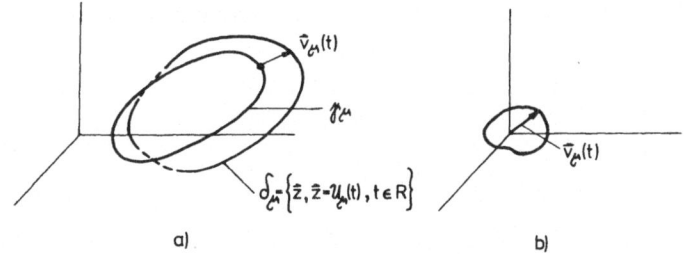

Figure C.16 Investigation of the stability of $Z_\mu(t)$. (a) U_μ—space, (b) v_μ—space.

procedure based on the use of a Poincaré map; a similar procedure has been used in certain numerical techniques. First, we introduce several notions necessary for the definition of a Poincaré map.

Let M be a smooth manifold (for a definition, see, e.g., [1.22]). Let us denote by Diff(M) a group of diffeomorphisms of a manifold M into itself. A group multiplication is the composition of maps in the usual way. Flow on the manifold is a one-parameter group of diffeomorphisms, which forms a sub-group Diff(M) and is determined by a homomorphic map

$$h: R \to \text{Diff}(M),$$

where R is an additive group of real numbers. For $t \in R^1$ let us set $h(t) = g^t$; hence, $g^t: M \to M$ is a diffeomorphism for fixed t and, furthermore,

$$g^{t+s} = g^s \circ g^t, \quad g^0 = \text{identity}, s, t \in R^1.$$

We shall also use the notation $g(t, \mathbf{x})$ instead of $g^t(\mathbf{x})$. The flow g^t determines a vector field $\mathbf{v}(\mathbf{x})$ on M:

$$\mathbf{v}(\mathbf{x}) = \lim_{t \to 0} \frac{g^t(\mathbf{x}) - g^0(\mathbf{x})}{t} = \frac{d}{dt}(g^t(\mathbf{x}))\bigg|_{t=0}. \tag{C90}$$

Conversely, if a vector field $\mathbf{v}(\mathbf{x})$ is given on M, it determines (under certain conditions) a flow g^t which satisfies (C90).

An orbit of a flow is a set of points

$$\gamma = \{\mathbf{m} \in M, \mathbf{m} = g^t(\mathbf{x}); t \in R^1\}, \tag{C91}$$

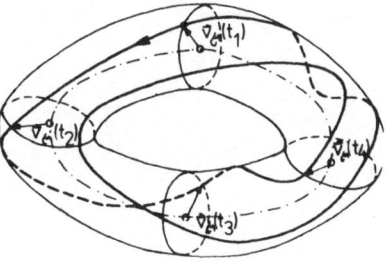

Figure C.17 Subharmonic 2ω periodic solution $U_\mu(t)$.

i.e., $\mathbf{x} \in M$ is fixed and t varies. The orbit is a curve on M and is a trajectory of a differential equation

$$\dot{\mathbf{x}} = \mathbf{v}(\mathbf{x}) \tag{C92}$$

on a manifold M which forms a phase space of this equation. If (C92) has periodic solutions, $g^{t+\omega}(\mathbf{x}) = g^t(\mathbf{x})$, $t \in R^1$ and the orbit (C91) is closed ($\mathbf{x}$ is an initial state of the solution of (C92), $g^0(\mathbf{x}) = \mathbf{x}$).

Otherwise, to solve differential equation (C92) entails finding a flow g^t (which generates the vector field $\mathbf{v}(\mathbf{x})$ by (C90)); the orbits of the flow g^t are trajectories of a given equation (i.e., integral curves of a vector field $\mathbf{v}(\mathbf{x})$).

In Fig. C.18, γ is a closed orbit of the flow g^t. Let us choose $\mathbf{m} \in \gamma$ and let S be a transversal "cross section" of the orbit at the point $\mathbf{m}$. The orbits of the flow g^t then "define" a map of S into itself in the following way: we take a point $\mathbf{x} \in S$ and denote $\gamma^{\mathbf{x}}$ an orbit of a flow g^t passing through $\mathbf{x}$. Let us denote the first intersection of the orbit $\gamma^{\mathbf{x}}$ (starting at $\mathbf{x}$) with S as $P(\mathbf{x})$: $P(\mathbf{x})$ is the value of Poincaré map at the point $\mathbf{x}$. Properties of a Poincaré map (i.e., whether $P(\mathbf{x})$ is contractive or expansive, etc.) determine the behavior of orbits close to γ (i.e., whether γ is an attracting or repelling set, etc). The point $\mathbf{m}$ is a fixed point of the Poincaré map, and there is a mutual one-to-one relationship between fixed points of the Poincaré map and closed orbits of the flow g^t. Let us express the above notions in a more exact way. Let a flow g^t generating a vector field $\mathbf{v}(\mathbf{x})$ (by relation (C90)) be given on a manifold M. Let γ be a closed orbit of the flow g^t; $\gamma = \{\mathbf{z} \in M, \mathbf{z} = g^t(\mathbf{m}); t \in R^1\}$, and let ω be the smallest positive number for which

$$g^{t+\omega}(\mathbf{m}) = g^t(\mathbf{m}), \qquad \text{for all} \quad t \in R^1;$$

then ω is a period of the orbit γ. Let S be a transversal cross section at the point $\mathbf{m} \in \gamma$ to γ, i.e., the submanifold of codimension 1.

A Poincaré map (of the orbit γ) is a map $P: W_0 \to W_1$, where W_0, W_1 are open neighborhoods (in S) of the point $\mathbf{m} \in S$,

$$P(\mathbf{x}) = g^{\tau(\mathbf{x})}(\mathbf{x}) = g(\tau(\mathbf{x}), \mathbf{x}), \tag{C93}$$

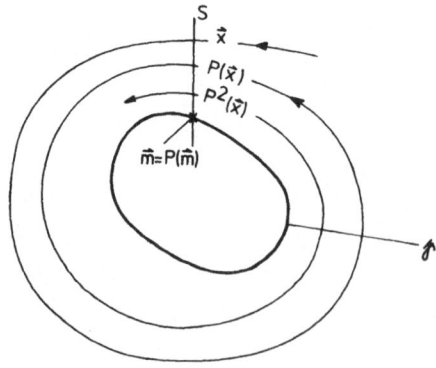

Figure C.18 Construction of a Poincaré map.

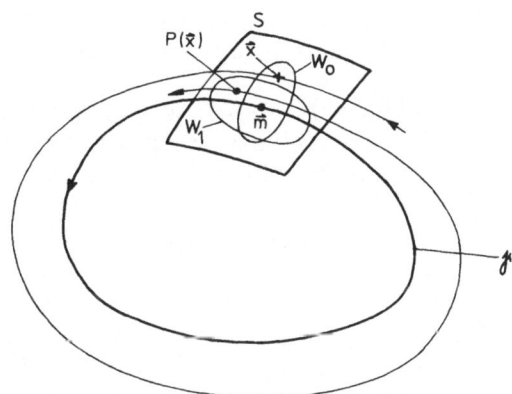

Figure C.19 Construction of a
Poincaré map.

where $\tau: W_0 \to R^1$, $\tau(\mathbf{m}) = \omega$ is such that for $t \in (0, \tau(\mathbf{x}))$, $g^t(\mathbf{x}) \notin S$ (see Fig.
C.19). The condition $\tau(\mathbf{m}) = \omega$ requires that $P(\mathbf{m}) = \mathbf{m}$, since $P(\mathbf{m}) =$
$g^{\tau(\mathbf{m})}(\mathbf{m}) = g^{\omega}(\mathbf{m}) = g^0(\mathbf{m}) = \mathbf{m}$. The presence of the function τ in (C93)
expresses the fact that for $\mathbf{x} \in S$, $\mathbf{x} \neq \mathbf{m}$ the point $g^{\omega}(\mathbf{x})$ need not lie on S but
"in the front of" or "behind" S.

The smoothness of a Poincaré map depends on the smoothness of the flow
g^t. It can be proved that, for a finite-dimensional problem, $P(\mathbf{x})$ is a local
diffeomorphism, and we shall deal solely with this case.

The matrix $DP(\mathbf{m})$, i.e., the linearization of the Poincaré map at the point
$\mathbf{m} \in \gamma$, is called a monodromy operator.

Let us consider system (C67) in a form such that the right-hand side does
not depend explicitly on time (autonomous system):

$$\dot{\mathbf{z}} = \mathbf{F}(\mu, \mathbf{z}) = B_\mu \mathbf{z} + \mathbf{f}_\mu(\mathbf{z}). \tag{C94}$$

Let (C94) have a periodic solution $\mathbf{Z}_\mu(t)$ with a corresponding closed orbit
γ_μ with a Poincaré map $P_\mu(\mathbf{x})$ in a phase space. Then the following theorem
holds:

*Let the eigenvalues $\rho(\mu)$ of the matrix $DP_\mu(\mathbf{m})$ lie inside the unit circle;
then the solution $\mathbf{Z}_\mu(t)$ is asymptotically stable (γ_μ is an attracting closed
orbit) (see [2.16]). The eigenvalues of the matrix $DP(\mathbf{m})$ are multipliers of
the equations in variations:*

$$\dot{\mathbf{v}}_\mu(t) = \mathbf{F}'_\mathbf{z}(\mu, \mathbf{Z}_\mu(t))\mathbf{v}_\mu(t);$$

in addition, another multiplier equal to one exists.

Let $P_\mu(\mathbf{x})$ be a k-parametric system of local diffeomorphisms, $\mathbf{\mu} \in U \subset R^k$,
where U is a certain neighborhood of the point $\mathbf{0} \in R^k$ and $\mathbf{x} \in V \subset R^d$, where
V is a certain neighborhood of the point $\mathbf{0} \in R^d$.

The Shoshitaishvili Theorem (cf. Sect. C.5) modified for diffeomorphisms,
can be used for the study of k-parametric systems.

We shall begin with the simplest types of bifurcations, i.e., the types which occur in one-parametric systems of diffeomorphisms P_μ, $\mu \in R^1$.

EXAMPLE 1: For $\mu < 0$, all multipliers lie inside the unit circle, and for $\mu = 0$, just one multiplier is equal to 1 (others remain inside the unit circle). From the modified Shoshitaishvilli Theorem, it follows that the bifurcation problem is, in principle, one-dimensional, i.e., it suffices to study a one-parameter system of diffeomorphisms in a certain neighborhood of $0 \in R^1$, for example,

$$P_\mu(x) = x + x^2 + \mu. \tag{C95}$$

(a) For $\mu = 0$ we have $P_0(x) = x + x^2$. The diffeomorphism P_0 has one fixed point $x = 0$, which is a solution of the equation

$$P_0(x) = x.$$

In this case, the flow $g_\mu^t|_{\mu=0}$ for which this diffeomorphism is a Poincaré map has one closed orbitally unstable orbit (cf., Fig. C.20). As depicted in Fig. C.20, iterations of a Poincaré map for $x > 0$ move away from γ, and they approach γ for $x < 0$. Hence, the orbits of the corresponding plane flow, passing through the points from a certain left neighborhood of zero, are attracted by γ, while the orbits going through points from the right neighborhood of zero are repelled by γ.

(b) $\mu < 0 \Rightarrow P_\mu(x)$ has two fixed points, which are solutions of the equation $P_\mu(x) = x$,

$$x + x^2 + \mu = x \Rightarrow x_1 = +\sqrt{-\mu}, \quad x_2 = -\sqrt{-\mu}.$$

Hence,

$$P'_\mu(x_1) = 1 + 2\sqrt{-\mu} = \rho_1(\mu), \tag{C96}$$
$$P'_\mu(x_2) = 1 - 2\sqrt{-\mu} = \rho_2(\mu), \tag{C97}$$

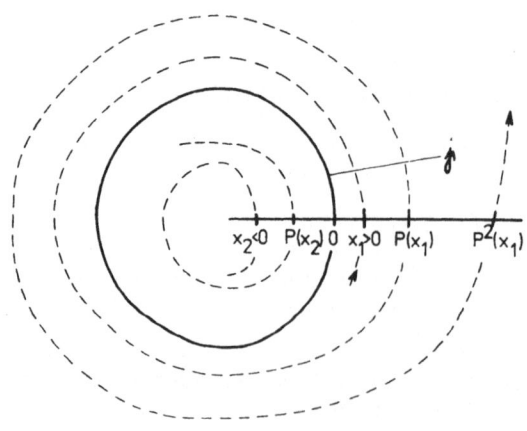

Figure C.20 Poincaré map with unstable orbit.
$P(x_1) = x_1 + x_1^2 > x_1$,
$P^2(x_1) = x_1 + 2x_1^2 + 2x_1^3 + x_1^4 > x_1 + x_1^2$,
$P(x_2) = x_2 + x_2^2 > x_2$.

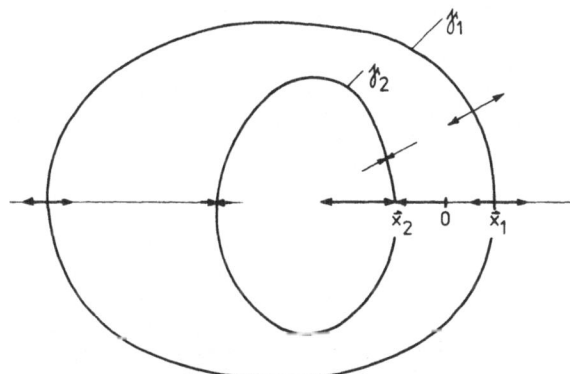

Figure C.21 Stable (γ_2) and
unstable (γ_1) periodic orbit.

where $' = d/dx$. For $\mu < 0$, the flow g^t has two closed orbits, γ_1 passing through x_1 and γ_2 passing through x_2. The orbit γ_1 is unstable and the orbit γ_2 is stable (see Fig. C.21).

(c) $\mu > 0$; $P_\mu(x)$ does not have a fixed point since the equation $P_\mu(x) = x$ does not have a solution; the corresponding flow g^t does not have a closed orbit.

When the parameter crosses the critical value from the "left to right," the periodic solution disappears; when the parameter crosses from the "right to left," two periodic solutions (one stable, the other unstable) arise. When we follow the multipliers of the orbit γ_2, we see that for $\mu \to 0-$, $\rho(\mu) \to 1-$, but does not exceed 1 (it will only "touch it").

EXAMPLE 2: Let us discuss the case where a multiplier is equal to -1 for a certain value of μ. We shall study a one-parametric system of Poincaré maps:

$$P_\mu(x) = (\mu - 1)x + x^3. \tag{C98}$$

From (C98), it follows that $P_\mu(0) = 0$ for $\mu < 0$, $\mu = 0$, and $\mu > 0$, i.e., the fixed point 0 of a Poincaré map P does not change with the varying parameter, but the stability of the fixed point changes.

However, the equation $P_\mu^2(x) = x$ has for $\mu < 0$, $|\mu| \ll 1$, two nonzero solutions, $x_1(\mu)$ and $x_2(\mu)$, in the neighborhood of zero, i.e., the second iteration of the Poincaré map has two new fixed points. This corresponds to the appearance of a closed orbit with a double period (see Fig. C.22).

The systems of diffeomorphisms (C95) and (C98) are versal deformations of corresponding differeomorphism $P_0(x) = x^2 + x$ and $P_0(x) = -x + x^3$, respectively. As for vector fields, a theory of normal forms of maps in the neighborhood of fixed points and a theory of k-parametric systems of diffeomorphisms can be built up, as described in [1.3].

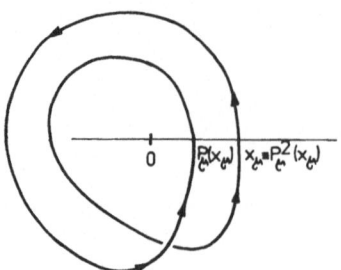

Figure C.22 Poincaré map for a periodic orbit with double period.

EXAMPLE 3: As a last example, we shall discuss the case where a pair of multipliers crosses the unit circle. Let a one-parameter system of diffeomorphisms P_μ in the neighborhood of $0 \in R^n$ be chosen such that $DP_\mu(0)$ has two eigenvalues $\rho(\mu)$, $\bar\rho(\mu)$ with $|\rho(\mu)| < 1$ for $\mu < 0$ and $|\rho(\mu)| > 1$ for $\mu > 0$. Let the other eigenvalues of $DP_\mu(0)$ lie inside the unit circle. Applying the Shoshitaishvili Theorem, we find that the bifurcation problem is two dimensional, hence, $P_\mu : R^2 \to R^2$, $P_\mu(0) = 0$, and below we shall denote $z = x + iy$.

We shall find a normal form of the diffeomorphism $P_\mu(z)$ in the neighborhood of the fixed point 0. First, we introduce several basic notions. Let us have a map

$$\mathbf{f}: \mathbb{C}^n \to \mathbb{C}^n, \qquad \mathbf{f}(\mathbf{x}) = A\mathbf{x} + \cdots \text{(i.e., } \mathbf{f}(0) = 0\text{)}, \tag{C99}$$

$$A = D\mathbf{f}(0).$$

Let $\lambda_1, \ldots, \lambda_n$ be eigenvalues of the matrix A. We say that the eigenvalues are at resonance if

$$\lambda_i = \lambda_1^{m_1} \cdot \lambda_2^{m_2} \cdots \lambda_n^{m_n} \tag{C100}$$

for some $i \in (1, 2, \ldots, n)$, $m_j \geq 0$; $\sum_{j=1}^{n} m_j \geq 2$.

A resonance monomial corresponding to resonance relation (C100) takes the form $x_1^{m_1} \cdot \ldots \cdot x_n^{m_n} \cdot \mathbf{e}_i$, where $\mathbf{e}_i$ is an eigenvector of the matrix A corresponding to λ_i. According to Poincaré–Dulac Theorem, by a (formal) change of variables $\mathbf{x} = \mathbf{z} + \varphi(\mathbf{z})$, we can transform the series in (C99) into the form

$$\mathbf{f}(\mathbf{z}) = A\mathbf{z} + \mathbf{w}(\mathbf{z}), \tag{C101}$$

where $\mathbf{w}(\mathbf{z})$ contains resonance monomials only. Let us apply the Poincaré–Dulac Theorem to Example 3, i.e., the map $P_0(\mathbf{z})$, where now $z \in C^1$. According to the above assumption, $DP_0(0)$ has the eigenvalues, $\rho_1 = e^{i\omega_0}$, $\rho_2 = e^{-i\omega_0}$. The following resonance relations hold:

$$\rho_1 = \rho_1^{k+1} \cdot \rho_2^k, \tag{C102}$$

$$\rho_2 = \rho_1^k \cdot \rho_2^{k+1}, \quad k \in \mathbb{N}, \tag{C103}$$

as $\rho_1 \rho_2 = 1$. A resonance monomial $z^{k+1} \bar{z}^k e_1$ corresponds to (C102) and $z^k \bar{z}^{k+1} e_2$ to (C103). In addition to the resonance relations (C102) and (C103), other relations may arise in the case

$$\omega_0 = \frac{2\pi}{m}, \qquad \text{for some} \quad m \in \mathbb{N}, \qquad m \geq 3 \tag{C104}$$

since $\rho_1^m = 1$ and $\rho_2^m = 1$, then there follow resonance relations

$$\begin{array}{ll} \rho_1 = \rho_1^{m+l+1} \rho_2^l, & \rho_1 = \rho_1^{l+1} \rho_2^{m+l} \\ \rho_2 = \rho_1^{m+l} \rho_2^{l+1}, & \rho_2 = \rho_1^l \rho_2^{m+l+1} \end{array} \qquad \text{for} \quad l = 0, 1, 2, \ldots \tag{C105}$$

and corresponding monomials

$$z^{m+l+1} \bar{z}^l e_1, \quad z^{l+1} \bar{z}^{m+l} e_1, \quad z^{m+l} \bar{z}^{l+1} e_2, \quad z^l \bar{z}^{m+l+1} e_2 \quad l = 0, 1, 2, \ldots \tag{C106}$$

The normal form is

$$\begin{aligned} P_0(z) = \rho_1 z &+ (b_0 z^{m+1} + b_1 z^{m+2} \bar{z} + \cdots) + (c_0 z \bar{z}^m + c_1 z^2 \bar{z}^{m+1} + \cdots) \\ &+ a_1 z^2 \bar{z} + a_2 z^3 \bar{z}^2 + \cdots. \end{aligned} \tag{C107}$$

If

$$\omega_0 \neq \frac{2\pi}{m} \qquad \text{for} \quad m = 1, 2, 3, 4 \tag{C108}$$

(which means that $\rho_1^m \neq 1$ for $m = 1, 2, 3, 4$) we can rewrite (C107) as

$$P_0(z) = \rho z + a_1 z^2 \bar{z} + O(|z|^5),$$

and for $P_\mu(z)$ (any μ), the normal form is

$$P_\mu(z) = \rho(\mu) \cdot z + a_1(\mu) z \cdot |z|^2 + O(|z|^5), \tag{C109}$$

where $\rho(0) = \exp(i\omega_0)$, $a_1(0) \neq 0$.

Let as assume that

$$\left. \frac{d|\rho(\mu)|}{d\mu} \right|_{\mu=0} > 0. \tag{C110}$$

A transformation of coordinates can be used to transform $P_\mu(z)$ into

$$P_\mu(z) = (1 + \mu)z + a_1(\mu) \cdot z|z|^2 + O(|z|^5), \tag{C111}$$

and in polar coordinates, $z = r \exp(i\varphi)$ (see [1.3]), we have

$$P_\mu(r, \varphi) = [(1 + \mu)r - g(\mu)r^3; \varphi + f_1(\mu) + f_2(\mu)r^2)] + O(|r|^5). \tag{C112}$$

If the term $O(|r|^5)$ in (C112) is omitted, we obtain a map

$$\tilde{P}_\mu(r, \varphi) = [(1 + \mu)r - g(\mu)r^3, \varphi + f_1(\mu) + f_2(\mu) \cdot r^2] = [R, \Phi], \tag{C113}$$

where $P_\mu(z) = w = R \exp(i\Phi)$.

We can see that the new radius R depends only on r and not on φ. If

$$g(0) > 0, \tag{C114}$$

the map P has an invariant circle with radius r_0 determined by

$$(1 + \mu)r_0 - g(\mu)r_0^3 = r_0 \Rightarrow r_0 = \frac{\mu}{g(\mu)}. \tag{C115}$$

This means that the flow corresponding to the Poincaré map $\tilde{P}_\mu$ has an invariant torus formed by orbits passing through the points of this invariant circle.

If we consider the omitted term $O(|r|^5)$, we can also show that the invariant circle will exist; however, the structure will be more complicated. In the case of a full map P_μ, the invariant circle is divided by nodes and saddle points of the diffeomorphism P_μ (see Fig. C.23). To describe the behavior of orbits, we have to use the rotation number [1.3] of the restriction P_μ on an invariant circle. If the rotation number is rational, P_μ has a finite number of periodic points on the invariant circle above (nodes and saddle points of the diffeomorphism P_μ). These periodic points of P_μ give rise to closed orbits of a corresponding flow; they describe subharmonic solutions (the orbits "circulate" around the original orbit several times before they close upon themselves). If the rotation number is irrational, the orbits densely fill the given invariant torus (an ergodic case).

Kozjakin [C.21] has shown that in a general one-parametric case, where the limit cycle loses its stability without strong resonance, we observe the creation and annihilation of a large number (infinite number) of oscillations with very large periods in the neighborhood of the critical parameter value.

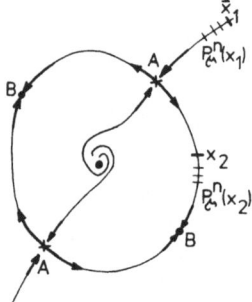

Figure C.23 Saddles (A) and nodes (B) of the diffeomorphism P_μ.

References

Chapter 1

[1.1] R. Thom: *Structural Stability and Morphogenesis*. Benjamin, Reading, MA, 1975.

[1.2] M. Golubitsky, V. Guillemin: *Stable Mappings and Their Singularities*. Springer-Verlag, Berlin Heidelberg New York, 1973.

[1.3] V. I. Arnold: *Additional Chapters of the Theory of Ordinary Differential Equations*. Nauka, Moscow, 1978 (in Russian).
V. I. Arnold, A. N. Varchenko, S. M. Gusein-Zade: Singularities of differentiable mappings, classification of critical points, caustics and wave fronts. Nauka, Moscow, 1982 (in Russian).

[1.4] V. I. Arnold: Lectures on bifurcations in versal families. Russ. Math. Surveys **27** (1972), 54–123.

[1.5] M. T. Thompson, G. W. Hunt: *A General Theory of Elastic Stability*. Wiley, London, 1973.

[1.6] J. J. Stoker, *Nonlinear Elasticity*. Gordon and Breach, New York, 1968.

[1.7] T. von Kármán: Festigkeitsprobleme im Maschinenbau, in *Encyk. der Math. Wissen.*, Vol. IV. Teubner, Leipzig, 1910. pp. 348–352.

[1.8] K. O. Friedrichs, J. J. Stoker: Amer. J. Math. **63** (1941), 839–888.

[1.9] H. B. Keller, J. B. Keller, E. L. Reiss: Quart. Appl. Math. **20** (1962), 55–65.

[1.10] H. S. Berger, P. Fife: Comm. on Pure and Appl. Math. **20** (1967), 687–719.

[1.11] M. S. Berger, P. Fife: Comm. Pure Appl. Math. **21** (1968), 227–241.

[1.12] L. S. Cheo, E. L. Reiss: Quart. Appl. Math. **31** (1973), 75–91.

[1.13] L. S. Cheo, E. L. Reiss: SIAM J. Appl. Math. **26** (1974), 490–495.

[1.14] L. Bauer, E. L. Reiss: J. Soc. Indust. Appl. Math. **13** (1965), 603–626.

[1.15] B. J. Matkowsky, L. J. Putnick: Internat. J. Non-Linear Mech. **9** (1974), 89–103.

[1.16] G. H. Knightly, D. Sather: Arch. Rat. Mech. Anal. **54** (1974), 356–372.

[1.17] M. Ojalvo, F. H. Hull: J. Eng. Mech. Div., ASCE **84**, EM 3 (1958), 1718-1–1718-20.

[1.18] M. Stein: Loads and Deformations of Buckled Rectangular Plates. NASA Technical Report R-40 (1959).

[1.19] G. J. Stroebel, W. H. Warner: J. Elast. **3** (1973), 185–202.

[1.20] L. Bauer, E. L. Reiss: Numerical bifurcation and secondary bifurcation: a case history, in *Numerical Solutions of Partial Differential Equations*, III. Academic, New York, 1976. pp. 443–466.

[1.21] O. R. J. Chillingworth: The catastrophe of a buckling beam, in *Warwick Dynamical Systems*, edited by A. K. Manning. Lecture Notes in Mathematics, No. 468. Springer-Verlag, Berlin Heidelberg New York, 1974.

[1.22] O. R. J. Chillingworth: *Differential Topology with a View to Applications*. Pitman, London, 1976.

[1.23] E. H. Dowell: *Aeroelasticity of Plates and Shells*. Noordhoff, Leyden, 1975.

[1.24] P. Holmes, J. Marsden: Bifurcations to divergence and flutter in flow-induced oscillations: An infinite dimensional analysis. Preprint, University of California, Berkeley, 1978.
P. Holmes, J. Marsden: Domains of stability in a wind-induced oscillation problem. J. Appl. Mech. **46** (1979).

[1.25] R. E. Huigol: Quart. Appl. Math. **36** (1978), 85–94.

[1.26] L. M. Sweet, J. A. Sivak: J. Dyn. Sys. Meas. and Control **101** (1979), 238–246.

[1.27] L. M. Sweet: Experimentally derived reformulation of the wheelset nonlinear hunting problem, in *New Approaches to Nonlinear Problems in Dynamics*, edited by P. J. Holmes. SIAM, Philadelphia, 1980.

[1.28] R. H. Plant: Z. Angew. Math. Phys. **30** (1979), 548–552.

[1.29] E. C. Zeeman: Euler buckling, in *Structural Stability, the Theory of Catastrophes and Applications in the Sciences*. Lecture Notes in Mathematics, No. 525. Springer-Verlag, Berlin Heidelberg New York, 1976. pp. 373–395.

[1.30] M. Golubitsky, D. Schaeffer: A theory for imperfect bifurcation via singularity theory, MRC Technical Summary Report No. 1844. University of Wisconsin, Madison, WI, 1978.

[1.31] O. A. Ladyzhenskaya: *The Mathematical Theory of Viscous Incompressible Flow*, 2nd ed. Gordon and Breach, New York, 1969.

[1.32] J. O. Hinze: *Turbulence*, 2nd ed. McGraw-Hill, New York, 1975.

[1.33] D. D. Joseph: *Stability of Fluid Motions*. Vols. I and II. Springer Tracts in Natural Philosophy, Vols. 27 and 28. Springer-Verlag, Berlin Heidelberg New York, 1976.

[1.34] H. W. Liepman: American Scientist **67** (1979), 221–228.

[1.35] D. H. Constantinescu, L. Michel, L. A. Radicati: Spontaneous symmetry breaking and bifurcations from the Maclaurin and Jacobi sequences. Preprint, 1977.
S. Chandrasekhar: *Hydrodynamic and Hydromagnetic Stability*. Oxford Univ. Press, 1961.
A. Wintner: *The Analytical Foundation of Celestial Mechanics*. Princeton Univ. Press, 1947.

[1.36] G. I. Taylor: Phil. Trans. Roy. Soc. London Ser. A, **223** (1923), 289–343.

[1.37] K. Kirchgässner, H. Kielhöfer: Rocky Mountains J. of Math. **3** (1973), 275–318.

[1.38] A. L. Hodgkin, A. F. Huxley: J. Physiol. **117** (1952), 500–544.

[1.39] A. S. Scott: Rev. Mod. Phys. **47** (1975), 487–533.

[1.40] R. Fitzhugh: Mathematical models of excitation and propagation in nerve, in *Biological Engineering*, edited by H. P. Schwan. McGraw-Hill, New York, 1969. pp. 1–85.

[1.41] J. S. Nagumo, S. Arimoto, S. Yoshizawa: Proc. IRE **50** (1962), 2061–2070.

[1.42] J. Rinzel: *Simple Model Equations for Active Nerve Conduction and Passive Neuronal Integration*. Lectures on Mathematics in the Life Sciences, 8, edited by S. A. Levin. American Math. Soc., Providence, RI, 1976.

[1.43] J. Rinzel, J. B. Keller: Biophys. J. **13** (1973), 1313–1337.

[1.44] K. Maginn: J. Math. Biology **6** (1978), 49–57.

[1.45] B. Kashef, R. Bellman: Math. Biosciences **19** (1974), 1–8.

[1.46] B. Hassard: J. Theor. Biol. **71** (1978), 401–420.

[1.47] R. G. Casten, H. Cohen, P. Lagerstrom: Quart. of Appl. Math. **32** (1975), 365–402.

[1.48] H. R. Wilson, J. D. Cowan: Kybernetik **13** (1973), 55–80.

[1.49] R. Lefever, R. Garay: A mathematical model of the immune surveillance against cancer. in Theoretical Immunology, G. I. Bell, A. S. Pevelson, G. Pimbley, eds., New York, 1978.

[1.50] G. H. Pimbley: Arch. Rat. Mech. Anal. **64** (1977), 169–192.

[1.51] R. M. May: *Stability and Complexity in Model Ecosystems*. Princeton University Press, Princeton, NJ, 1973.

[1.52] J. Maynard Smith: *Models in Ecology*. Cambridge University Press, Cambridge, 1974.

[1.53] M. Tansky: *Structure, Stability and Efficiency of Ecosystem*. Progress in Theoretical Biology, Vol. 4. Academic, New York, 1976.

[1.54] R. M. May, G. F. Oster: The American Naturalist **110** (1976), 573–599.

[1.55] M. Mimura, Y. Nishiura, M. Yamaguti: Some diffusive prey and predator systems and their bifurcation problem. Preprint, Dept. of Mathematics, Kyoto University, 1977.

[1.56] E. E. Selkov: Biofizika **15** (1976), 1065–1076 (in Russian).

[1.57] U. F. Franck, E. Wicke (editors): *Proceedings of the Symposium Kinetics of Physico-chemical Oscillations*, Vols. I–III. Aachen, 1979.

[1.58] Ch. F. Walter: Kinetic and thermodynamic aspects of biological and biochemical control mechanisms, in *Biochemical Mechanisms of Eukaryotic Cells*, edited by E. Kun, S. Grisolia. Wiley, New York, 1972. Chapter 11.

[1.59] B. Hess: Oscillations in homogeneous systems, in *Proceedings of the Symposium Kinetics of Physicochemical Oscillations*, Vol. III, edited by U. F. Franck, E. Wicke. Aachen, 1979. pp. 606–620.

[1.60] A. Boiteux, B. Hess: Spatial dissipative structures in yeast extracts, in *Proceedings of the Symposium Kinetics of Physicochemical Oscillations*, Vol. II, edited by U. F. Franck, E. Wicke. Aachen, 1979. pp. 370–378.

[1.61] L. Glass, M. C. Mackey: Pathological conditions resulting from instabilities in physiological control systems, in *Bifurcation Theory and Applications in Scientific Disciplines*, edited by O. Gurel and O. E. Rössler. Annals of the New York Academy of Science, 1978.

[1.62] M. Kubíček, M. Marek: J. Chem. Phys. **67** (1977), 1997–2006.

[1.63] V. Hlaváček, M. Kubíček, J. Sinkule: J. Theor. Biology **36** (1974), 283–290.

[1.64] M. Kubíček, M. Marek, P. Husták, V. Rýzler: *Bifurcation, Multiplicity and Stability in Reaction-Diffusion Systems* Proceedings of the 5th Symposium Computers in Chemical Engineering, Vysoké Tatry, October 5–9, 1977. pp. 903–939.

[1.65] P. C. Fife: *Mathematical Aspects of Reacting and Diffusing Systems*. Lecture Notes in Biomathematics, Vol. 28, edited by S. Levin. Springer-Verlag, Berlin Heidelberg New York, 1979.

[1.66] H. G. Othmer, L. E. Scriven: I & EC Fundamentals **8** (1969), 303–313.

[1.67] W. H. Ray: Bifurcation phenomena in chemically reacting systems, in *Applications of Bifurcation Theory*. Academic, New York San Francisco London, 1977. pp. 285–315.

[1.68] A. Uppal, W. H. Ray, A. B. Poore: Chem. Eng. Sci. **31** (1976), 205–214.

[1.69] A. B. Poore: Arch. Rat. Mech. Analysis **52** (1973), 358–371.

[1.70] A. B. Poore: Arch. Rat. Mech. Analysis **60** (1976), 371–393.

[1.71] A. Uppal, W. H. Ray, A. B. Poore: Chem. Engng. Sci. **29** (1974), 967–985.

[1.72] M. Golubitsky, B. L. Keyfitz: SIAM J. Math. Analysis **11** (1980), 316–339.

[1.73] R. P. Goldstein, N. R. Amundson: Chem. Eng. Sci. **20** (1965), 195–236; 449–476; 477–500; 501–527.

[1.74] R. A. Schmitz: Multiplicity, stability and sensitivity of states in chemically reacting systems—a review. *Advances in Chemistry Ser.* **148** (1975), 156–211.

[1.75] R. K. Bhat, W. F. Schneider: Ber. Bunsen Ges. Phys. Chem. **80** (1976), 1153–1160.

[1.76] R. Aris: *The Mathematical Theory of Diffusion and Reaction in Permeable Catalysts*, Vols. I and II. Clarendon, Oxford, 1975.

[1.77] M. Kubíček, H. Hofmann, V. Hlaváček: Chem. Engng Sci. **34** (1979), 593–600.

[1.78] N. G. Karanth, R. Hughes: Cat. Rev. Sci. Eng. **9** (1974), 169–208.

[1.79] Y. Villermaux, P. Trambouze (editors): Chemical Reaction Engineering—Nice, Vols. I and II. Proceedings of the International Symposium on Chemical Reaction Engineering. Chem. Eng. Sci. **35**, No. 1, 2 (1980).

[1.80] V. W. Weekman, D. Luss (editors): Chemical Reaction Engineering Reviews—Houston, ACS Symposium Series No. 72 (1978).

[1.81] V. Hlaváček, H. Hofmann, M. Kubíček: Chem. Engng. Sci. **26** (1971), 1629–1634.

[1.82] G. Iooss, D. D. Joseph: Elementary Stability and Bifurcation Theory. Springer-Verlag, New York, 1980.

[1.83] L. Bauer, H. B. Keller, E. L. Reiss: SIAM Rev. **17** (1975), 101–122.

[1.84] D. Henry: Geometric Theory of Semilinear Parabolic Equations. Springer, 1982.

[1.85] V. Hlaváček, M. Kubíček, J. Jelínek: Chem. Engng. Sci. **25** (1970), 1441–1461.

[1.86] D. D. Perlmutter: *Stability of Chemical Reactors*, Prentice-Hall, Englewood Cliffs, NJ, 1972.

[1.87] V. Hlaváček, M. Kubíček, K. Višňák: Chem. Engng. Sci. **27** (1972), 719–742.

[1.88] R. J. Field, E. Körös, R. M. Noyes: J. Amer. Chem. Soc. **94** (1972), 8649–8661.

[1.89] R. M. Noyes: Mechanisms of chemical oscillators in *Synergetics, Far from Equilibrium*, edited by A. Pacault, C. Vidal. Springer-Verlag, Berlin Heidelberg New York, 1979. pp. 34–42.

[1.90] M. Kubíček, H. Hofmann, V. Hlaváček, J. Sinkule: Chem. Engng. Sci. **35** (1980), 987–996.

[1.91] I. Prigogine, R. Lefever: J. Chem. Phys. **48** (1968), 1695–1701.

[1.92] I. Schreiber: Masters Thesis, Prague Institute of Chemical Technology, Prague, 1979.

[1.93] I. Schreiber, M. Kubíček, M. Marek: On coupled cells, in *New Approaches to Nonlinear Problems in Dynamics*, edited by P. J. Holmes. SIAM, Philadelphia, 1980, pp. 496–508.

[1.94] S. P. Graef, J. F. Andrews: J. Water Pol. Contr. Fed. **46** (1974), 666–683.
[1.95] G. Nicolis, I. Prigogine: *Self-Organization in Nonequilibrium Systems.* Wiley, New York, 1977.
[1.96] T. Erneux, M. Herschkowitz-Kaufman: J. Chem. Phys. **66** (1977), 248–257.
[1.97] A. Gierer, M. Meinhardt: Kybernetik **12** (1972), 30–41.
[1.98] G. N. Lance, M. H. Rogers: Proc. Roy. Soc. **A266** (1962), 109.
[1.99] G. L. Mellov, P. J. Chapple, V. K. Stokes: J. Fluid. Mech. **31** (1968), 95–112.
[1.100] C. E. Pearson: J. Fluid Mech. **21** (1965), 623–633.
[1.101] V. Hlaváček, M. Kubíček, M. Marek: J. Catalysis **15** (1969), 17–30.
[1.102] V. Hlaváček, M. Kubíček, M. Marek: J. Catalysis **15** (1969), 31–42.

Chapter 2

[2.1] J. M. Ortega, W. C. Rheinboldt: *Iterative Solution of Nonlinear Equations in Several Variables.* Academic Press, New York London, 1970.
[2.2] W. E. Milne: *Numerical Solution of Differential Equations.* Wiley, New York London, 1953.
[2.3] P. Henrici: *Discrete Variable Methods in Ordinary Differential Equations.* Wiley, New York, 1962.
[2.4] M. Kubíček: ACM Trans. on Math. Software **2** (1976), 98–107.
[2.5] D. K. Faddejev, V. N. Faddejeva: *Numerical Methods of Linear Algebra* (Fizmatgiz, Moscow, 1960), translated by R. Williams. Freeman, San Francisco, 1963.
[2.6] A. Ralston: *A First Course in Numerical Analysis.* McGraw-Hill, New York St. Louis San Francisco Toronto London Sydney, 1965.
[2.7] I. Stakgold: SIAM Rev. **13** (1971), 289–332.
[2.8] P. Glansdorf, I. Prigogine: *Thermodynamic Theory of Structure, Stability and Fluctuations.* Wiley-Interscience, London, New York, Sydney, Toronto, 1971.
[2.9] M. Kubíček: Appl. Math. Comput. **1** (1975), 341–352.
[2.10] R. Seydel: Numer. Math. **32** (1979), 51–68.
[2.11] M. Kubíček, M. Marek: Appl. Math. Comput. **5** (1979), 253–264.
[2.12] M. Kubíček, A. Klíč: Direction of branches bifurcating at a bifurcation point. Determination of starting points for a continuation algorithm. Appl. Math. Comput. **9** (1983).
[2.13] A. Kurosh: *Higher Algebra.* Mir, Moscow, 1972 (in Russian).
[2.14] M. Kubíček, I. Stuchl, M. Marek: Isolas in solution diagrams, J. Comp. Phys. **48** (1982), 106–116.
[2.15] D. Dellwo, H. B. Keller, B. J. Matkowsky, E. L. Reiss: On the Birth of Isolas (to be published).
[2.16] J. E. Marsden, M. McCracken: *The Hopf Bifurcation and its Applications.* Springer-Verlag, Berlin Heidelberg New York, 1976.
[2.17] B. D. Hassard: Numerical evaluation of Hopf bifurcation formulae. Preprint, State University of New York at Buffalo, 1979.
[2.18] M. Kubíček: SIAM J. Appl. Math. **38** (1980), 103–107.
[2.19] M. Holodniok, M. Kubíček: New algorithms for evaluation of complex bifurcation points in ordinary differential equations, a comparative numerical study. Appl. Math. Comput. (to be published).
[2.20] A. D. Jepson: Ph.D. Thesis, Department of Mathematics, California Institute of Technology, 1981.
[2.21] E. Fehlberg: Computing **4** (1969), 93–106.
[2.22] F. T. Krogh: Changing Stepsize in the Integration of Differential Equations Using Modified Divided Differences. Preprint, Jet Propulsion Laboratory, California Institute of Technology, Pasadena, 1973.
[2.23] C. W. Gear: Communs. of the ACM **14** (1971), 185–190.
[2.24] G. Dahlquist: Math. Scand. **4** (1956), 33–53.
[2.25] L. Lapidus, J. H. Seinfeld: *Numerical Solution of Ordinary Differential Equations.* Academic, New York, 1971.
[2.26] G. D. Byrne, A. C. Hindmarsh: ACM Trans. Math. Software **1** (1975), 971–988.

[2.27] M. L. Michelsen: Semiimplicit Runge–Kutta methods for stiff systems. Tech. Rpt. Danmarks Tekniske Hojskole, Instituttet for Kemiteknik, Lyngby 1976. Chem. Engng J. **14** (1977), 107–112.

[2.28] W. Linniger, R. A. Willoughby: SIAM J. Num. Anal. **6** (1969), 47–59.

[2.29] M. Kubíček, K. Višňák: Chem. Engng. Commun. **1** (1974), 291–296.

[2.30] M. Kubíček: Sci. Papers of the Inst. of Chem. Technol. Prague, **K12** (1977), 11–16.

[2.31] K. Višňák, M. Kubíček: J. Inst. Math. Applic. **21** (1978), 251–264.

[2.32] U. A. Mitropolskii, D. J. Martyniuk: *Periodic and Quasiperiodic Oscillations of the System with Time-Delay.* High School Publ. House, Kiev, 1979 (in Russian).

[2.33] E. A. Coddington, N. Levinson: *Theory of Ordinary Differential Equations.* McGraw-Hill, New York, 1955.

[2.34] L. O. Chua, P. M. Lin: *Computer Aided Analysis of Electronic Circuits. Algorithms and Computational Techniques.* Prentice-Hall, Englewood Cliffs, N.J., 1975.

[2.35] M. Holodniok, M. Kubíček: DERPER—An algorithm for continuation of periodic solutions in ordinary differential equations. J. Computational Physics (to be published).

[2.36] M. Holodniok, M. Kubíček: Algorithms for determination of period doubling bifurcation points in ordinary differential equations. Preprints TUM, Institut für Mathematik, Technical University München, 1983.

[2.37] E. N. Lorenz: J. Atm. Sci. **20** (1963), 130–141.

[2.38] D. Ruelle, F. Takens: Commun. Math. Phys. **20** (1971), 167–192.

[2.39] R. M. May: Nature **261** (1976), 459–467.

[2.40] P. Collet, J.-P. Eckmann: *Iterated Maps on the Interval as Dynamical Systems.* Birkhauser, Boston, 1980.

[2.41] M. J. Feigenbaum: J. Stat. Phys. **19** (1978), 25–52. J. Stat. Phys. **21** (1979), 669–705.

[2.42] N. Metropolis, M. L. Stein, P. R. Stein: J. Comb. Theory **A15** (1973), 25–44.

[2.43] P. Manneville, Y. Pomeau: Physica **1D** (1980), 219–226.

[2.44] J. L. Kaplan, J. A. Yorke: Commun. Math. Phys. **67** (1979), 93.

[2.45] M. Henon: Commun. Math. Phys. **50** (1976), 69.

[2.46] C. Simó: J. Stat. Phys. **21** (1979), 465.

[2.47] P. Collet, J.-P. Eckmann, H. Koch: J. Stat. Phys. **25** (1981), 1–14.

[2.48] C. Grebogi, E. Ott, J. A. Yorke: Preprint, Univ. of Maryland, 1982.

[2.49] L. Glas, C. Graves, G. A. Petrillo, M. C. Mackey: J. Theor. Biol. **86** (1980), 455–475.

[2.50] D. G. Aronson, M. A. Chory, G. R. Hall, R. P. McGehee: Commun. Math. Phys. **83** (1982), 303–354.

[2.51] M. J. Feigenbaum, L. P. Kadanoff, S. J. Shenker: Physica **5D**, (1982), 370–386.

[2.52] S. Ostlund, D. Rand, J. Sethna, E. Siggia: Preprint, NSF-ITP-82-80, University of California, Santa Barbara, 1982.

[2.53] V. Franceschini: J. Stat. Phys. **22** (1980), 397–407.

[2.54] V. Franceschini, C. Tebaldi: J. Stat. Phys. **21** (1979), 707.

[2.55] D. A. Russel, E. Ott: Phys. Fluids **24** (1981), 1976.

[2.56] I. Schreiber, M. Marek: Physica **5D** (1982), 258–272.

[2.57] I. Schreiber, M. Marek: Phys. Lett. **91A** (1982), 263–266.

[2.58] J. H. Curry: Commun. Math. Phys. **60** (1978), 193–204.

[2.59] V. Franceschini: Physica D (submitted 1982).

[2.60] A. Arneodo, P. Coullet, C. Tresser: Phys. Lett. **81A** (1981), 197–201.

[2.61] Z. Kurtanjek, M. Sheintuch, D. Luss: in *Kinetics of Physicochemical Oscillations.* Edited by V. Franck, E. Wicke. Aachen, 1979, Vol. 1, pp. 191–199.

[2.62] I. Schreiber, M. Kubíček, M. Marek: in *New Approaches to Nonlinear Problems in Dynamics.* Edited by P. J. Holmes. SIAM, Philadelphia, 1980, p. 496.

[2.63] J. C. Roux, J. S. Turner, W. D. McCormick, H. L. Swinney: in *Nonlinear Problems: Present and Future.* North-Holland, Amsterdam, 1982, pp. 409–422.

[2.64] H. L. Swinney: in Order and Chaos (to appear in Physica D, 1983).

[2.65] V. I. Oseledec: Trans. Moscow. Math. Soc. **19** (1968), 197.

[2.66] R. J. Higgins: Am. J. Phys. **44** (1976), 766.

[2.67] G. Benettin, C. Froeschle, J. P. Scheidecker: Phys. Rev. **A19** (1979), 2454–2460.

[2.68] I. Shimada, T. Nagashima: Prog. Theor. Phys. **61** (1979), 1605–1615.

[2.69] R. Shaw: Z. Naturforsch. **36a** (1981), 80.

[2.70] D. Farmer, J. P. Crutchfield, H. Froehling, N. H. Packard, R. Shaw: Ann. N.Y. Acad.
        Sci. **357** (1980), 453.
[2.71] I. Schreiber, M. Kubiček, M. Marek: Z. Naturforsch. Ç (submitted).
[2.72] M. Holodniok, M. Kubíček, M. Marek: Technical report, Technische Universität
        München, TUM-M8217, 1982.
[2.73] C. T. Sparrow: *The Lorenz Equations: Bifurcations, Chaos and Strange Attractors.*
        Springer-Verlag, New York Heidelberg Berlin, 1982.
[2.74] A. J. Lichtenberg, M. A. Lieberman: *Regular and Stochastic Motion* (Springer-Verlag,
        New York Heidelberg Berlin, 1982).
[2.75] R. Helleman: in *Fundamental Problems in Statistical Mechanics*, Vol. 5, Edited by E. G. D.
        Cohen. North Holland, Amsterdam, 1980, pp. 165–233.
[2.76] I. Schreiber, M. Marek: *Chaotic Behavior in Deterministic Systems.* Academia, Prague,
        1983 (in Czech).

Chapter 3

[3.1]   J. Villadsen, M. L. Michelsen: *Solution of Differential Equation Models by Polynomial
        Approximation.* Prentice-Hall, Englewood Cliffs, NJ, 1978.
[3.2]   B. A. Finlayson: *The Method of Weighted Residuals and Variational Principles.* Academic
        Press, New York, 1972.
[3.3]   D. Gottleib, S. A. Orszag: *Numerical Analysis of Spectral Methods: Theory and Applica-
        tions.* NSF-CBMS Monograph 26, SIAM, Philadelphia, 1978.
[3.4]   L. Fox: *The Numerical Solution of Two-Point Boundary Problems in ODE.* Oxford Uni-
        versity Press, London, 1957.
[3.5]   H. B. Keller: *Numerical Methods for Two-Point Boundary Value Problems.* Blaisdell,
        Waltham, MA, 1968.
[3.6]   S. M. Roberts, J. S. Shipman: *Two-Point Boundary Value Problems: Shooting Methods.*
        American Elsevier, New York, 1972.
[3.7]   M. Kubíček, V. Hlaváček: *Numerical Solution of Nonlinear Boundary Value Problems with
        Applications.* Prentice-Hall, Englewood Cliffs, NJ, 1983.
[3.8]   M. Holodniok, M. Kubíček, V. Hlaváček: J. Fluid Mech. **81** (1977), 689–699.
[3.9]   V. Pereyra: PASVA 3: An adaptive finite difference FORTRAN program for first order
        nonlinear, ordinary boundary problems, in *Codes for Boundary-Value Problems in
        Ordinary Differential Equations.* Lecture Notes in Computer Science. Springer-Verlag,
        Berlin Heidelberg New York, 1979. pp. 67–88.
[3.10] D. D. Morison, J. D. Riley, J. F. Zancanaro: Communs. ACM **5** (1962), 613–614.
[3.11] H. J. Pesch, P. Rentrop: ZAMM **58** (1978), 23–28.
[3.12] D. W. Decker, H. B. Keller: Solution branching, A constructive technique, in *New
        Approaches to Nonlinear Problems in Dynamics*, edited by P. J. Holmes. SIAM, Phila-
        delphia, 1980. pp. 53–69.
[3.13] M. Kubíček, V. Hlaváček: Chem. Engng. Sci. **26** (1971), 705–709.
[3.14] M. Kubíček, M. Holodniok, V. Hlaváček: Computers and Fluids **4** (1976), 59–64.
[3.15] M. Kubíček, V. Hlaváček: Chem. Engng. Sci. **28** (1972), 1049–1052.
[3.16] M. Kubíček, V. Hlaváček: J. Inst. Math. Appl. **12** (1973), 287–293.
[3.17] M. Kubíček, V. Hlaváček: Chem. Engng. Sci. **27** (1972), 743–750, 2095–2098.
[3.18] M. Kubíček, V. Rýzler, M. Marek: Bioph. Chem. **8** (1978), 235–246.
[3.19] M. Holodniok, M. Kubíček, V. Hlaváček: J. Fluid Mech. 108 (1981), 227–240.
[3.20] M. Schlehöferová, M. Holodniok, M. Kubíček: Sci. Papers of the Prague Inst. of Chem.
        Technology (to be published).
[3.21] M. S. Klamkin: SIAM Rev. 4, 43 (1962).
[3.22] M. S. Klamkin: J. Math. Anal. Appl. **32**, 308 (1970).
[3.23] P. Husták, M. Kubíček, I. Marek, M. Marek: Bifurcation in reaction-diffusion systems, in
        Theory of Nonlinear Operators. Akademie Verlag, Berlin, 1978. pp. 117–128.
[3.24] M. Kubíček, V. Hlaváček: Chem. Engng. Sci. **29** (1974), 1645–1649.
[3.25] M. Kubíček, V. Hlaváček: Chem. Engng. Sci. **30** (1975), 1439–1440.
[3.26] M. Kubíček: Communs of the ACM **16** (1973), 760–761.
[3.27] M. Kubíček: Sci. Papers of the Prague Inst. of Chem. Technol. **K12** (1977), 5–9.

[3.28] H. B. Kell , R. K.-H. Szeto: Calculation of flows between rotating disks (preprint).
[3.29] M. Kubíček M. Holodniok: J. Comput. Physics (to be published).
[3.30] V. Hlaváček, M. Kubíček, J. Caha: Chem. Engng. Sci. **27** (1971), 1743–1752.
[3.31] K. F. Jensen, W. H. Ray: Chem. Engng. Sci. **37** (1982), 199–222.
[3.32] J. Peterson, K. A. Overholser, R. F. Heinemann: Chem. Engng. Sci. **36** (1981), 628–638.
[3.33] R. F. Heinemann, A. B. Poore: Chem. Engng. Sci. **36** (1981), 1411–1419.
[3.34] M. Kubíček, M. Holodniok: Chem. Engng. Sci. **38** (1983) in press.
[3.35] G. E. Forsythe, W. R. Wasow: *Finite-Difference Methods for Partial Differential Equations*. Wiley, New York London, 1960.
[3.36] A. R. Mitchel: *Computational Methods in Partial Differential Equations*. Wiley, London, 1969.
[3.37] L. Fox (editor): *Numerical Solution of Ordinary and Partial Differential Equations*. Pergamon, New York Oxford London Paris, 1962.
[3.38] R. D. Richtmyer. *Difference Methods for Initial-Value Problems*. Interscience, New York, 1956.
[3.39] G. Eigenberger, J. B. Butt: Chem. Engng. Sci. **31** (1976), 681–691.
[3.40] V. Hlaváček, M. Kubíček: Chem. Engng. Sci. **25** (1970), 1527–1536.

Chapter 4

[4.1] A. D. Wentzel, M. I. Freidlin: *Fluctuations in Dynamic Systems—Effects of Small Random Perturbations*. Nauka, Moscow, 1979 (in Russian).
[4.2] M. Marek, M. Kubíček: Bull. of Math. Biology **43** (1981), 259–270.
[4.3] P. Raschman: Masters Thesis, Prague Institute of Chemical Technology, Prague, 1979.
[4.4] P. Raschman, M. Kubíček, M. Marek: Waves in distributed chemical systems: experiments and computations, in *New Approaches to Nonlinear Problems in Dynamics*, edited by P. J. Holmes. SIAM, Philadelphia, 1980. pp. 271–288.

Chapter 5

[5.1] H. B. Keller: Numerical solution of bifurcation and nonlinear eigenvalue problems, in *Applications of Bifurcation Theory*, edited by P. H. Rabinowitz. Academic, New York, 1977.
[5.2] H. B. Keller: Constructive methods for bifurcation and nonlinear eigenvalue problems, in Proceedings of the 3rd International Symposium on Computing Method in Applied Science and Engineering, Versailles, 1977.
[5.3] D. W. Decker: Some topics in bifurcation theory, PhD. Thesis. California Institute of Technology, Pasadena, 1978.
[5.4] R. Szeto: The flow between rotating coaxial disks, PhD. Thesis. California Institute of Technology, Pasadena, 1978.
[5.5] W. C. Rheinboldt: SIAM J. Numer. Anal. **17** (1980), 221–237.
[5.6] C. Den Heijer, W. C. Rheinboldt: SIAM J. Numer. Anal. **18** (1981), 925–947.
[5.7] W. C. Rheinboldt, J. V. Burkardt: A Program for a Locally-Parametrized Continuation Process. Tech. Rept. ICMA-81-30, Department of Mathematic and Statistics, University of Pittsburgh, PA 1981.
[5.8] R. Menzel, H. Schwetlick: Computing **16** (1976), 187–199.
[5.9] H. Schwetlick: Effective methods for computing turning points of curves implicitly defined by nonlinear equations. Preprint, Sekt. Math., Univ. Halle, German Democratic Republic, 1981.
[5.10] E. Riks: J. Appl. Mech., **39** (1972), 1060–65.
[5.11] R. A. Brown, L. E. Scriven, W. J. Silliman: Computer-aided analysis of nonlinear problems in transport phenomena, in *New Approaches to Nonlinear Problems in Dynamics*, edited by P. J. Holmes. SIAM, Philadelphia, 1980. pp. 281–308.
[5.12] R. K. Mehra, J. V. Carrol: Global Stability and Control Analysis of Aircraft at High Angles of Attack. Report ONR-CR 215-248-1, 1977.

[5.13] R. K. Mehra, J. V. Carroll: Global Stability and Control Analysis of Aircraft at High Angles of Attack. Report ONR-CR 215-248-2, 1978.
[5.14] R. K. Mehra, J. V. Carroll: Global Stability and Control Analysis of Aircraft at High Angles of Attack. Report ONR-CR 215-248-3, 1979.
[5.15] R. B. Simpson: SIAM J. Numer. Anal. 12 (1975), 439–451.
[5.16] J. P. Abbot: J. Comp. Appl. Math. 4 (1978), 19–26.
[5.17] W. C. Rheinboldt: SIAM J. Numer. Anal. 15 (1978), 1–11.
[5.18] G. Moore, A. Spence: SIAM J. Numer. Anal. 17 (1980), 567–576.
[5.19] G. Moore: Numer. Funct. Anal. and Optimiz. 2 (1980), 441–472.
[5.20] G. Moore, A. Spence: IMA J. of Numer. Analysis 1 (1981), 23–38.
[5.21] B. D. Hassard, N. D. Kazarinoff, Y. H. Wan: *Theory and Applications of the Hopf Bifurcation.* Cambridge University Press, Cambridge 1981.
[5.22] M. Golubitsky, D. Schaeffer: A singularity theory approach to steady state bifurcation theory, in *Nonlinear Partial Differential Equations in Engineering and Applied Science.* Marcel Dekker, New York, 1979.
[5.23] M. Kubíček, M. Holodniok: Computation of periodic solutions in distributed parameter systems (in preparation).
[5.24] K. P. Hadeler: Numer. Math. 34 (1980), 457–467.

Appendix C

[C.1] A. A. Andronov, E. A. Leontovich, I. I. Gordon, A. G. Maier: *Theory of Bifurcations of Dynamic Systems on a Plane.* Wiley, New York, 1973.
[C.2] A. A. Andronov, A. A. Vitt, S. E. Chaikin: *Theory of Oscillations.* Fizmatgiz, Moscow, 1959 (in Russian).
[C.3] N. Minorsky: *Nonlinear Oscillations.* Van Nostrand, Princeton, NJ, 1962.
[C.4] N. N. Bogoliubov, U. A. Mitropolskii: *Asymptotic Methods in the Theory of Nonlinear Oscillations.* Nauka, Moscow, 1974 (in Russian).
[C.5] G. Hayashi: *Nonlinear Oscillations in Physical Systems.* McGraw-Hill, New York, 1965.
[C.6] J. K. Hale: *Ordinary Differential Equations.* McGraw-Hill, New York, 1969.
[C.7] S. Smale: Bull. Amer. Math. Soc. 73 (1967), 747–817.
[C.8] J. Moser: *Stable and Random Motions in Dynamical Systems.* Princeton University Press, Princeton, NJ, 1973.
[C.9] Z. Nitecki: *Differentiable Dynamics.* MIT Press, Cambridge, MA, 1971.
[C.10] G. Iooss: *Topics in Bifurcation of Maps and Applications.* Lecture Notes, Math. Studies. North Holland, Amsterdam, 1979.
[C.11] P. J. Holmes: Bifurcation in coupled nonlinear oscillators with applications to flutter and divergence. Preprint, 1979.
[C.12] M. W. Hirsh, C. C. Pugh, M. Shub: *Invariant Manifolds.* Lecture Notes in Mathematics, No. 583. Springer-Verlag, Heidelberg Berlin New York, 1977.
[C.13] B. D. Hassard, Y. H. Wan: J. Math. Anal. Appl. 63 (1978), 297–312.
[C.14] A. D. Brjuno: Trudy MMO 25 (1971), 119–262 (in Russian).
[C.15] F. Takens: Singularities of vector fields, in Publications of the Institut des Haute Etudes Scientifiques, Bures-sur-Yvette, Paris 43 (1974), 47–100.
[C.16] S. Sternberg: Amer. J. Math. 80 (1958), 623–631; 81 (1959), 578–604.
[C.17] G. Iooss: *Direct Bifurcation of a Steady Solution of the Navier-Stokes Equations into an Invariant Torus*, Lecture Notes in Math. Vol. 565. (Springer-Verlag, Berlin, 1975. pp. 69–84.
[C.18] J. Guckenheimer: On a codimension two singularity. Preprint, University of California at Santa Cruz, 1979.
[C.19] W. F. Langford: SIAM J. Appl. Math. 37 (1979), 22–48.
[C.20] R. I. Bogdanov: Funct. Anal. Appl. 9 (2) (1975), 144–145.
[C.21] V. S. Kozjakin: DAN USSR, 232 (1977), 25–27 (in Russian).
[C.22] P. Holmes: Bifurcation in coupled oscillators with applications to flutter and divergence. Preprint, Cornell University, Ithaca, 1979.

Index